;eneral Motors
GMC Acadia
Buick Enclave
Saturn Outlook
& Chevrolet
Traverse
Automotive
Repair
Manual

by Jeff Killingsworth, Robert Maddox and John H Haynes

Member of the Guild of Motoring Writers

Models covered:

GMC Acadia - 2007 through 2016
GMC Acadia LTD - 2017
Buick Enclave - 2008 through 2017
Saturn Outlook - 2007 through 2010
Chevrolet Traverse - 2009 through 2017

(38001-5AA3)

J H Haynes & Co. Ltd.
Haynes North America, Inc

www.haynes.com

Acknowledgements

Technical writers who contributed to this project include Trip Aiken, Demian Hurst, Peter Sessler, Brian Godfrey, Dennis Gibb, Ben King and Scott "Gonzo" Weaver.

A book in the Haynes Automotive Repair Manual Series

ISBN-10: 1-62092-336-X
ISBN-13: 978-1-62092-336-8

Library of Congress Control Number: 2018960955

Disclaimer

There are risks associated with automotive repairs. The ability to make repairs depends on individual skill, experience and proper tools. Individuals should act with due care and acknowledge and assume the risk of making automotive repairs. While every attempt is made to ensure that the information in this manual is correct, no liability can be accepted by the authors or publishers for loss, damage or injury caused by any errors in, or omissions from, the information given.

Contents

Introductory pages

About this manual	0-5
Introduction	0-5
Vehicle identification numbers	0-6
Recall information	0-7
Buying parts	0-11
Maintenance techniques, tools and working facilities	0-11
Jacking and towing	0-18
Booster battery (jump) starting	0-19
Automotive chemicals and lubricants	0-20
Conversion factors	0-21
Fraction/decimal/millimeter equivalents	0-22
Safety first!	0-23
Troubleshooting	0-24

Chapter 1
Tune-up and routine maintenance — **1-1**

Chapter 2 Part A
3.6L V6 engine — **2A-1**

Chapter 2 Part B
General engine overhaul procedures — **2B-1**

Chapter 3
Cooling, heating and air conditioning systems — **3-1**

Chapter 4
Fuel and exhaust systems — **4-1**

Chapter 5
Engine electrical systems — **5-1**

Chapter 6
Emissions and engine control systems — **6-1**

Chapter 7 Part A
Automatic transaxle — **7A-1**

Chapter 7 Part B
Transfer case — **7B-1**

Chapter 8
Driveline — **8-1**

Chapter 9
Brakes — **9-1**

Chapter 10
Suspension and steering systems — **10-1**

Chapter 11
Body — **11-1**

Chapter 12
Chassis electrical system — **12-1**

Wiring diagrams — **12-21**

Index — **IND-1**

Haynes mechanic and photographer with a 2007 GMC Acadia

About this manual

Its purpose

The purpose of this manual is to provide comprehensive, useful and accessible automotive repair information, to help you get the best value from your vehicle. It can do so in several ways. It can help you decide what work must be done, even if you choose to have it done by a dealer service department or a repair shop; it provides information and procedures for routine maintenance and servicing; and it offers diagnostic and repair procedures to follow when trouble occurs.

We hope you use the manual to tackle the work yourself. For many simpler jobs, doing it yourself may be quicker than arranging an appointment to get the vehicle into a shop and making the trips to leave it and pick it up. More importantly, a lot of money can be saved by avoiding the expense the shop must pass on to you to cover its labor and overhead costs. An added benefit is the sense of satisfaction and accomplishment that you feel after doing the job yourself. However, this manual is not a substitute for a professional certified technician or mechanic. There are risks associated with automotive repairs. The ability to make repairs on a vehicle depends on individual skill, experience, and proper tools. Individuals should act with due care and acknowledge and assume the risk of performing automotive repairs.

Using the manual

The manual is divided into Chapters. Each Chapter is divided into numbered Sections, which are headed in bold type between horizontal lines. Each Section consists of consecutively numbered paragraphs.

The reference numbers used in illustration captions pinpoint the pertinent Section and the Step within that Section. That is, illustration 3.2 means the illustration refers to Section 3 and Step (or paragraph) 2 within that Section.

Procedures, once described in the text, are not normally repeated. When it's necessary to refer to another Chapter, the reference will be given as Chapter and Section number. Cross references given without use of the word "Chapter" apply to Sections and/or paragraphs in the same Chapter. For example, "see Section 8" means in the same Chapter. References to the left or right side of the vehicle assume you are sitting in the driver's seat, facing forward.

This repair manual is produced by a third party and is not associated with an individual car manufacturer. If there is any doubt or discrepancy between this manual and the owner's manual or the factory service manual, please refer to factory service manual or seek assistance from a professional certified technician or mechanic. Even though we have prepared this manual with extreme care, neither the publisher nor the author can accept responsibility for any errors in, or omissions from, the information given.

NOTE

A **Note** provides information necessary to properly complete a procedure or information which will make the procedure easier to understand.

CAUTION

A **Caution** provides a special procedure or special steps which must be taken while completing the procedure where the Caution is found. Not heeding a Caution can result in damage to the assembly being worked on.

WARNING

A **Warning** provides a special procedure or special steps which must be taken while completing the procedure where the Warning is found. Not heeding a Warning can result in personal injury.

Introduction

This manual covers the GMC Acadia, Buick Enclave, Chevrolet Traverse and Saturn Outlook. These vehicles feature 3.6L V6 engines. The engine drives the front wheels via independent driveaxles. On AWD models, the rear wheels are also propelled by way of a transfer case, driveshaft, torque tube, rear differential and two rear driveaxles.

Both Front Wheel Drive (FWD) and All Wheel Drive (AWD) models are equipped with a fully electronic, six-speed (6T70/6T75) automatic transaxle. AWD models are equipped with a bolt-on Getrag 790 transfer case.

Suspension is independent at all four wheels, with MacPherson struts used at the front and coil springs with shock absorbers at the rear. The rack-and-pinion steering unit is mounted on the subframe.

The brakes are disc at the front and rear, with power assist standard. All models are equipped with an Anti-Lock Brake System (ABS).

Vehicle identification numbers

Modifications are a continuing and unpublicized process in vehicle manufacturing. Since spare parts manuals and lists are compiled on a numerical basis, the individual vehicle numbers are essential to correctly identify the component required.

Vehicle Identification Number (VIN)

This very important identification number is stamped on a plate attached to the dashboard inside the windshield on the driver's side of the vehicle (see illustration). It can also be found on the certification label located on the driver's side door post. The VIN also appears on the Vehicle Certificate of Title and Registration. It contains information such as where and when the vehicle was manufactured, the model year and the body style.

On the vehicles covered by this manual, the model year codes are:

7	2007
8	2008
9	2009
A	2010
B	2011
C	2012
D	2013
E	2014
F	2015
G	2016
H	2017

On the vehicles covered by this manual, the engine codes are:

- 7 ....3.6L V6 (LY7) engine with Sequential Fuel Injection (SFI) - 2008 and earlier models
- D ....3.6L V6 (LLT) engine with high-pressure direct fuel injection - 2009 and later models

Certification label

The certification label is attached to the end of the driver's door (see illustration). The label contains the name of the manufacturer, the month and year of production, the Gross Vehicle Weight Rating (GVWR), the Gross Axle Weight Rating (GAWR) and the certification statement.

Engine number

The engine identification number is stamped or etched on left end (driver's side) end of the engine block, near the transaxle, on the radiator side of the engine (see illustration).

Transaxle number

The transaxle identification number has important information, such as the transaxle type and build date, on a label that is attached to the left end of the transaxle (see illustration). It is also etched into the case on the front edge of the transaxle bellhousing, near the engine identification number.

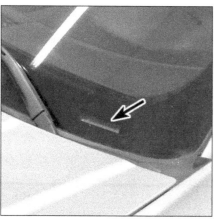

3.2 The Vehicle Identification Number (VIN) is located on a plate on top of the dash (visible through the windshield)

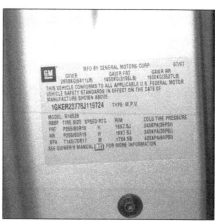

3.5 The Certification label is located on the end of the driver's door

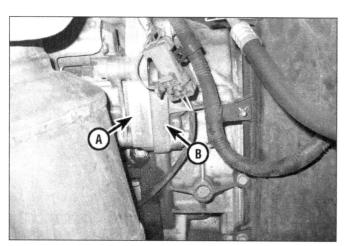

3.6 Location of the engine identification number (A) and the transaxle identifcation number (B) (shown from the front with the radiator removed)

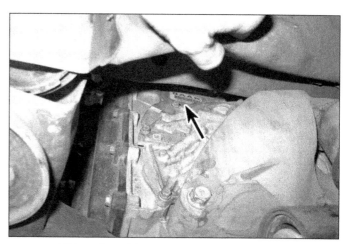

3.7 Location of the transaxle identification label

Recall information

Vehicle recalls are carried out by the manufacturer in the rare event of a possible safety-related defect. The vehicle's registered owner is contacted at the address on file at the Department of Motor Vehicles and given the details of the recall. Remedial work is carried out free of charge at a dealer service department.

If you are the new owner of a used vehicle which was subject to a recall and you want to be sure that the work has been carried out, it's best to contact a dealer service department and ask about your individual vehicle - you'll need to furnish them your Vehicle Identification Number (VIN).

The table below is based on information provided by the National Highway Traffic Safety Administration (NHTSA), the body which oversees vehicle recalls in the United States. The recall database is updated constantly. For the latest information on vehicle recalls, check the NHTSA website at www. nhtsa.gov, www.safercar.gov, or call the NHTSA hotline at 1-888-327-4236.

Recall date	Recall campaign number	Model(s) affected	Concern
Feb 21, 2007	07V062000	2007 GMC Acadia and Saturn Outlook	On certain vehicles, the sensing and diagnostic module (SDM), which controls the function of front airbags, may not operate properly. As a result, the front airbags may fail to deploy in a frontal crash. Also, the airbag warning lamp on the instrument panel may fail to provide warning that the system is inoperative. In the event of a crash, this condition could increase the risk of injury to occupants in the front seat.
Aug 14, 2008	08V410000	2008 Buick Enclave; 2007 and 2008 GMC Acadia and Saturn Outlook	On some models, if a buildup of snow or ice on the windshield or on the wipers restricts the movement of the wiper arm, the windshield wiper linkage may become detached from the motor shaft and the wipers may become inoperative. If this were to occur, driver visibility could be reduced, which could result in a vehicle crash.
Aug 28, 2008	08V441000	2008 Buick Enclave; 2007 and 2008 GMC Acadia and Saturn Outlook	On certain models, a short circuit on the printed circuit board for the washer fluid heater may overheat the control-circuit ground wire. This may cause other electrical features to malfunction, create an odor, or cause smoke, increasing the risk of a fire.
Nov 24, 2008	08V615000	2009 Buick Enclave, GMC Acadia, Chevrolet Traverse and Saturn Outlook	Some models may have been built with a safety belt buckle in the second or third row that is missing a rivet. In a vehicle crash, if the rivet is missing, the buckle may separate from the mounting strap, increasing the risk of injury to the passenger.
Mar 6, 2009	09V073000	2009 Buick Enclave, GMC Acadia and Saturn Outlook	On some models, the transmission shift cable adjustment clip may not be fully engaged. If the clip is not fully engaged, the shift lever and the actual position of the transmission gear may not match. With this condition, the driver could move the shifter to "Park" and remove the ignition key, but the transmission gear may not be in "Park." The driver may not be able to restart the vehicle and the vehicle could roll away after the driver has exited the vehicle, resulting in a possible crash.

Recall date	Recall campaign number	Model(s) affected	Concern
May 6, 2009	09V153000	2009 Chevrolet Traverse	Some models may have a parking brake cable link (connector) that is not to specification. This connector may fracture when the parking brake pedal is depressed. If the transmission shift lever is in the Park position, there will be no unintended vehicle movement. However, if the keys are left in the ignition, the transmission shift lever is in any position other than park, and the vehicle is parked on a sufficient slope, it could result in unintended vehicle movement. This could result in a crash.
June 4, 2010	10V240000	2008 and 2009 Buick Enclave; 2007, 2008 and 2009 GMC Acadia and Saturn Outlook; 2009 Chevrolet Traverse	On certain models, a recall was implemented in 2008 to add a fuse to the control circuit harness to address the potential consequences of a printed circuit board (PCB) electrical short. However, there have been new reports of thermal incidents on HWFS modules after this improvement was installed. These incidents resulted from a new failure mode attributed to the device's thermal protection feature. The significance varies from minor distortion to considerable melting of the plastic around the HWFS fluid chamber. It is possible for the heated washer module to ignite and a fire may occur.
Aug 12, 2010	10V375000	2009 and 2010 Buick Enclave, GMC Acadia, Chevrolet Traverse, Saturn Outlook	On some models, the second row seat side trim shield restricts the upward rotation of the safety belt buckle when the seat back is returned to a seating position after being folded flat. If the buckle contacts the seat frame, additional effort is required to return the seat to a seating position. If sufficient force is applied, the buckle cover could be pushed down the strap, exposing and partially depressing the red release button. The safety belt may not restrain the occupant as intended during a crash, increasing the risk of injury to the seat occupant.
Apr 9, 2012	12V151000	2011 and 2012 Buick Enclave, GMC Acadia, Chevrolet Traverse	On some models, snow or ice buildup on the windshield or on the wiper may restrict the movement of the wiper arm, causing the wiper arm to loosen and cause the wiper to become inoperative. If this occurs, driver visibility could be reduced, increasing the risk of a crash.
Feb 21, 2014	14V092000	2014 Acadia, Enclave and Traverse	On some models, the transmission shift cable adjuster clip may disengage from the transmission shift lever. If the shift cable disengages from the transmission shift lever, a driver may be unable to shift gear positions and the indicated shift position may not represent the gear position the vehicle is in. Should a disengagement occur while the vehicle is being driven, when the driver goes to stop and park the vehicle, the driver may be able to shift the lever to the "PARK" position, but the vehicle transmission may not be in the "PARK" gear position. If the vehicle is not in the "PARK" position there is a risk the vehicle will roll away as the driver and other occupants exit the vehicle or anytime thereafter. A vehicle rollaway increases the risk of injury to exiting occupants and bystanders.

Recall date	Recall campaign number	Model(s) affected	Concern
March 17, 2014	14V118000	2009, 2010, 2011, 2012, 2013 Chevrolet Traverse, GMC Acadia and Buick Enclave, 2008, 2009, 2010 Saturn Outlook	In the affected vehicles, increased resistance in the driver and passenger seat mounted side impact airbag (SIAB) wiring harnesses may result in the SIAB and seat belt pretensioners not deploying in the event of a crash. Failure of the side impact airbags and seat belt pretensioners to deploy in a crash increases the risk of injury to the driver and front seat occupant.
Apr 30, 2014	14V223000	Enclave, Acadia and Traverse (built March 25, 2013 - August 15, 2013)	On some models, the Powertrain Control Module (PCM) software may cause the fuel level information to become inaccurate and the vehicle may stall without warning with an empty fuel tank. GM will reprogram the PCM with accurate parameters to correct the instrument cluster fuel display.
May 20, 2014	14V266000	Buick Enclave (built April 8, 2008 - May 14, 2014), GMC Acadia (built April 9, 2008 - May 14, 2014), Chevrolet Traverse (built June 6, 2008 - May 14, 2014) and Saturn Outlook (built April 14, 2008 - March 18, 2010).	On some models, a flexible steel cable that connects the seat belt to the front seat positions may fatigue and become weak over time, causing seat belt failure.
Oct 1, 2014	14V614000	2014 Acadia, Enclave	On some models, the chassis electronic module or vehicle control module (VCM) may short causing stalling or rough running. This problem may be caused by internal contamination or water damage.
Jan 28, 2015	15V044000	Buick Enclave (built Dec. 9, 2014 - January 14, 2015), GMC Acadia (built Dec.9, 2014 - January 16, 2015), Chevrolet Traverse (built December 9, 2014 - January 20, 2015) with Goodyear P255/65 R18 Fortera HL tires	Models with these Goodyear Fortera HL tires may experience tread cracking and eventual failure. Tire cracking could lead to loss of tire air pressure and vehicle handling problems.
June 30, 2015	15V415000	2009, 2010, 2011, 2012, 2013 Chevrolet Traverse, 2008, 2009, 2010, 2011, 2012 Buick Enclave, 2007, 2008, 2009, 2010, 2011, 2012 GMC Acadia, 2007, 2008, 2009, 2010 Saturn Outlook	The affected vehicles, equipped with the power liftgate option, have gas struts that hold the power liftgate up when open. These struts may prematurely wear and the open liftgate may suddenly fall. If the open liftgate unexpectedly falls, it may strike a person, increasing their risk of injury.
Sept 29, 2015	15V609000	2016 Chevrolet Traverse, GMC Acadia, Buick Enclave	On certain models, due to a manufacturing defect, the windshield wiper motor may overheat when used. A wiper motor that overheats increases the risk of a fire.
Dec 11, 2015	15V833000	2016 Chevrolet Traverse, GMC Acadia, Buick Enclave	On certain models, the third row left lower seat frame may have welds that are not in the correct location. If the seat frame is not welded in the correct places, the seat may not perform as intended, and in the event of a crash, the seat occupant could be at an increased risk of injury.

Notes

Buying parts

Replacement parts are available from many sources, which generally fall into one of two categories - authorized dealer parts departments and independent retail auto parts stores. Our advice concerning these parts is as follows:

Retail auto parts stores: Good auto parts stores will stock frequently needed components which wear out relatively fast, such as clutch components, exhaust systems, brake parts, tune-up parts, etc. These stores often supply new or reconditioned parts on an exchange basis, which can save a considerable amount of money. Discount auto parts stores are often very good places to buy materials and parts needed for general vehicle maintenance such as oil, grease, filters, spark plugs, belts, touch-up paint, bulbs, etc. They also usually sell tools and general accessories, have convenient hours, charge lower prices and can often be found not far from home.

Authorized dealer parts department: This is the best source for parts which are unique to the vehicle and not generally available elsewhere (such as major engine parts, transmission parts, trim pieces, etc.).

Warranty information: If the vehicle is still covered under warranty, be sure that any replacement parts purchased - regardless of the source - do not invalidate the warranty!

To be sure of obtaining the correct parts, have engine and chassis numbers available and, if possible, take the old parts along for positive identification.

Maintenance techniques, tools and working facilities

Maintenance techniques

There are a number of techniques involved in maintenance and repair that will be referred to throughout this manual. Application of these techniques will enable the home mechanic to be more efficient, better organized and capable of performing the various tasks properly, which will ensure that the repair job is thorough and complete.

Fasteners

Fasteners are nuts, bolts, studs and screws used to hold two or more parts together. There are a few things to keep in mind when working with fasteners. Almost all of them use a locking device of some type, either a lockwasher, locknut, locking tab or thread adhesive. All threaded fasteners should be clean and straight, with undamaged threads and undamaged corners on the hex head where the wrench fits. Develop the habit of replacing all damaged nuts and bolts with new ones. Special locknuts with nylon or fiber inserts can only be used once. If they are removed, they lose their locking ability and must be replaced with new ones.

Rusted nuts and bolts should be treated with a penetrating fluid to ease removal and prevent breakage. Some mechanics use turpentine in a spout-type oil can, which works quite well. After applying the rust penetrant, let it work for a few minutes before trying to loosen the nut or bolt. Badly rusted fasteners may have to be chiseled or sawed off or removed with a special nut breaker, available at tool stores.

If a bolt or stud breaks off in an assembly, it can be drilled and removed with a special tool commonly available for this purpose. Most automotive machine shops can perform this task, as well as other repair procedures, such as the repair of threaded holes that have been stripped out.

Flat washers and lockwashers, when removed from an assembly, should always be replaced exactly as removed. Replace any damaged washers with new ones. Never use a lockwasher on any soft metal surface (such as aluminum), thin sheet metal or plastic.

Fastener sizes

For a number of reasons, automobile manufacturers are making wider and wider use of metric fasteners. Therefore, it is important to be able to tell the difference between standard (sometimes called U.S. or SAE) and metric hardware, since they cannot be interchanged.

All bolts, whether standard or metric, are sized according to diameter, thread pitch and length. For example, a standard 1/2 - 13 x 1 bolt is 1/2 inch in diameter, has 13 threads per inch and is 1 inch long. An M12 - 1.75 x 25 metric bolt is 12 mm in diameter, has a thread pitch of 1.75 mm (the distance between threads) and is 25 mm long. The two bolts are nearly identical, and easily confused, but they are not interchangeable.

In addition to the differences in diameter, thread pitch and length, metric and standard bolts can also be distinguished by examining the bolt heads. To begin with, the distance across the flats on a standard bolt head is measured in inches, while the same dimension on a metric bolt is sized in millimeters (the same is true for nuts). As a result, a standard wrench should not be used on a metric bolt and a metric wrench should not be used on a standard bolt. Also, most standard bolts have slashes radiating out from the center of the head to denote the grade or strength of the bolt, which is an indication of the amount of torque that can be applied to it. The greater the number of slashes, the greater the strength of the bolt. Grades 0 through 5 are commonly used on automobiles. Metric bolts have a property class (grade) number, rather than a slash, molded into their heads to indicate bolt strength. In this case, the higher the number, the stronger the bolt. Property class numbers 8.8, 9.8 and 10.9 are commonly used on automobiles.

Strength markings can also be used to distinguish standard hex nuts from metric hex nuts. Many standard nuts have dots stamped into one side, while metric nuts are marked with a number. The greater the number of dots, or the higher the number, the greater the strength of the nut.

Metric studs are also marked on their ends according to property class (grade). Larger studs are numbered (the same as metric bolts), while smaller studs carry a geometric code to denote grade.

It should be noted that many fasteners, especially Grades 0 through 2, have no distinguishing marks on them. When such is the case, the only way to determine whether it is standard or metric is to measure the thread pitch or compare it to a known fastener of the same size.

Standard fasteners are often referred to as SAE, as opposed to metric. However, it should be noted that SAE technically refers to a non-metric fine thread fastener only. Coarse thread non-metric fasteners are referred to as USS sizes.

Since fasteners of the same size (both standard and metric) may have different strength ratings, be sure to reinstall any bolts, studs or nuts removed from your vehicle in their original locations. Also, when replacing a fastener with a new one, make sure that the new one has a strength rating equal to or greater than the original.

Tightening sequences and procedures

Most threaded fasteners should be tightened to a specific torque value (torque is the twisting force applied to a threaded component such as a nut or bolt). Overtightening the fastener can weaken it and cause it to break, while undertightening can cause it to eventually come loose. Bolts, screws and studs, depending on the material they are made of and their thread diameters, have specific torque values, many of which are noted in the Specifications at the beginning of each Chapter. Be sure to follow the torque recommendations closely. For fasteners not assigned a

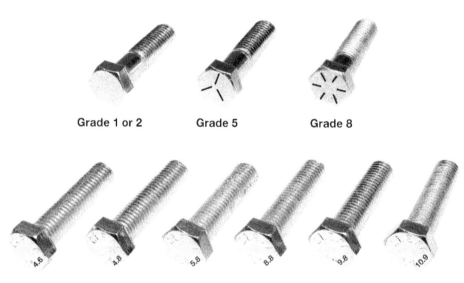

Grade 1 or 2   Grade 5   Grade 8

Bolt strength marking (standard/SAE/USS; bottom - metric)

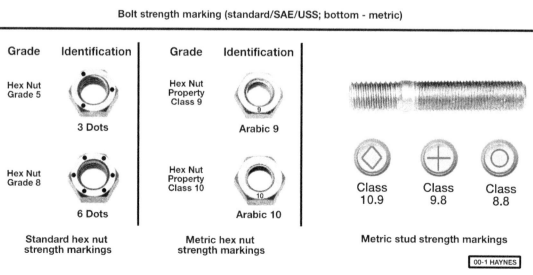

Grade	Identification
Hex Nut Grade 5	3 Dots
Hex Nut Grade 8	6 Dots

Standard hex nut strength markings

Grade	Identification
Hex Nut Property Class 9	Arabic 9
Hex Nut Property Class 10	Arabic 10

Metric hex nut strength markings

Class 10.9   Class 9.8   Class 8.8

Metric stud strength markings

00-1 HAYNES

specific torque, a general torque value chart is presented here as a guide. These torque values are for dry (unlubricated) fasteners threaded into steel or cast iron (not aluminum). As was previously mentioned, the size and grade of a fastener determine the amount of torque that can safely be applied to it. The figures listed here are approximate for Grade 2 and Grade 3 fasteners. Higher grades can tolerate higher torque values.

Fasteners laid out in a pattern, such as cylinder head bolts, oil pan bolts, differential cover bolts, etc., must be loosened or tightened in sequence to avoid warping the component. This sequence will normally be shown in the appropriate Chapter. If a specific pattern is not given, the following procedures can be used to prevent warping.

Initially, the bolts or nuts should be assembled finger-tight only. Next, they should be tightened one full turn each, in a crisscross or diagonal pattern. After each one has been tightened one full turn, return to the first one and tighten them all one-half turn, following the same pattern. Finally, tighten each of them one-quarter turn at a time until each fastener has been tightened to the proper torque. To loosen and remove the fasteners, the procedure would be reversed.

Component disassembly

Component disassembly should be done with care and purpose to help ensure that

Metric thread sizes	Ft-lbs	Nm
M-6	6 to 9	9 to 12
M-8	14 to 21	19 to 28
M-10	28 to 40	38 to 54
M-12	50 to 71	68 to 96
M-14	80 to 140	109 to 154

Pipe thread sizes		
1/8	5 to 8	7 to 10
1/4	12 to 18	17 to 24
3/8	22 to 33	30 to 44
1/2	25 to 35	34 to 47

U.S. thread sizes		
1/4 - 20	6 to 9	9 to 12
5/16 - 18	12 to 18	17 to 24
5/16 - 24	14 to 20	19 to 27
3/8 - 16	22 to 32	30 to 43
3/8 - 24	27 to 38	37 to 51
7/16 - 14	40 to 55	55 to 74
7/16 - 20	40 to 60	55 to 81
1/2 - 13	55 to 80	75 to 108

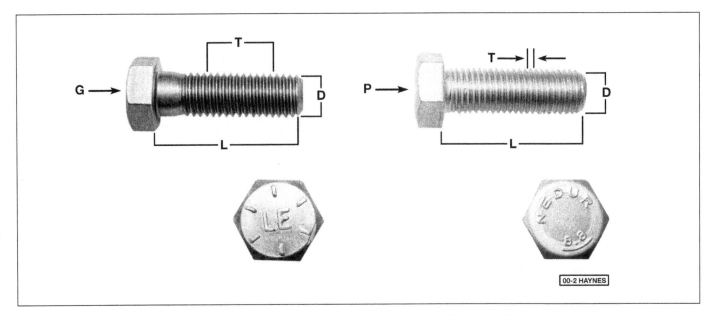

Standard (SAE and USS) bolt dimensions/grade marks

G	Grade marks (bolt strength)
L	Length (in inches)
T	Thread pitch (number of threads per inch)
D	Nominal diameter (in inches)

Metric bolt dimensions/grade marks

P	Property class (bolt strength)
L	Length (in millimeters)
T	Thread pitch (distance between threads in millimeters)
D	Diameter

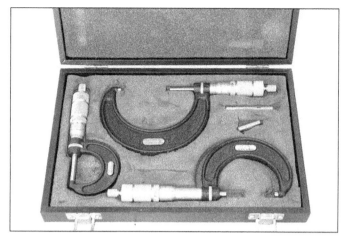

Micrometer set

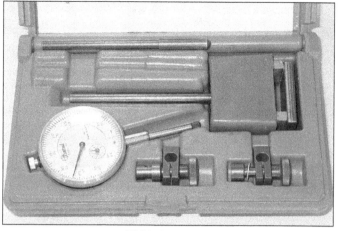

Dial indicator set

the parts go back together properly. Always keep track of the sequence in which parts are removed. Make note of special characteristics or marks on parts that can be installed more than one way, such as a grooved thrust washer on a shaft. It is a good idea to lay the disassembled parts out on a clean surface in the order that they were removed. It may also be helpful to make sketches or take instant photos of components before removal.

When removing fasteners from a component, keep track of their locations. Sometimes threading a bolt back in a part, or putting the washers and nut back on a stud, can prevent mix-ups later. If nuts and bolts cannot be returned to their original locations, they should be kept in a compartmented box or a series of small boxes. A cupcake or muffin tin is ideal for this purpose, since each cavity can hold the bolts and nuts from a particular area (i.e. oil pan bolts, valve cover bolts, engine mount bolts, etc.). A pan of this type is especially helpful when working on assemblies with very small parts, such as the carburetor, alternator, valve train or interior dash and trim pieces. The cavities can be marked with paint or tape to identify the contents.

Whenever wiring looms, harnesses or connectors are separated, it is a good idea to identify the two halves with numbered pieces of masking tape so they can be easily reconnected.

Gasket sealing surfaces

Throughout any vehicle, gaskets are used to seal the mating surfaces between two parts and keep lubricants, fluids, vacuum or pressure contained in an assembly.

Many times these gaskets are coated with a liquid or paste-type gasket sealing compound before assembly. Age, heat and pressure can sometimes cause the two parts to stick together so tightly that they are very difficult to separate. Often, the assembly can be loosened by striking it with a soft-face hammer near the mating surfaces. A regular hammer can be used if a block of wood is placed between the hammer and the part. Do

not hammer on cast parts or parts that could be easily damaged. With any particularly stubborn part, always recheck to make sure that every fastener has been removed.

Avoid using a screwdriver or bar to pry apart an assembly, as they can easily mar the gasket sealing surfaces of the parts, which must remain smooth. If prying is absolutely necessary, use an old broom handle, but keep in mind that extra clean up will be necessary if the wood splinters.

After the parts are separated, the old gasket must be carefully scraped off and the gasket surfaces cleaned. Stubborn gasket material can be soaked with rust penetrant or treated with a special chemical to soften it so it can be easily scraped off. **Caution:** *Never use gasket removal solutions or caustic chemicals on plastic or other composite components.* A scraper can be fashioned from a piece of copper tubing by flattening and sharpening one end. Copper is recommended because it is usually softer than the surfaces to be scraped, which reduces the chance of gouging the part. Some gaskets can be removed with a wire brush, but regardless of the method used, the mating surfaces must be left clean and smooth. If for some reason the gasket surface is gouged, then a gasket sealer thick enough to fill scratches will have to be used during reassembly of the components. For most applications, a non-drying (or semi-drying) gasket sealer should be used.

Hose removal tips

Warning: *If the vehicle is equipped with air conditioning, do not disconnect any of the A/C hoses without first having the system depressurized by a dealer service department or a service station.*

Hose removal precautions closely parallel gasket removal precautions. Avoid scratching or gouging the surface that the hose mates against or the connection may leak. This is especially true for radiator hoses. Because of various chemical reactions, the rubber in hoses can bond itself to the metal spigot that the hose fits over. To remove

a hose, first loosen the hose clamps that secure it to the spigot. Then, with slip-joint pliers, grab the hose at the clamp and rotate it around the spigot. Work it back and forth until it is completely free, then pull it off. Silicone or other lubricants will ease removal if they can be applied between the hose and the outside of the spigot. Apply the same lubricant to the inside of the hose and the outside of the spigot to simplify installation.

As a last resort (and if the hose is to be replaced with a new one anyway), the rubber can be slit with a knife and the hose peeled from the spigot. If this must be done, be careful that the metal connection is not damaged.

If a hose clamp is broken or damaged, do not reuse it. Wire-type clamps usually weaken with age, so it is a good idea to replace them with screw-type clamps whenever a hose is removed.

Tools

A selection of good tools is a basic requirement for anyone who plans to maintain and repair his or her own vehicle. For the owner who has few tools, the initial investment might seem high, but when compared to the spiraling costs of professional auto maintenance and repair, it is a wise one.

To help the owner decide which tools are needed to perform the tasks detailed in this manual, the following tool lists are offered: *Maintenance and minor repair, Repair/overhaul* and *Special.*

The newcomer to practical mechanics should start off with the *maintenance and minor repair* tool kit, which is adequate for the simpler jobs performed on a vehicle. Then, as confidence and experience grow, the owner can tackle more difficult tasks, buying additional tools as they are needed. Eventually the basic kit will be expanded into the *repair and overhaul* tool set. Over a period of time, the experienced do-it-yourselfer will assemble a tool set complete enough for most repair and overhaul procedures and will add tools from the special category when it is felt that the expense is justified by the frequency of use.

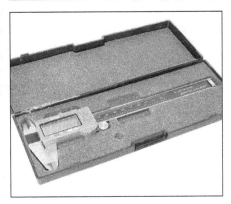

Dial caliper

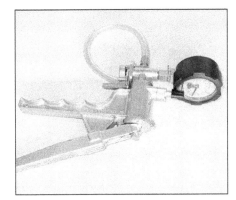

Hand-operated vacuum pump

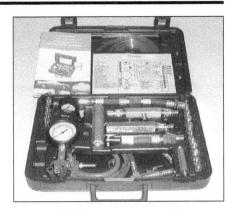

Fuel pressure gauge set

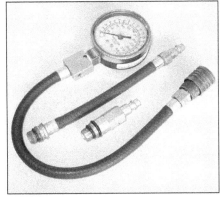

Compression gauge with spark plug
hole adapter

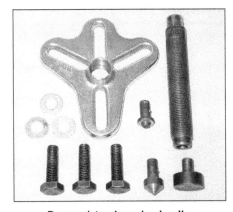

Damper/steering wheel puller

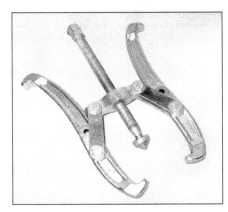

General purpose puller

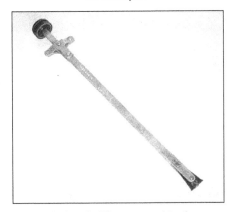

Hydraulic lifter removal tool

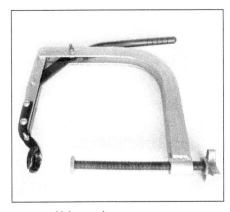

Valve spring compressor

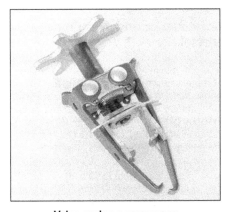

Valve spring compressor

Ridge reamer

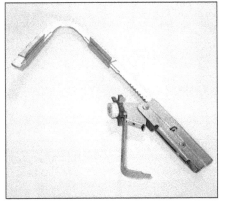

Piston ring groove cleaning tool

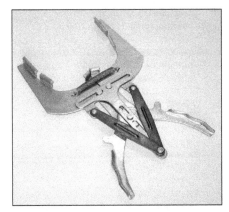

Ring removal/installation tool

Ring compressor

Cylinder hone

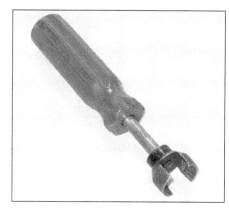

Brake hold-down spring tool

Torque angle gauge

Clutch plate alignment tool

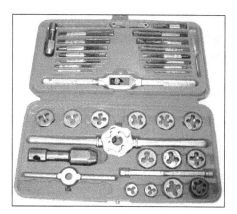

Tap and die set

Maintenance and minor repair tool kit

The tools in this list should be considered the minimum required for performance of routine maintenance, servicing and minor repair work. We recommend the purchase of combination wrenches (box-end and open-end combined in one wrench). While more expensive than open end wrenches, they offer the advantages of both types of wrench.

Combination wrench set (1/4-inch to 1 inch or 6 mm to 19 mm)
Adjustable wrench, 8 inch
Spark plug wrench with rubber insert
Spark plug gap adjusting tool
Feeler gauge set
Brake bleeder wrench
Standard screwdriver (5/16-inch x 6 inch)
Phillips screwdriver (No. 2 x 6 inch)
Combination pliers - 6 inch
Hacksaw and assortment of blades
Tire pressure gauge
Grease gun
Oil can
Fine emery cloth
Wire brush
Battery post and cable cleaning tool
Oil filter wrench
Funnel (medium size)
Safety goggles
Jackstands (2)
Drain pan

Note: *If basic tune-ups are going to be part of routine maintenance, it will be necessary to purchase a good quality stroboscopic timing light and combination tachometer/dwell meter. Although they are included in the list of special tools, it is mentioned here because they are absolutely necessary for tuning most vehicles properly.*

Repair and overhaul tool set

These tools are essential for anyone who plans to perform major repairs and are in addition to those in the maintenance and minor repair tool kit. Included is a comprehensive set of sockets which, though expensive, are invaluable because of their versatility, especially when various extensions and drives are available. We recommend the 1/2-inch drive over the 3/8-inch drive. Although the larger drive is bulky and more expensive, it has the capacity of accepting a very wide range of large sockets. Ideally, however, the mechanic should have a 3/8-inch drive set and a 1/2-inch drive set.

Socket set(s)
Reversible ratchet
Extension - 10 inch
Universal joint
Torque wrench (same size drive as sockets)
Ball peen hammer - 8 ounce
Soft-face hammer (plastic/rubber)
Standard screwdriver (1/4-inch x 6 inch)

Standard screwdriver (stubby - 5/16-inch)
Phillips screwdriver (No. 3 x 8 inch)
Phillips screwdriver (stubby - No. 2)
Pliers - vise grip
Pliers - lineman's
Pliers - needle nose
Pliers - snap-ring (internal and external)
Cold chisel - 1/2-inch
Scribe
Scraper (made from flattened copper tubing)
Centerpunch
Pin punches (1/16, 1/8, 3/16-inch)
Steel rule/straightedge - 12 inch
Allen wrench set (1/8 to 3/8-inch or 4 mm to 10 mm)
A selection of files
Wire brush (large)
Jackstands (second set)
Jack (scissor or hydraulic type)

Note: *Another tool which is often useful is an electric drill with a chuck capacity of 3/8-inch and a set of good quality drill bits.*

Special tools

The tools in this list include those which are not used regularly, are expensive to buy, or which need to be used in accordance with their manufacturer's instructions. Unless these tools will be used frequently, it is not very economical to purchase many of them. A consideration would be to split the cost and use between yourself and a friend or friends. In addition, most of these tools can be obtained from a tool rental shop on a temporary basis.

This list primarily contains only those tools and instruments widely available to the public, and not those special tools produced by the vehicle manufacturer for distribution to dealer service departments. Occasionally, references to the manufacturer's special tools are included in the text of this manual. Generally, an alternative method of doing the job without the special tool is offered. However, sometimes there is no alternative to their use. Where this is the case, and the tool cannot be purchased or borrowed, the work should be turned over to the dealer service department or an automotive repair shop.

Valve spring compressor
Piston ring groove cleaning tool
Piston ring compressor
Piston ring installation tool
Cylinder compression gauge
Cylinder ridge reamer
Cylinder surfacing hone
Cylinder bore gauge
Micrometers and/or dial calipers
Hydraulic lifter removal tool
Balljoint separator
Universal-type puller
Impact screwdriver
Dial indicator set
Stroboscopic timing light (inductive
  pick-up)
Hand operated vacuum/pressure pump
Tachometer/dwell meter
Universal electrical multimeter
Cable hoist
Brake spring removal and installation
  tools
Floor jack

Buying tools

For the do-it-yourselfer who is just starting to get involved in vehicle maintenance and repair, there are a number of options available when purchasing tools. If maintenance and minor repair is the extent of the work to be done, the purchase of individual tools is satisfactory. If, on the other hand, extensive work is planned, it would be a good idea to purchase a modest tool set from one of the large retail chain stores. A set can usually be bought at a substantial savings over the individual tool prices, and they often come with a tool box. As additional tools are needed, add-on sets, individual tools and a larger tool box can be purchased to expand the tool selection. Building a tool set gradually allows the cost of the tools to be spread over a longer period of time and gives the mechanic the freedom to choose only those tools that will actually be used.

Tool stores will often be the only source of some of the special tools that are needed, but regardless of where tools are bought, try to avoid cheap ones, especially when buying screwdrivers and sockets, because they won't last very long. The expense involved in replacing cheap tools will eventually be greater than the initial cost of quality tools.

Care and maintenance of tools

Good tools are expensive, so it makes sense to treat them with respect. Keep them clean and in usable condition and store them properly when not in use. Always wipe off any dirt, grease or metal chips before putting them away. Never leave tools lying around in the work area. Upon completion of a job, always check closely under the hood for tools that may have been left there so they won't get lost during a test drive.

Some tools, such as screwdrivers, pliers, wrenches and sockets, can be hung on a panel mounted on the garage or workshop wall, while others should be kept in a tool box or tray. Measuring instruments, gauges, meters, etc. must be carefully stored where they cannot be damaged by weather or impact from other tools.

When tools are used with care and stored properly, they will last a very long time. Even with the best of care, though, tools will wear out if used frequently. When a tool is damaged or worn out, replace it. Subsequent jobs will be safer and more enjoyable if you do.

How to repair damaged threads

Sometimes, the internal threads of a nut or bolt hole can become stripped, usually from overtightening. Stripping threads is an all-too-common occurrence, especially when working with aluminum parts, because aluminum is so soft that it easily strips out.

Usually, external or internal threads are only partially stripped. After they've been cleaned up with a tap or die, they'll still work. Sometimes, however, threads are badly damaged. When this happens, you've got three choices:

1) *Drill and tap the hole to the next suitable oversize and install a larger diameter bolt, screw or stud.*
2) *Drill and tap the hole to accept a threaded plug, then drill and tap the plug to the original screw size. You can also buy a plug already threaded to the original size. Then you simply drill a hole to the specified size, then run the threaded plug into the hole with a bolt and jam nut. Once the plug is fully seated, remove the jam nut and bolt.*
3) *The third method uses a patented thread repair kit like Heli-Coil or Slimsert. These easy-to-use kits are designed to repair*

damaged threads in straight-through holes and blind holes. Both are available as kits which can handle a variety of sizes and thread patterns. Drill the hole, then tap it with the special included tap. Install the Heli-Coil and the hole is back to its original diameter and thread pitch.

Regardless of which method you use, be sure to proceed calmly and carefully. A little impatience or carelessness during one of these relatively simple procedures can ruin your whole day's work and cost you a bundle if you wreck an expensive part.

Working facilities

Not to be overlooked when discussing tools is the workshop. If anything more than routine maintenance is to be carried out, some sort of suitable work area is essential.

It is understood, and appreciated, that many home mechanics do not have a good workshop or garage available, and end up removing an engine or doing major repairs outside. It is recommended, however, that the overhaul or repair be completed under the cover of a roof.

A clean, flat workbench or table of comfortable working height is an absolute necessity. The workbench should be equipped with a vise that has a jaw opening of at least four inches.

As mentioned previously, some clean, dry storage space is also required for tools, as well as the lubricants, fluids, cleaning solvents, etc. which soon become necessary.

Sometimes waste oil and fluids, drained from the engine or cooling system during normal maintenance or repairs, present a disposal problem. To avoid pouring them on the ground or into a sewage system, pour the used fluids into large containers, seal them with caps and take them to an authorized disposal site or recycling center. Plastic jugs, such as old antifreeze containers, are ideal for this purpose.

Always keep a supply of old newspapers and clean rags available. Old towels are excellent for mopping up spills. Many mechanics use rolls of paper towels for most work because they are readily available and disposable. To help keep the area under the vehicle clean, a large cardboard box can be cut open and flattened to protect the garage or shop floor.

Whenever working over a painted surface, such as when leaning over a fender to service something under the hood, always cover it with an old blanket or bedspread to protect the finish. Vinyl covered pads, made especially for this purpose, are available at auto parts stores.

Jacking and towing

Jacking

The jack supplied with the vehicle should only be used for raising the vehicle for changing a tire or placing jackstands under the frame.

Warning: *Never crawl under the vehicle or start the engine when the jack is being used as the only means of support.*

All vehicles are supplied with a scissors-type jack. When jacking the vehicle, it should be engaged with the rocker panel flange **(see illustration)**.

The vehicle should be on level ground with the wheels blocked and the transmission in Park (automatic). Pry off the hub cap (if equipped) using the tapered end of the lug wrench. Loosen the lug nuts one-half turn and leave them in place until the wheel is raised off the ground.

Place the jack under the side of the vehicle in the indicated position. Use the supplied wrench to turn the jackscrew clockwise until the wheel is raised off the ground. Remove the lug nuts, pull off the wheel and replace it with the spare.

With the beveled side in, reinstall the lug nuts and tighten them until snug. Lower the vehicle by turning the jackscrew counterclockwise. Remove the jack and tighten the nuts in a diagonal pattern to the torque listed in the Chapter 1 Specifications. If a torque wrench is not available, have the torque checked by a service station as soon as possible. Replace the hubcap by placing it in position and using the heel of your hand or a rubber mallet to seat it.

Towing

These models can be towed with all four wheels on the ground, from the front; this is known as "Dinghy towing." The transmission must be in Neutral, and the ignition key in the ACC position.If the vehicle is being towed more than a few hours the the engine should be run for five minutes to ensure the proper lubrication of the transaxle. If the engine can't be run at least five minutes a day the vehicle should not be towed by this method.

Note: *To prevent the battery from draining, remove the BATT1 (50-amp) fuse from the underhood fuse/relay box.*

Front wheel drive (FWD) models can also be towed with the front wheels on a dolly. All-wheel drive models can't be towed with the front wheels on a dolly; all four wheels must be off the ground (unless it is dinghy towed). A sling-type tow truck cannot be used, as body damage will result. The best way to tow the vehicle is with a flat-bed car carrier.

Caution: *Don't tow a FWD vehicle with the front tires on the ground if the compact spare tire is installed on one of the front wheels - the transaxle could be damaged.*

In an emergency, the vehicle can be towed a short distance with a cable or chain attached to one of the towing eyelets located under the front or rear bumpers. The driver must remain in the vehicle to operate the steering and brakes (remember that power steering and power brakes will not work with the engine off).

7.2 The jack fits over the rocker panel flange (there are two jacking points on each side of the vehicle)

Booster battery (jump) starting

Observe these precautions when using a booster battery to start a vehicle:

a) *Before connecting the booster battery, make sure the ignition switch is in the Off position.*
b) *Turn off the lights, heater and other electrical loads.*
c) *Your eyes should be shielded. Safety goggles are a good idea.*
d) *Make sure the booster battery is the same voltage as the dead one in the vehicle.*

e) *The two vehicles MUST NOT TOUCH each other!*
f) *Make sure the transaxle is in Park (automatic).*
g) *If the booster battery is not a maintenance-free type, remove the vent caps and lay a cloth over the vent holes.*

Connect the red jumper lead between the positive (+) terminals of the two batteries **(see illustration)**.

Note: *These vehicles are equipped with a remote positive terminal located on the side of the underhood fuse/relay box, to make jumper cable connection easier* **(see illustration)**.

Connect one end of the black jumper cable to the negative (-) terminal of the booster battery, or a good grounding point, such as a bracket on the engine. The other end of this cable should be connected to a good ground on the vehicle to be started, such as a bolt or bracket on the body.

Start the engine using the booster battery, then, with the engine running at idle speed, disconnect the jumper cables in the reverse order of connection.

8.2a Make the booster battery connections in the numerical order shown (note that the negative cable of the booster battery is NOT attached to the negative terminal of the dead battery)

8.2b The remote positive terminal is located on the underhood fuse/relay box. To access it, remove the cover

Automotive chemicals and lubricants

A number of automotive chemicals and lubricants are available for use during vehicle maintenance and repair. They include a wide variety of products ranging from cleaning solvents and degreasers to lubricants and protective sprays for rubber, plastic and vinyl.

Cleaners

Carburetor cleaner and choke cleaner is a strong solvent for gum, varnish and carbon. Most carburetor cleaners leave a dry-type lubricant film which will not harden or gum up. Because of this film it is not recommended for use on electrical components.

Brake system cleaner is used to remove brake dust, grease and brake fluid from the brake system, where clean surfaces are absolutely necessary. It leaves no residue and often eliminates brake squeal caused by contaminants.

Electrical cleaner removes oxidation, corrosion and carbon deposits from electrical contacts, restoring full current flow. It can also be used to clean spark plugs, carburetor jets, voltage regulators and other parts where an oil-free surface is desired.

Demoisturants remove water and moisture from electrical components such as alternators, voltage regulators, electrical connectors and fuse blocks. They are non-conductive and non-corrosive.

Degreasers are heavy-duty solvents used to remove grease from the outside of the engine and from chassis components. They can be sprayed or brushed on and, depending on the type, are rinsed off either with water or solvent.

Lubricants

Motor oil is the lubricant formulated for use in engines. It normally contains a wide variety of additives to prevent corrosion and reduce foaming and wear. Motor oil comes in various weights (viscosity ratings) from 0 to 50. The recommended weight of the oil depends on the season, temperature and the demands on the engine. Light oil is used in cold climates and under light load conditions. Heavy oil is used in hot climates and where high loads are encountered. Multi-viscosity oils are designed to have characteristics of both light and heavy oils and are available in a number of weights from 0W-20 to 20W-50.

Gear oil is designed to be used in differentials, manual transmissions and other areas where high-temperature lubrication is required.

Chassis and wheel bearing grease is a heavy grease used where increased loads and friction are encountered, such as for wheel bearings, balljoints, tie-rod ends and universal joints.

High-temperature wheel bearing grease is designed to withstand the extreme temperatures encountered by wheel bearings in disc brake equipped vehicles. It usually contains molybdenum disulfide (moly), which is a dry-type lubricant.

White grease is a heavy grease for metal-to-metal applications where water is a problem. White grease stays soft under both low and high temperatures (usually from -100 to +190-degrees F), and will not wash off or dilute in the presence of water.

Assembly lube is a special extreme pressure lubricant, usually containing moly, used to lubricate high-load parts (such as main and rod bearings and cam lobes) for initial start-up of a new engine. The assembly lube lubricates the parts without being squeezed out or washed away until the engine oiling system begins to function.

Silicone lubricants are used to protect rubber, plastic, vinyl and nylon parts.

Graphite lubricants are used where oils cannot be used due to contamination problems, such as in locks. The dry graphite will lubricate metal parts while remaining uncontaminated by dirt, water, oil or acids. It is electrically conductive and will not foul electrical contacts in locks such as the ignition switch.

Moly penetrants loosen and lubricate frozen, rusted and corroded fasteners and prevent future rusting or freezing.

Heat-sink grease is a special electrically non-conductive grease that is used for mounting electronic ignition modules where it is essential that heat is transferred away from the module.

Sealants

RTV sealant is one of the most widely used gasket compounds. Made from silicone, RTV is air curing, it seals, bonds, waterproofs, fills surface irregularities, remains flexible, doesn't shrink, is relatively easy to remove, and is used as a supplementary sealer with almost all low and medium temperature gaskets.

Anaerobic sealant is much like RTV in that it can be used either to seal gaskets or to form gaskets by itself. It remains flexible, is solvent resistant and fills surface imperfections. The difference between an anaerobic sealant and an RTV-type sealant is in the curing. RTV cures when exposed to air, while an anaerobic sealant cures only in the absence of air. This means that an anaerobic sealant cures only after the assembly of parts, sealing them together.

Thread and pipe sealant is used for sealing hydraulic and pneumatic fittings and vacuum lines. It is usually made from a Teflon compound, and comes in a spray, a paint-on liquid and as a wrap-around tape.

Chemicals

Anti-seize compound prevents seizing, galling, cold welding, rust and corrosion in fasteners. High-temperature ant-seize, usually made with copper and graphite lubricants, is used for exhaust system and exhaust manifold bolts.

Anaerobic locking compounds are used to keep fasteners from vibrating or working loose and cure only after installation, in the absence of air. Medium strength locking compound is used for small nuts, bolts and screws that may be removed later. High-strength locking compound is for large nuts, bolts and studs which aren't removed on a regular basis.

Oil additives range from viscosity index improvers to chemical treatments that claim to reduce internal engine friction. It should be noted that most oil manufacturers caution against using additives with their oils.

Gas additives perform several functions, depending on their chemical makeup. They usually contain solvents that help dissolve gum and varnish that build up on carburetor, fuel injection and intake parts. They also serve to break down carbon deposits that form on the inside surfaces of the combustion chambers. Some additives contain upper cylinder lubricants for valves and piston rings, and others contain chemicals to remove condensation from the gas tank.

Miscellaneous

Brake fluid is specially formulated hydraulic fluid that can withstand the heat and pressure encountered in brake systems. Care must be taken so this fluid does not come in contact with painted surfaces or plastics. An opened container should always be resealed to prevent contamination by water or dirt.

Weatherstrip adhesive is used to bond weatherstripping around doors, windows and trunk lids. It is sometimes used to attach trim pieces.

Undercoating is a petroleum-based, tar-like substance that is designed to protect metal surfaces on the underside of the vehicle from corrosion. It also acts as a sound-deadening agent by insulating the bottom of the vehicle.

Waxes and polishes are used to help protect painted and plated surfaces from the weather. Different types of paint may require the use of different types of wax and polish. Some polishes utilize a chemical or abrasive cleaner to help remove the top layer of oxidized (dull) paint on older vehicles. In recent years many non-wax polishes that contain a wide variety of chemicals such as polymers and silicones have been introduced. These non-wax polishes are usually easier to apply and last longer than conventional waxes and polishes.

Conversion factors

Length (distance)
Inches (in)	X	25.4	= Millimeters (mm)	X	0.0394	= Inches (in)
Feet (ft)	X	0.305	= Meters (m)	X	3.281	= Feet (ft)
Miles	X	1.609	= Kilometers (km)	X	0.621	= Miles

Length (distance)
Inches (in)   X 25.4   = Millimeters (mm)   X 0.0394 = Inches (in)
Feet (ft)   X 0.305 = Meters (m)   X 3.281   = Feet (ft)
Miles   X 1.609 = Kilometers (km)   X 0.621   = Miles

Volume (capacity)
Cubic inches (cu in; in^3)   X 16.387 = Cubic centimeters (cc; cm^3)   X 0.061 = Cubic inches (cu in; in^3)
Imperial pints (Imp pt)   X 0.568 = Liters (l)   X 1.76   = Imperial pints (Imp pt)
Imperial quarts (Imp qt)   X 1.137 = Liters (l)   X 0.88   = Imperial quarts (Imp qt)
Imperial quarts (Imp qt)   X 1.201 = US quarts (US qt)   X 0.833 = Imperial quarts (Imp qt)
US quarts (US qt)   X 0.946 = Liters (l)   X 1.057 = US quarts (US qt)
Imperial gallons (Imp gal)   X 4.546 = Liters (l)   X 0.22   = Imperial gallons (Imp gal)
Imperial gallons (Imp gal)   X 1.201 = US gallons (US gal)   X 0.833 = Imperial gallons (Imp gal)
US gallons (US gal)   X 3.785 = Liters (l)   X 0.264 = US gallons (US gal)

Mass (weight)
Ounces (oz)   X 28.35 = Grams (g)   X 0.035 = Ounces (oz)
Pounds (lb)   X 0.454 = Kilograms (kg)   X 2.205 = Pounds (lb)

Force
Ounces-force (ozf; oz)   X 0.278 = Newtons (N)   X 3.6   = Ounces-force (ozf; oz)
Pounds-force (lbf; lb)   X 4.448 = Newtons (N)   X 0.225 = Pounds-force (lbf; lb)
Newtons (N)   X 0.1   = Kilograms-force (kgf; kg)   X 9.81   = Newtons (N)

Pressure
Pounds-force per square inch (psi; lbf/in^2; lb/in^2)   X 0.070 = Kilograms-force per square centimeter (kgf/cm^2; kg/cm^2)   X 14.223 = Pounds-force per square inch (psi; lbf/in^2; lb/in^2)
Pounds-force per square inch (psi; lbf/in^2; lb/in^2)   X 0.068 = Atmospheres (atm)   X 14.696 = Pounds-force per square inch (psi; lbf/in^2; lb/in^2)
Pounds-force per square inch (psi; lbf/in^2; lb/in^2)   X 0.069 = Bars   X 14.5 = Pounds-force per square inch (psi; lbf/in^2; lb/in^2)
Pounds-force per square inch (psi; lbf/in^2; lb/in^2)   X 6.895 = Kilopascals (kPa)   X 0.145 = Pounds-force per square inch (psi; lbf/in^2; lb/in^2)
Kilopascals (kPa)   X 0.01 = Kilograms-force per square centimeter (kgf/cm^2; kg/cm^2)   X 98.1 = Kilopascals (kPa)

Torque (moment of force)
Pounds-force inches (lbf in; lb in)   X 1.152 = Kilograms-force centimeter (kgf cm; kg cm)   X 0.868 = Pounds-force inches (lbf in; lb in)
Pounds-force inches (lbf in; lb in)   X 0.113 = Newton meters (Nm)   X 8.85 = Pounds-force inches (lbf in; lb in)
Pounds-force inches (lbf in; lb in)   X 0.083 = Pounds-force feet (lbf ft; lb ft)   X 12 = Pounds-force inches (lbf in; lb in)
Pounds-force feet (lbf ft; lb ft)   X 0.138 = Kilograms-force meters (kgf m; kg m)   X 7.233 = Pounds-force feet (lbf ft; lb ft)
Pounds-force feet (lbf ft; lb ft)   X 1.356 = Newton meters (Nm)   X 0.738 = Pounds-force feet (lbf ft; lb ft)
Newton meters (Nm)   X 0.102 = Kilograms-force meters (kgf m; kg m)   X 9.804 = Newton meters (Nm)

Vacuum
Inches mercury (in. Hg)   X 3.377 = Kilopascals (kPa)   X 0.2961 = Inches mercury
Inches mercury (in. Hg)   X 25.4 = Millimeters mercury (mm Hg)   X 0.0394 = Inches mercury

Power
Horsepower (hp)   X 745.7 = Watts (W)   X 0.0013 = Horsepower (hp)

Velocity (speed)
Miles per hour (miles/hr; mph)   X 1.609 = Kilometers per hour (km/hr; kph)   X 0.621 = Miles per hour (miles/hr; mph)

Fuel consumption*
Miles per gallon, Imperial (mpg)   X 0.354 = Kilometers per liter (km/l)   X 2.825 = Miles per gallon, Imperial (mpg)
Miles per gallon, US (mpg)   X 0.425 = Kilometers per liter (km/l)   X 2.352 = Miles per gallon, US (mpg)

Temperature
Degrees Fahrenheit = (°C x 1.8) + 32

Degrees Celsius (Degrees Centigrade; °C) = (°F - 32) x 0.56

*It is common practice to convert from miles per gallon (mpg) to liters/100 kilometers (l/100km), where mpg (Imperial) x l/100 km = 282 and mpg (US) x l/100 km = 235

DECIMALS to MILLIMETERS

Decimal	mm	Decimal	mm
0.001	0.0254	0.500	12.7000
0.002	0.0508	0.510	12.9540
0.003	0.0762	0.520	13.2080
0.004	0.1016	0.530	13.4620
0.005	0.1270	0.540	13.7160
0.006	0.1524	0.550	13.9700
0.007	0.1778	0.560	14.2240
0.008	0.2032	0.570	14.4780
0.009	0.2286	0.580	14.7320
		0.590	14.9860
0.010	0.2540		
0.020	0.5080		
0.030	0.7620		
0.040	1.0160	0.600	15.2400
0.050	1.2700	0.610	15.4940
0.060	1.5240	0.620	15.7480
0.070	1.7780	0.630	16.0020
0.080	2.0320	0.640	16.2560
0.090	2.2860	0.650	16.5100
		0.660	16.7640
0.100	2.5400	0.670	17.0180
0.110	2.7940	0.680	17.2720
0.120	3.0480	0.690	17.5260
0.130	3.3020		
0.140	3.5560		
0.150	3.8100	0.700	17.7800
0.160	4.0640	0.710	18.0340
0.170	4.3180	0.720	18.2880
0.180	4.5720	0.730	18.5420
0.190	4.8260	0.740	18.7960
		0.750	19.0500
0.200	5.0800	0.760	19.3040
0.210	5.3340	0.770	19.5580
0.220	5.5880	0.780	19.8120
0.230	5.8420	0.790	20.0660
0.240	6.0960		
0.250	6.3500		
0.260	6.6040	0.800	20.3200
0.270	6.8580	0.810	20.5740
0.280	7.1120	0.820	21.8280
0.290	7.3660	0.830	21.0820
		0.840	21.3360
0.300	7.6200	0.850	21.5900
0.310	7.8740	0.860	21.8440
0.320	8.1280	0.870	22.0980
0.330	8.3820	0.880	22.3520
0.340	8.6360	0.890	22.6060
0.350	8.8900		
0.360	9.1440		
0.370	9.3980		
0.380	9.6520		
0.390	9.9060	0.900	22.8600
0.400	10.1600	0.910	23.1140
0.410	10.4140	0.920	23.3680
0.420	10.6680	0.930	23.6220
0.430	10.9220	0.940	23.8760
0.440	11.1760	0.950	24.1300
0.450	11.4300	0.960	24.3840
0.460	11.6840	0.970	24.6380
0.470	11.9380	0.980	24.8920
0.480	12.1920	0.990	25.1460
0.490	12.4460	1.000	25.4000

FRACTIONS to DECIMALS to MILLIMETERS

Fraction	Decimal	mm	Fraction	Decimal	mm
1/64	0.0156	0.3969	33/64	0.5156	13.0969
1/32	0.0312	0.7938	17/32	0.5312	13.4938
3/64	0.0469	1.1906	35/64	0.5469	13.8906
1/16	0.0625	1.5875	9/16	0.5625	14.2875
5/64	0.0781	1.9844	37/64	0.5781	14.6844
3/32	0.0938	2.3812	19/32	0.5938	15.0812
7/64	0.1094	2.7781	39/64	0.6094	15.4781
1/8	0.1250	3.1750	5/8	0.6250	15.8750
9/64	0.1406	3.5719	41/64	0.6406	16.2719
5/32	0.1562	3.9688	21/32	0.6562	16.6688
11/64	0.1719	4.3656	43/64	0.6719	17.0656
3/16	0.1875	4.7625	11/16	0.6875	17.4625
13/64	0.2031	5.1594	45/64	0.7031	17.8594
7/32	0.2188	5.5562	23/32	0.7188	18.2562
15/64	0.2344	5.9531	47/64	0.7344	18.6531
1/4	0.2500	6.3500	3/4	0.7500	19.0500
17/64	0.2656	6.7469	49/64	0.7656	19.4469
9/32	0.2812	7.1438	25/32	0.7812	19.8438
19/64	0.2969	7.5406	51/64	0.7969	20.2406
5/16	0.3125	7.9375	13/16	0.8125	20.6375
21/64	0.3281	8.3344	53/64	0.8281	21.0344
11/32	0.3438	8.7312	27/32	0.8438	21.4312
23/64	0.3594	9.1281	55/64	0.8594	21.8281
3/8	0.3750	9.5250	7/8	0.8750	22.2250
25/64	0.3906	9.9219	57/64	0.8906	22.6219
13/32	0.4062	10.3188	29/32	0.9062	23.0188
27/64	0.4219	10.7156	59/64	0.9219	23.4156
7/16	0.4375	11.1125	15/16	0.9375	23.8125
29/64	0.4531	11.5094	61/64	0.9531	24.2094
15/32	0.4688	11.9062	31/32	0.9688	24.6062
31/64	0.4844	12.3031	63/64	0.9844	25.0031
1/2	0.5000	12.7000	1	1.0000	25.4000

Safety first!

Regardless of how enthusiastic you may be about getting on with the job at hand, take the time to ensure that your safety is not jeopardized. A moment's lack of attention can result in an accident, as can failure to observe certain simple safety precautions. The possibility of an accident will always exist, and the following points should not be considered a comprehensive list of all dangers. Rather, they are intended to make you aware of the risks and to encourage a safety conscious approach to all work you carry out on your vehicle.

Essential DOs and DON'Ts

DON'T rely on a jack when working under the vehicle. Always use approved jackstands to support the weight of the vehicle and place them under the recommended lift or support points.

DON'T attempt to loosen extremely tight fasteners (i.e. wheel lug nuts) while the vehicle is on a jack - it may fall.

DON'T start the engine without first making sure that the transmission is in Neutral (or Park where applicable) and the parking brake is set.

DON'T remove the radiator cap from a hot cooling system - let it cool or cover it with a cloth and release the pressure gradually.

DON'T attempt to drain the engine oil until you are sure it has cooled to the point that it will not burn you.

DON'T touch any part of the engine or exhaust system until it has cooled sufficiently to avoid burns.

DON'T siphon toxic liquids such as gasoline, antifreeze and brake fluid by mouth, or allow them to remain on your skin.

DON'T inhale brake lining dust - it is potentially hazardous (see *Asbestos* below).

DON'T allow spilled oil or grease to remain on the floor - wipe it up before someone slips on it.

DON'T use loose fitting wrenches or other tools which may slip and cause injury.

DON'T push on wrenches when loosening or tightening nuts or bolts. Always try to pull the wrench toward you. If the situation calls for pushing the wrench away, push with an open hand to avoid scraped knuckles if the wrench should slip.

DON'T attempt to lift a heavy component alone - get someone to help you.

DON'T *rush or take unsafe shortcuts to finish a job.*

DON'T allow children or animals in or around the vehicle while you are working on it.

DO wear eye protection when using power tools such as a drill, sander, bench grinder, etc. and when working under a vehicle.

DO keep loose clothing and long hair well out of the way of moving parts.

DO make sure that any hoist used has a safe working load rating adequate for the job.

DO get someone to check on you periodically when working alone on a vehicle.

DO carry out work in a logical sequence and make sure that everything is correctly assembled and tightened.

DO keep chemicals and fluids tightly capped and out of the reach of children and pets.

DO remember that your vehicle's safety affects that of yourself and others. If in doubt on any point, get professional advice.

Steering, suspension and brakes

These systems are essential to driving safety, so make sure you have a qualified shop or individual check your work. Also, compressed suspension springs can cause injury if released suddenly - be sure to use a spring compressor.

Airbags

Airbags are explosive devices that can **CAUSE** injury if they deploy while you're working on the vehicle. Follow the manufacturer's instructions to disable the airbag whenever you're working in the vicinity of airbag components.

Asbestos

Certain friction, insulating, sealing, and other products - such as brake linings, brake bands, clutch linings, torque converters, gaskets, etc. - may contain asbestos or other hazardous friction material. Extreme care must be taken to avoid inhalation of dust from such products, since it is hazardous to health. If in doubt, assume that they do contain asbestos.

Fire

Remember at all times that gasoline is highly flammable. Never smoke or have any kind of open flame around when working on a vehicle. But the risk does not end there. A spark caused by an electrical short circuit, by two metal surfaces contacting each other, or even by static electricity built up in your body under certain conditions, can ignite gasoline vapors, which in a confined space are highly explosive. Do not, under any circumstances, use gasoline for cleaning parts. Use an approved safety solvent.

Always disconnect the battery ground (-) cable at the battery before working on any part of the fuel system or electrical system. Never risk spilling fuel on a hot engine or exhaust component. It is strongly recommended that a fire extinguisher suitable for use on fuel and electrical fires be kept handy in the garage or workshop at all times. Never try to extinguish a fuel or electrical fire with water.

Fumes

Certain fumes are highly toxic and can quickly cause unconsciousness and even death if inhaled to any extent. Gasoline vapor falls into this category, as do the vapors from some cleaning solvents. Any draining or pouring of such volatile fluids should be done in a well ventilated area.

When using cleaning fluids and solvents, read the instructions on the container carefully. Never use materials from unmarked containers.

Never run the engine in an enclosed space, such as a garage. Exhaust fumes contain carbon monoxide, which is extremely poisonous. If you need to run the engine, always do so in the open air, or at least have the rear of the vehicle outside the work area.

The battery

Never create a spark or allow a bare light bulb near a battery. They normally give off a certain amount of hydrogen gas, which is highly explosive.

Always disconnect the battery ground (-) cable at the battery before working on the fuel or electrical systems.

If possible, loosen the filler caps or cover when charging the battery from an external source (this does not apply to sealed or maintenance-free batteries). Do not charge at an excessive rate or the battery may burst.

Take care when adding water to a non maintenance-free battery and when carrying a battery. The electrolyte, even when diluted, is very corrosive and should not be allowed to contact clothing or skin.

Always wear eye protection when cleaning the battery to prevent the caustic deposits from entering your eyes.

Household current

When using an electric power tool, inspection light, etc., which operates on household current, always make sure that the tool is correctly connected to its plug and that, where necessary, it is properly grounded. Do not use such items in damp conditions and, again, do not create a spark or apply excessive heat in the vicinity of fuel or fuel vapor.

Secondary ignition system voltage

A severe electric shock can result from touching certain parts of the ignition system (such as the spark plug wires) when the engine is running or being cranked, particularly if components are damp or the insulation is defective. In the case of an electronic ignition system, the secondary system voltage is much higher and could prove fatal.

Hydrofluoric acid

This extremely corrosive acid is formed when certain types of synthetic rubber, found in some O-rings, oil seals, fuel hoses, etc. are exposed to temperatures above 750-degrees F (400-degrees C). The rubber changes into a charred or sticky substance containing the acid. *Once formed, the acid remains dangerous for years. If it gets onto the skin, it may be necessary to amputate the limb concerned.*

When dealing with a vehicle which has suffered a fire, or with components salvaged from such a vehicle, wear protective gloves and discard them after use.

Troubleshooting

Contents

Symptom	Section

Engine and performance
Engine will not rotate when attempting to start............................ 1
Engine rotates but will not start 2
Engine hard to start when cold 3
Engine hard to start when hot.. 4
Starter motor noisy or excessively rough in engagement............. 5
Engine starts but stops immediately.. 6
Oil puddle under engine 7
Engine lopes while idling or idles erratically 8
Engine misses at idle speed.. 9
Engine misses throughout driving speed range 10
Engine stumbles on acceleration.. 11
Engine surges while holding accelerator steady 12
Engine stalls 13
Engine lacks power 14
Engine backfires 15
Pinging or knocking engine sounds during acceleration or uphill... 16
Engine runs with oil pressure light on...................................... 17
Engine continues to run after switching off............................... 18

Engine electrical system
Battery will not hold a charge 19
Voltage warning light fails to go out 20
Voltage warning light fails to come on when key is turned on 21

Fuel system
Excessive fuel consumption 22
Fuel leakage and/or fuel odor... 23

Cooling system
Overheating.. 24
Overcooling 25
External coolant leakage 26
Internal coolant leakage 27
Coolant loss... 28
Poor coolant circulation 29

Automatic transaxle
Fluid leakage 30
Transaxle fluid brown or has a burned smell............................. 31
General shift mechanism problems... 32
Engine will start in gears other than Park or Neutral 33
Transaxle slips, shifts roughly, is noisy or has no drive in
    forward or reverse gears 34

Driveaxles
Clicking noise in turns... 35
Knock or clunk when accelerating after coasting 36
Shudder or vibration during acceleration.................................. 37

Driveshaft
Leaks at front of driveshaft 38
Knock or clunk when transmission is under initial load
    (just after transmission is put into gear) 39
Metallic grating sound consistent with vehicle speed.................. 40
Vibration 41
Scraping noise.. 42

Rear differential
Noise - same when in drive as when vehicle is coasting.............. 43
Knocking sound when starting or shifting gears......................... 44
Noise when turning.. 45
Vibration 46
Oil leaks.. 47

Brakes
Vehicle pulls to one side during braking 48
Noise (grinding or high-pitched squeal) when the
    brakes are applied... 49
Brake roughness or chatter (pedal pulsates)............................. 50
Excessive pedal effort required to stop vehicle 51
Excessive brake pedal travel.. 52
Dragging brakes 53
Grabbing or uneven braking action 54
Brake pedal feels spongy when depressed............................... 55
Brake pedal travels to the floor with little resistance................... 56
Parking brake does not hold... 57

Suspension and steering systems
Vehicle pulls to one side 58
Abnormal or excessive tire wear 59
Wheel makes a thumping noise 60
Shimmy, shake or vibration 61
Hard steering 62
Steering wheel does not return to center position correctly.......... 63
Abnormal noise at the front end 64
Wander or poor steering stability 65
Erratic steering when braking... 66
Excessive pitching and/or rolling around corners or
    during braking 67
Suspension bottoms 68
Cupped tires 69
Excessive tire wear on outside edge 70
Excessive tire wear on inside edge 71
Tire tread worn in one place 72
Excessive play or looseness in steering system 73
Rattling or clicking noise in steering gear 74

This section provides an easy reference guide to the more common problems which may occur during the operation of your vehicle. Various symptoms and their possible causes are grouped under headings denoting components or systems, such as Engine, Cooling system, etc. They also refer to the Chapter and/or Section that deals with the problem.

Remember that successful troubleshooting isn't a mysterious art practiced only by professional mechanics. It's simply the result of knowledge combined with an intelligent, systematic approach to a problem. Always use a process of elimination, starting with the simplest solution and working through to the most complex - and never overlook the obvious. Anyone can run the gas tank dry or leave the lights on overnight, so don't assume that you're exempt from such oversights.

Finally, always establish a clear idea why a problem has occurred and take steps to ensure that it doesn't happen again. If the electrical system fails because of a poor connection, check all other connections in the system to make sure they don't fail as well. If a particular fuse continues to blow, find out why - don't just go on replacing fuses. Remember, failure of a small component can often be indicative of potential failure or incorrect functioning of a more important component or system.

Engine and performance

1 Engine will not rotate when attempting to start

1   Battery terminal connections loose or corroded (Chapter 1).
2   Battery discharged or faulty (Chapter 1).
3   Automatic transaxle not completely engaged in Park (Chapter 7A).
4   Broken, loose or disconnected wiring in the starting circuit (Chapters 5 and 12).
5   Starter motor pinion jammed in flywheel ring gear (Chapter 5).
6   Starter solenoid faulty (Chapter 5).
7   Starter motor faulty (Chapter 5).
8   Ignition switch faulty (Chapter 12).
9   Transmission range switch faulty (Chapter 6).
10   Starter pinion or driveplate teeth worn or broken (Chapter 5).
11   Body Control Module (BCM) or Powertrain Control Module (PCM) faulty (Chapter 6).

2 Engine rotates but will not start

1   Fuel tank empty.
2   Battery discharged (engine rotates slowly) (Chapter 1).
3   Battery terminal connections loose or corroded (Chapter 1).

4   Fuel not reaching fuel injection system (Chapter 4).
5   Ignition system problem (Chapter 5).
6   Worn, faulty or incorrectly gapped spark plugs (Chapter 1).

3 Engine hard to start when cold

1   Battery discharged or low (Chapter 1).
2   Fuel system malfunctioning (Chapter 4).
3   Emissions or engine control system malfunctioning (Chapter 6).

4 Engine hard to start when hot

1   Air filter clogged (Chapter 1).
2   Fuel not reaching the fuel injection system (Chapter 4).
3   Corroded battery connections, especially ground (Chapter 1).
4   Emissions or engine control system malfunctioning (Chapter 6).

5 Starter motor noisy or excessively rough in engagement

1   Pinion or driveplate gear teeth worn or broken (Chapter 5).
2   Starter motor mounting bolts loose or missing (Chapter 5).

6 Engine starts but stops immediately

1   Loose or faulty electrical connections at coil pack or alternator (Chapter 5).
2   Insufficient fuel reaching the fuel injectors (Chapter 4).
3   Vacuum leak at the gasket between the intake manifold/plenum and throttle body (Chapters 1 and 4).
4   Restricted exhaust system (most likely the catalytic converter) (Chapters 4 and 6).

7 Oil puddle under engine

1   Oil pan gasket and/or oil pan drain bolt seal leaking (Chapters 1 and 2A).
2   Oil pressure sending unit leaking (Chapter 2B).
3   Valve cover gaskets leaking (Chapter 2A).
4   Engine oil seals leaking (Chapter 2A).

8 Engine lopes while idling or idles erratically

1   Vacuum leakage (Chapter 2B).
2   Leaking EGR valve or plugged PCV system (Chapter 6).
3   Air filter clogged (Chapter 1).

4   Fuel pump not delivering sufficient fuel to the fuel injection system (Chapter 4).
5   Leaking head gasket (Chapter 2A).
6   Camshaft lobes worn (Chapter 2A).

9 Engine misses at idle speed

1   Spark plugs worn or not gapped properly (Chapter 1).
2   Faulty ignition coil(s) (Chapter 5).
3   Vacuum leaks (Chapters 1 and 4).
4   Uneven or low compression (Chapter 2B).

10 Engine misses throughout driving speed range

1   Fuel filter clogged and/or impurities in the fuel system (Chapters 1 and 4).
2   Low fuel output at the injector (Chapter 4).
3   Faulty or incorrectly gapped spark plugs (Chapter 1).
4   Faulty emission system components (Chapter 6).
5   Low or uneven cylinder compression pressures (Chapter 2B).
6   Weak or faulty ignition coil(s) (Chapter 5).
7   Vacuum leak in fuel injection system, intake manifold or vacuum hoses (Chapter 4).

11 Engine stumbles on acceleration

1   Spark plugs worn (Chapter 1).
2   Fuel injection system problem (Chapter 4).
3   Fuel filter clogged (Chapter 1).
4   Intake manifold air leak (Chapter).

12 Engine surges while holding accelerator steady

1   Intake air leak (Chapter 4).
2   Fuel pump faulty (Chapter 4).
3   Defective PCM (Chapter 6).

13 Engine stalls

1   Fuel filter clogged and/or water and impurities in the fuel system (Chapters 1 and 4).
2   Ignition system problem (Chapter 5).
3   Faulty emissions system components (Chapter 6).
4   Vacuum leak in the intake manifold or vacuum hoses (Chapter 4).

14 Engine lacks power

1   Faulty or incorrectly gapped spark plugs (Chapter 1).

2 Restricted exhaust system (most likely the catalytic converter) (Chapters 4 and 6).
3 Fuel injection system malfunctioning (Chapter 4).
4 Faulty coil(s) (Chapter 5).
5 Brakes binding (Chapter 9).
6 Automatic transaxle fluid level incorrect (Chapter 1).
7 Fuel filter clogged and/or impurities in the fuel system (Chapter 1).
8 Emission control system not functioning properly (Chapter 6).
9 Low or uneven cylinder compression pressures (Chapter 2B).

15 Engine backfires

1 Emissions system not functioning properly (Chapter 6).
2 Fuel injection system malfunctioning (Chapter 4).
3 Vacuum leak at fuel injectors, intake manifold or vacuum hoses (Chapter 4).
4 Valves worn or sticking (Chapter 2B).

16 Pinging or knocking engine sounds during acceleration or uphill

1 Incorrect grade of fuel.
2 Engine control system malfunctioning (Chapter 6).
3 Improper or damaged spark plugs (Chapter 1).
4 Faulty emissions system (Chapter 6).
5 Vacuum leak (Chapter 4).

17 Engine runs with oil pressure light on

1 Low oil level (Chapter 1).
2 Faulty oil pressure sender (Chapter 2B).
3 Worn engine bearings and/or oil pump (Chapter 2A).

18 Engine continues to run after switching off

Faulty ignition switch (Chapter 12).

Engine electrical system

19 Battery will not hold a charge

1 Drivebelt defective (Chapter 1).
2 Battery terminals loose or corroded (Chapter 1).
3 Alternator not charging properly (Chapter 5).

4 Loose, broken or faulty wiring in the charging circuit (Chapter 5).
5 Internally defective battery (Chapters 1 and 5).

20 Voltage warning light fails to go out

1 Faulty alternator or charging circuit (Chapter 5).
2 Alternator drivebelt defective or out of adjustment (Chapter 1).
3 Alternator voltage regulator inoperative (Chapter 5).

21 Voltage warning light fails to come on when key is turned on

1 Warning light bulb defective (Chapter 12).
2 Fault in the printed circuit, dash wiring or bulb holder (Chapter 12).

Fuel system

22 Excessive fuel consumption

1 Dirty or clogged air filter element (Chapter 1).
2 Emissions system not functioning properly (Chapter 6).
3 Fuel injection system malfunctioning (Chapter 4).
4 Low tire pressure or incorrect tire size (Chapter 1).

23 Fuel leakage and/or fuel odor

1 Leak in a fuel feed or vent line (Chapter 4).
2 Tank overfilled.
3 Evaporative emissions control canister defective (Chapters 1 and 6).
4 Fuel injector seals faulty (Chapter 4).

Cooling system

24 Overheating

1 Insufficient coolant in system (Chapter 1).
2 Drivebelt defective (Chapter 1).
3 Radiator core blocked or grille restricted (Chapter 3).
4 Thermostat faulty (Chapter 3).
5 Electric cooling fan not operating (Chapter 3).
6 Expansion tank cap not maintaining proper pressure (Chapter 3).
7 Blown head gasket (Chapter 2A).

25 Overcooling

Incorrect (opening temperature too low) or faulty thermostat (Chapter 3).

26 External coolant leakage

1 Deteriorated/damaged hoses or loose clamps (Chapters 1 and 3).
2 Water pump seal defective (Chapters 1 and 3).
3 Leakage from radiator core (Chapter 3).
4 Engine drain or water jacket core plugs leaking (Chapter).

27 Internal coolant leakage

1 Leaking cylinder head gasket (Chapter 2A).
2 Cracked cylinder bore or cylinder head (Chapter 2A).

28 Coolant loss

1 Too much coolant in system (Chapter 1).
2 Coolant boiling away because of overheating (Chapter 3).
3 Internal or external leakage (Chapter 3).
4 Faulty expansion tank cap (Chapter 3).

29 Poor coolant circulation

1 Inoperative water pump (Chapter 3).
2 Restriction in cooling system (Chapters 1 and 3).
3 Water pump drivebelt defective or out of adjustment (Chapter 1).
4 Thermostat sticking (Chapter 3).

Automatic transaxle

30 Fluid leakage

1 Automatic transmission fluid is a deep red color. Fluid leaks should not be confused with engine oil, which can easily be blown by airflow to the transaxle.
2 To pinpoint a leak, first remove all built-up dirt and grime from the transaxle housing with degreasing agents and/or steam cleaning. Drive the vehicle at low speeds so air flow will not blow the leak far from its source. Raise the vehicle and determine where the leak is coming from. Common areas of leakage are:

   a) *Fluid pan*
   b) *Fill plug (Chapter 1)*
   c) *Fluid cooler lines (Chapter 7A)*
   d) *Vehicle Speed Sensor (Chapter 6)*

31 Transaxle fluid brown or has a burned smell

Transaxle overheated. Change fluid (Chapter 1).

32 General shift mechanism problems

1 Chapter 7A deals with checking and adjusting the shift cable on automatic transaxles. Common problems which may be attributed to a poorly adjusted cable are:

a) *Engine starting in gears other than Park or Neutral.*
b) *Indicator on shifter pointing to a gear other than the one actually being used.*
c) *Vehicle moves when in Park.*

2 Refer to Chapter 7A for the shift cable adjustment procedure.

33 Engine will start in gears other than Park or Neutral

Transmission range switch malfunctioning (Chapter 6).

34 Transaxle slips, shifts roughly, is noisy or has no drive in forward or reverse gears

1 There are many probable causes for the above problems, but the home mechanic should be concerned with only one possibility - fluid level. Before taking the vehicle to a repair shop, check the level and condition of the fluid as described in Chapter 1.
2 Correct the fluid level as necessary or change the fluid and filter if needed. If the problem persists, have a professional diagnose the probable cause.

Driveaxles

35 Clicking noise in turns

Worn or damaged outer CV joint. Check for cut or damaged boots (Chapter 1). Repair as necessary (Chapter 8).

36 Knock or clunk when accelerating after coasting

Worn or damaged CV joint. Check for cut or damaged boots (Chapter 1). Repair as necessary (Chapter 8).

37 Shudder or vibration during acceleration

1 Worn or damaged CV joints. Repair or replace as necessary (Chapter 8).
2 Sticking inner joint assembly. Correct or replace as necessary (Chapter 8).

Driveshaft

38 Leaks at front of driveshaft

Defective transfer case seal.

39 Knock or clunk when transmission is under initial load (just after transmission is put into gear)

1 Loose or disconnected rear suspension components. Check all mounting bolts and bushings (Chapter 10).
2 Loose driveshaft bolts. Inspect all bolts and nuts and tighten them securely.
3 Worn or damaged universal joint bearings.
4 Worn sleeve yoke and mainshaft spline.

40 Metallic grating sound consistent with vehicle speed

1 Pronounced wear in the universal joint bearings. Replace driveshaft (Chapter 8).
2 Worn center support bearing. Replace driveshaft (Chapter 8).

41 Vibration

1 Before blaming the driveshaft, make sure the tires are perfectly balanced and perform the following test. Refer to Chapter 8 for more information.
2 Install a tachometer inside the vehicle to monitor engine speed as the vehicle is driven. Drive the vehicle and note the engine speed at which the vibration (roughness) is most pronounced. Now shift the transmission to a different gear and bring the engine speed to the same point.
3 If the vibration occurs at the same engine speed (rpm) regardless of which gear the transmission is in, the driveshaft is NOT at fault since the driveshaft speed varies.
4 If the vibration decreases or is eliminated when the transmission is in a different gear at the same engine speed, refer to the following probable causes:

a) *Bent or dented driveshaft. Inspect and replace as necessary.*

b) *Undercoating or built-up dirt, etc. on the driveshaft. Clean the shaft thoroughly.*
c) *Worn universal joint bearings. Replace the driveshaft.*
d) *Driveshaft and/or companion flange out of balance. Check for missing weights on the shaft. Remove driveshaft and reinstall 180-degrees from original position, then recheck. Have the driveshaft balanced if problem persists.*
e) *Loose driveshaft mounting bolts/nuts.*
f) *Worn center support bearing. Replace the driveshaft.*
g) *Worn transfer case rear bushing.*

42 Scraping noise

Make sure there is nothing, such as an exhaust heat shield or safety loop, rubbing on the driveshaft.

Rear differential

43 Noise - same when in drive as when vehicle is coasting

1 Road noise. No corrective action available.
2 Tire noise. Inspect tires and check tire pressures (Chapter 1).
3 Hub bearings worn or damaged (Chapter 8).
4 Insufficient differential oil (Chapter 1).
5 Defective differential.

44 Knocking sound when starting or shifting gears

Worn differential.

45 Noise when turning

Worn differential.

46 Vibration

See probable causes under *Driveshaft*. Proceed under the guidelines listed for the driveshaft. If the problem persists, check the rear hub bearings by raising the rear of the vehicle and spinning the wheels by hand. Listen for evidence of rough (noisy) bearings. Remove and inspect.

47 Oil leaks

1 Pinion oil seal damaged.
2 Driveaxle oil seals damaged.

3   Loose filler or drain plug on differential (Chapter 1).

4   Clogged or damaged breather on differential.

Brakes

48   Vehicle pulls to one side during braking

1   Incorrect tire pressures (Chapter 1).

2   Front end out of alignment (have the front end aligned).

3   Unmatched tires on same axle.

4   Restricted brake lines or hoses (Chapter 1).

5   Sticking caliper piston (Chapter 9).

6   Loose suspension parts (Chapter 10).

7   Contaminated brake pad material (Chapter 9).

49   Noise (grinding or high-pitched squeal) when the brakes are applied

Brake pads worn out. Replace the pads with new ones immediately (Chapter 9).

50   Brake roughness or chatter (pedal pulsates)

1   Excessive brake disc lateral runout (Chapter 9).

2   Parallelism of disc not within specifications (Chapter 9).

3   Uneven pad wear caused by caliper not sliding due to improper clearance or dirt (Chapter 9).

4   Defective brake disc (Chapter 9).

51   Excessive pedal effort required to stop vehicle

1   Malfunctioning power brake booster (Chapter 9).

2   Partial system failure (Chapter 9).

3   Excessively worn pads (Chapter 9).

4   One or more caliper pistons seized or sticking (Chapter 9).

5   Brake pads contaminated with oil or grease (Chapter 9).

6   New pads installed and not yet seated. It will take a while for the new material to seat.

52   Excessive brake pedal travel

1   Partial brake system failure (Chapter 9).

2   Insufficient fluid in master cylinder (Chapters 1 and 9).

3   Air trapped in system (Chapter 9).

4   Faulty master cylinder (Chapter 9).

53   Dragging brakes

1   Master cylinder pistons not returning correctly (Chapter 9).

2   Restricted brake lines or hoses (Chapters 1 and 9).

3   Incorrect parking brake adjustment (Chapter 9).

4   Defective brake calipers (Chapter 9).

54   Grabbing or uneven braking action

1   Malfunction of proportioning valve (Chapter).

2   Malfunction of power brake booster unit (Chapter 9).

3   Binding brake pedal mechanism (Chapter 9).

4   Contaminated brake linings (Chapter 9).

55   Brake pedal feels spongy when depressed

1   Air in hydraulic lines (Chapter 9).

2   Master cylinder mounting bolts loose (Chapter 9).

3   Master cylinder defective (Chapter 9).

56   Brake pedal travels to the floor with little resistance

Little or no fluid in the master cylinder reservoir caused by leaking caliper, or loose, damaged or disconnected brake lines (Chapter 9).

57   Parking brake does not hold

Parking brake cables improperly adjusted (Chapter 9).

Suspension and steering systems

58   Vehicle pulls to one side

1   Mismatched or uneven tires.

2   Broken or sagging springs (Chapter 10).

3   Wheel alignment incorrect (Chapter 10).

4   Front brakes dragging (Chapter 9).

59   Abnormal or excessive tire wear

1   Front wheel alignment incorrect (Chapter 10).

2   Sagging or broken springs (Chapter 10).

3   Tire out-of-balance (Chapter 10).

4   Worn strut or shock absorber (Chapter 10).

5   Overloaded vehicle.

6   Tires not rotated regularly.

60   Wheel makes a thumping noise

1   Blister or bump on tire (Chapter 1).

2   Improper strut or shock absorber action (Chapter 10).

61   Shimmy, shake or vibration

1   Tire or wheel out-of-balance or out-of-round (Chapter 10).

2   Loose or worn wheel bearings (Chapter 10).

3   Worn tie-rod ends (Chapter 10).

4   Worn balljoints (Chapter 10).

5   Excessive wheel runout (Chapter 10).

6   Blister or bump on tire (Chapter 1).

62   Hard steering

1   Lack of lubrication at balljoints, tie-rod ends and steering gear assembly (Chapter 10).

2   Front wheel alignment incorrect (Chapter 10).

3   Low tire pressure (Chapter 1).

63   Steering wheel does not return to center position correctly

1   Lack of lubrication at balljoints and tie-rod ends (Chapter 10).

2   Binding in steering column (Chapter 10).

3   Defective rack-and-pinion assembly (Chapter 10).

4   Front wheel alignment problem (Chapter 10).

64   Abnormal noise at the front end

1   Worn balljoints and tie-rod ends (Chapter 10).

2   Loose upper strut mount (Chapter 10).

3   Worn tie-rod ends (Chapter 10).

4   Loose stabilizer bar (Chapter 10).

5   Loose wheel lug nuts (Chapter 1).

6   Loose suspension bolts (Chapter 10).

65   Wander or poor steering stability

1   Mismatched or uneven tires (Chapter 10).

2   Worn balljoints or tie-rod ends (Chapters 1 and 10).

3   Broken or sagging springs (Chapter 10).

4   Front wheel alignment incorrect.

5   Worn steering gear clamp bushing (Chapter 10).

6   Wheel bearings worn (Chapter 10).

66 Erratic steering when braking

1   Wheel bearings worn (Chapter 9).
2   Broken or sagging springs (Chapter 10).
3   Leaking caliper (Chapter 9).
4   Warped brake discs (Chapter 9).
5   Worn steering gear clamp bushing (Chapter 10).
6   Wheel alignment incorrect.

67 Excessive pitching and/or rolling around corners or during braking

1   Loose stabilizer bar (Chapter 10).
2   Worn struts/shock absorbers or mounts (Chapter 10).
3   Broken or sagging springs (Chapter 10).
4   Overloaded vehicle.

68 Suspension bottoms

1   Overloaded vehicle.
2   Worn struts or shock absorbers (Chapter 10).
3   Incorrect, broken or sagging springs (Chapter 10).

69 Cupped tires

1   Front wheel alignment incorrect (Chapter 10).
2   Worn struts or shock absorbers (Chapter 10).
3   Hub bearings worn (Chapter 10).
4   Excessive tire or wheel runout (Chapter 10).
5   Worn balljoints (Chapter 10).

70 Excessive tire wear on outside edge

1   Inflation pressures incorrect (Chapter 1).
2   Excessive speed in turns.
3   Wheel alignment incorrect (excessive toe-in or positive camber). Have professionally aligned.
4   Suspension arm bent (Chapter 10).

71 Excessive tire wear on inside edge

1   Inflation pressures incorrect (Chapter 1).
2   Wheel alignment incorrect (toe-out or excessive negative camber). Have professionally aligned.

3   Loose or damaged steering components (Chapter 10).

72 Tire tread worn in one place

1   Tires out-of-balance.
2   Damaged or buckled wheel. Inspect and replace if necessary.
3   Defective tire (Chapter 1).

73 Excessive play or looseness in steering system

1   Hub bearings worn (Chapter 10).
2   Tie-rod end loose or worn (Chapter 10).
3   Steering gear loose (Chapter 10).

74 Rattling or clicking noise in steering gear

1   Steering gear mounting bolts loose (Chapter 10).
2   Steering gear defective (Chapter 10).

Notes

Chapter 1
Tune-up and routine maintenance

Contents

	Section			Section
Air filter replacement	23		Fluid level checks	4
Automatic transaxle fluid change	26		Fuel system check	18
Automatic transaxle fluid level check	7		Introduction	2
Battery check, maintenance and charging	11		Maintenance schedule	1
Brake fluid change	22		Power steering fluid level check	6
Brake system check	19		Seat belt check	9
Cabin air filter - replacement	30		Spark plug replacement	24
Cooling system check	14		Suspension and steering check	25
Cooling system servicing (draining, flushing and refilling)	29		Underhood hose check and replacement	13
Differential lubricant change (AWD models)	28		Tire and tire pressure checks	5
Differential lubricant level check (AWD models)	16		Tire rotation	15
Driveaxle boot check	17		Transfer case lubricant change (AWD models)	27
Drivebelt and tensioner check and replacement	12		Transfer case lubricant level check (AWD models)	21
Engine oil and filter change	8		Tune-up general information	3
Exhaust system check	20		Wiper blade inspection and replacement	10

Specifications

Recommended lubricants and fluids

Note: *Listed here are manufacturer recommendations at the time this manual was written. Manufacturers occasionally upgrade their fluid and lubricant specifications, so refer to your vehicle owner's manual or check with your local auto parts store for current recommendations.*

Engine oil type	API "certified for gasoline engines"
Engine oil viscosity	5W-30
Fuel	Unleaded fuel, 87-octane minimum
Automatic transaxle fluid	DEXRON-VI automatic transmission fluid
Transfer case	SAE 75W-90 synthetic gear oil
Differential fluid (rear, AWD models)	SAE 75W-90 synthetic gear oil
Power steering fluid	
2007 through 2013 models	GM power steering fluid, or equivalent
2014 and later models	DEXRON-VI automatic transmission fluid
Brake fluid	DOT 3 brake fluid
Engine coolant	50/50 mixture of DEX-COOL coolant and demineralized water
Parking brake mechanism grease	White lithium-based grease NLGI no. 2
Chassis lubrication grease	NLGI no. 2 GC or GC-LB chassis grease
Hood, door and trunk hinge lubricant	Lubriplate lubricant aerosol spray
Door hinge and check spring grease	NLGI no. 2 multi-purpose grease
Key lock cylinder lubricant	Graphite spray
Hood latch assembly lubricant	Lubriplate lubricant aerosol spray
Door latch lubricant	NLGI no. 2 multi-purpose grease or equivalent

Capacities*

Engine oil (including filter)		
2011 and earlier models..	5.5 quarts	5.2 liters
2012 and later models...	6.0 quarts	5.7 liters
Automatic transaxle, 6-speed (6T70/75) (drain and refill)	4.2 to 6.3 quarts	4 to 6 liters

Note: *This is an initial-fill specification. Be sure to check the fluid level (see Section 7).*

Differential (rear, AWD models)		
2008 and earlier models..	2.1 quarts	2 liters
2009 and 2010 models...	0.85 quart	0.80 liter
2011 through 2013 models...	1.06 quarts	1 liter
2014 and later models...	0.85 quarts	0.80 liter
Transfer case - Getrag 760 (AWD models)		
2009 and earlier models..	0.85 quart	0.80 liter
2010 and later models...	1.06 quarts	1 liter
Cooling system		
2008 and earlier models..	12.2 quarts	11.5 liters
2009 and later models...	11.4 quarts	10.8 liters

* All capacities are approximate. Add as necessary to bring to appropriate level

Brakes

Disc brake pad wear limit	3/32 inch	2.38 mm
Parking brake shoe wear limit	1/16 inch	1.59 mm

Ignition system

FRONT OF VEHICLE

38027-2B-SPECS HAYNES

Cylinder locations

Spark plug type		
2008 and earlier models..	AC Delco 41-107	
2009 and later models...	AC Delco 41-109	
Spark plug gap	0.043 inch	1.09 mm
Firing order	1-2-3-4-5-6	

Torque specifications

Ft-lbs (unless otherwise indicated)   **Nm**

Note: *One foot-pound (ft-lb) of torque is equivalent to 12 inch-pounds (in-lbs) of torque. Torque values below approximately 15 foot-pounds are expressed in inch-pounds, because most foot-pound torque wrenches are not accurate at these smaller values.*

Engine oil drain plug..	18	24
Automatic transaxle drain plug	106 in-lbs	12
Drivebelt tensioner bolt...	18	24
Drivebelt idler pulley bolt		
2007 through 2013 models	37	50
2014 and later models..	43	58
Differential drain/fill plugs (rear, AWD models)	26	35
Transfer case drain/fill plugs...	29	39
Spark plugs		
2007 through 2013 models	156 in-lbs	19
2014 and later models..	132 in-lbs	18
Wheel lug nuts..	140	190

Typical engine compartment underside components

1 Engine oil drain plug	4 Brake caliper	7 Inner driveaxle boot
2 Automatic transaxle drain plug	5 Strut/coil spring assembly	8 Tie-rod end
3 Steering gear	6 Outer driveaxle boot	9 Balljoint

Typical rear underside components

1 Shock absorber
2 Upper control arm
3 Lower control arm
4 Exhaust system hanger

5 Rear muffler/resonator
6 Muffler
7 Fuel tank

Engine compartment components - 3.6L (SFI)

1	Brake fluid reservoir	4	Transaxle fill tube cap/dipstick	7	Coolant reservoir
2	Air filter housing	5	Engine oil dipstick	8	Underhood fuse/relay box
3	Windshield washer fluid reservoir	6	Engine oil filler cap	9	Power steering fluid reservoir

Engine compartment components – 3.6L Direct injection (VVT)

1	Brake fluid reservoir	4	Transaxle filler cap/dipstick (not visible)	7	Coolant reservoir
2	Air filter housing	5	Engine oil dipstick	8	Underhood fuse/relay box
3	Windshield washer fluid reservoir	6	Engine oil filler cap	9	Power steering fluid reservoir

1 Maintenance schedule

The following maintenance intervals are based on the assumption that the vehicle owner will be doing the maintenance or service work, as opposed to having a dealer service department do the work. These are the minimum maintenance intervals recommended by the factory for vehicles that are driven daily. If you wish to keep your vehicle in peak condition at all times, you may wish to perform some of these procedures even more often. Because frequent maintenance enhances the efficiency, performance and resale value of your vehicle, we encourage you to do so. If you drive in dusty areas, tow a trailer, idle or drive at low speeds for extended periods or drive for short distances (less than four miles) in below freezing temperatures, shorter intervals are also recommended.

When the vehicle is new, follow the maintenance schedule to the letter, record the maintenance performed in your owners manual and keep all receipts to protect the new vehicle warranty. In many cases the initial maintenance check is done at no cost to the owner (check with your dealer service department for more information).

Every 250 miles or weekly, whichever comes first

Check the engine oil level (Section 4)
Check the coolant level (Section 4)
Check the windshield washer fluid level (Section 4)
Check the brake fluid level (Section 4)
Check the tires and tire pressures (Section 5)

Every 3,000 miles or 3 months, whichever comes first

Note: *All items listed above, plus...*
Check the power steering fluid level (Section 6)
Change the engine oil and filter (Section 8)

Every 6,000 miles or 6 months, whichever comes first

Note: *All items listed above, plus...*
Check the seat belts (Section 9)
Inspect the windshield wiper blades (Section 10)
Check and service the battery (Section 11)
Check the engine drivebelt (Section 12)
Inspect underhood hoses (Section 13)
Check the cooling system (Section 14)
Rotate the tires (Section 15)
Check the differential lubricant level (Section 16)

Every 15,000 miles or 12 months, whichever comes first

Note: *All items listed above, plus...*
Check the driveaxle boots (Section 17)
Check the fuel system (Section 18)
Check the brake system (Section 19)*
Check the exhaust system (Section 20)
Check the transfer case lubricant level (AWD models) (Section 21)

Every 30,000 miles or 30 months, whichever comes first

Note: *All items listed above, plus...*
Change the brake fluid (Section 22)
Replace the air filter (Section 23)
Replace the spark plugs (non-platinum or iridium type) (Section 24)
Check the steering and suspension components (Section 25)
Change the automatic transaxle fluid and filter (Section 26)**

Every 60,000 miles or 48 months, whichever comes first

Change the transfer case lubricant (Section 27)*
Change the differential lubricant (Section 28)*

Every 100,000 miles or 60 months, whichever comes first

Replace the spark plugs (platinum or iridium type) (Section 24)
Service the cooling system (drain, flush and refill) (Section 29)
This item is affected by severe operating conditions, as described below. If the vehicle is operated under severe conditions, perform all maintenance indicated with an asterisk (*) at half the indicated intervals. Severe conditions exist if you mainly operate the vehicle . . .
In dusty areas
Towing a trailer
Idling for extended periods
Driving at low speeds when outside temperatures remain below freezing, and most trips are less than four miles long
** *Perform this procedure at half the recommended interval if operated under one or more of the following conditions:*
In heavy city traffic where the outside temperature regularly reaches 90-degrees F or higher
In hilly or mountainous terrain
Frequent trailer towing
If the vehicle has been driven through deep water

2   Introduction

1   This Chapter is designed to help the home mechanic maintain the GMC Acadia, Saturn Outlook, Buick Enclave and Chevrolet Traverse with the goals of maximum performance, economy, safety and reliability in mind.

2   Included is a master maintenance schedule, followed by procedures dealing specifically with each item on the schedule. Visual checks, adjustments, component replacement and other helpful items are included. Refer to the accompanying illustrations of the engine compartment and the underside of the vehicle for the locations of various components.

3   Servicing your vehicle in accordance with the mileage/time maintenance schedule and the step-by-step procedures will result in a planned maintenance program that should produce a long and reliable service life. Keep in mind that it's a comprehensive plan, so maintaining some items but not others at the specified intervals will produce the same results.

4   As you service your vehicle, you will discover that many of the procedures can - and should - be grouped together because of the nature of the particular procedure you're performing or because of the close proximity of two otherwise unrelated components to one another.

5   For example, if the vehicle is raised for chassis lubrication, you should inspect the exhaust, suspension, steering and fuel systems while you're under the vehicle. When you're rotating the tires, it makes good sense to check the brakes since the wheels are already removed. Finally, let's suppose you have to borrow or rent a torque wrench. Even if you only need it to tighten the spark plugs, you might as well check the torque of as many critical fasteners as time allows.

6   The first step in this maintenance program is to prepare you before the actual work begins. Read through all the procedures you're planning to do, then gather up all the parts and tools needed. If it looks like you might run into problems during a particular job, seek advice from a mechanic or an experienced do-it-yourselfer.

Owner's Manual and VECI label information

7   Your vehicle Owner's Manual was written for your year and model and contains very specific information on component locations, specifications, fuse ratings, part numbers, etc. The Owner's Manual is an important resource for the do-it-yourselfer to have; if one was not supplied with your vehicle, it can generally be ordered from a dealer parts department.

8   Among other important information, the Vehicle Emissions Control Information (VECI) label contains specifications and procedures for tune-up adjustments (if applicable) and, in some instances, spark plugs (see Chapter 6 for more information on the VECI label). The information on this label is the exact mainte-nance data recommended by the manufacturer. This data often varies by intended operating altitude, local emissions regulations, month of manufacture, etc.

9   This Chapter contains procedural details, safety information and more ambitious maintenance intervals than you might find in the manufacturer's literature. However, you may also find procedures and specifications in your Owner's Manual or VECI label that differ with what's printed here. In these cases, the Owner's Manual or VECI label can be considered correct, since it is specific to your particular vehicle.

3   Tune-up general information

1   The term tune-up is used in this manual to represent a combination of individual operations rather than one specific procedure that will maintain a gasoline engine in proper tune.

2   If, from the time the vehicle is new, the routine maintenance schedule is followed closely and frequent checks are made of fluid levels and high wear items, as suggested throughout this manual, the engine will be kept in relatively good running condition and the need for additional work will be minimized.

3   More likely than not, however, there may be times when the engine is running poorly due to lack of regular maintenance. This is even more likely if a used vehicle, which has not received regular and frequent maintenance checks, is purchased. In such cases, an engine tune-up will be needed outside of the regular routine maintenance intervals.

4   The first step in any tune-up or diagnostic procedure to help correct a poor running engine is a cylinder compression check. Compression check (see Chapter 2B) will help determine the condition of internal engine components and should be used as a guide for tune-up and repair procedures. If, for instance, the compression check indicates serious internal engine wear, a conventional tune-up won't improve the performance of the engine and would be a waste of time and money. Because of its importance, someone should do the compression check with the right equipment and the knowledge to use it properly.

5   The following procedures are those most often needed to bring a generally poor running engine back into a proper state of tune.

Minor tune-up

Check all engine-related fluids (Section 4)
Clean, inspect and test the battery
    (Section 11)
Check the drivebelt (Section 12)
Check all underhood hoses (Section 13)
Check the cooling system (Section 14)
Check the air filter (Section 23)

Major tune-up

Note: *All items listed under Minor tune-up, plus...*

4.2 Engine oil dipstick (A) and oil filler cap (B). To prevent dirt from contaminating the engine, always make sure the area around the cap is clean before removing it

Replace the air filter (Section 23)
Replace the spark plugs (Section 24)
Check the ignition system (Chapter 5)
Check the charging system (Chapter 5)

4   Fluid level checks (every 250 miles or weekly)

Note: *The following are fluid level checks to be done on a 250 mile or weekly basis. Additional fluid level checks can be found in specific maintenance procedures that follow. Regardless of intervals, be alert to fluid leaks under the vehicle, which would indicate a fault to be corrected immediately.*

1   Fluids are an essential part of the lubrication, cooling, brake and windshield washer systems. Because the fluids gradually become depleted and/or contaminated during normal operation of the vehicle, they must be periodically replenished. See *Recommended lubricants and fluids* in this Chapter's Specifications before adding fluid to any of the following components.

Note: *The vehicle must be on level ground when fluid levels are checked.*

Engine oil

2   The engine oil level is checked with a dipstick that extends through a tube and into the oil pan at the bottom of the engine **(see illustration)**.

3   The oil level should be checked before the vehicle has been driven, or about 5 minutes after the engine has been shut off. If the oil is checked immediately after driving the vehicle, some of the oil will remain in the upper engine components, resulting in an inaccurate reading on the dipstick.

4   Pull the dipstick out of the tube and wipe all the oil from the end with a clean rag or paper towel. Insert the clean dipstick all the way back into the tube, then pull it out again. Note the oil at the end of the dipstick. Add oil as necessary

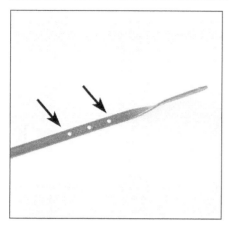

4.4 The oil level must be maintained within the cross-hatched zone at all times - it takes about one quart of oil to raise the level from the bottom of the zone to the top of the zone

4.8 The coolant reservoir is located on the right side of the engine compartment

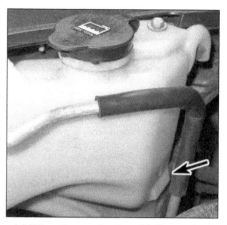

4.9 When the engine is cold, the coolant level must be up to the COLD FILL mark

to keep the level within the cross-hatched zone on the dipstick **(see illustration)**.

5   Do not overfill the engine by adding too much oil since this may result in oil-fouled spark plugs, oil leaks or oil seal failures.

6   Oil is added to the engine after unscrewing a cap from the valve cover. A funnel will help to reduce spills.

7   Checking the oil level is an important preventive maintenance step. A consistently low oil level indicates oil leakage through damaged seals, defective gaskets or past worn rings or valve guides. If the oil looks milky or has water droplets in it, the cylinder head gasket(s) may be blown or the head(s) or block may be cracked. The engine should be checked immediately. The condition of the oil should also be checked. Whenever you check the oil level, slide your thumb and index finger up the dipstick before wiping off the oil. If you see small dirt or metal particles clinging to the dipstick, the oil should be changed (see Section 8).

Engine coolant

Warning: *Do not allow antifreeze to come in contact with your skin or painted surfaces of the vehicle. Rinse off spills immediately with plenty of water. Antifreeze is highly toxic if ingested. Never leave antifreeze lying around in an open container or in puddles on the floor; children and pets are attracted by its sweet smell and may drink it. Check with local authorities on disposing of used antifreeze. Many communities have collection centers that will see that antifreeze is disposed of safely.*

Caution: *Never mix green-colored ethylene glycol antifreeze and orange-colored DEX-COOL silicate-free coolant because doing so will destroy the efficiency of the DEX-COOL coolant, which is designed to last for 100,000 miles or five years.*

Note: *Non-toxic antifreeze is now manufactured and available at auto parts stores, but even this type should be disposed of properly.*

8   All vehicles covered by this manual are equipped with a coolant reservoir, located at

the right side of the engine compartment **(see illustration)** and connected by a hose to the radiator filler neck.

9   The coolant level in the reservoir should be checked regularly. When the engine is cold, the coolant level should be at or slightly above the COLD FILL mark **(see illustration)**. If it isn't, add coolant to the tank.

Warning: *Never unscrew the expansion tank cap when the engine is warm. Only remove the cap when the engine and cooling system is cool.*

10   To add coolant, open the cap and add a 50/50 mixture of DEX-COOL coolant and water until the level is up to the COLD FILL mark (see the **Caution** at the beginning of this Section).

11   Drive the vehicle and recheck the coolant level. If only a small amount of coolant is required to bring the system up to the proper level, water can be used. However, repeated additions of water will dilute the antifreeze and water solution. In order to maintain the proper ratio of antifreeze and water, always top up the coolant level with the correct mixture. An empty plastic milk jug or bleach bottle makes an excellent container for mixing coolant. Do not use rust inhibitors or additives.

12   If the coolant level drops consistently, there may be a leak in the system. Inspect the radiator, hoses, filler cap, drain plugs and water pump (see Section 14). If no leaks are noted, have the radiator cap tested by a service station.

13   Check the condition of the coolant as well. It should be relatively clear. If it is brown or rust colored, the system should be drained, flushed and refilled. Even if the coolant appears to be normal, the corrosion inhibitors wear out, so it must be replaced at the specified intervals. If the system is filled with standard green coolant/water, it must be flushed and replaced more frequently than if the original DEX-COOL coolant is retained.

Windshield washer fluid

14   Fluid for the windshield washer system is located in a plastic reservoir in the left side

4.14 The windshield washer fluid tank is located at the left front corner of the engine compartment

of the engine compartment **(see illustration)**.

15   In milder climates, plain water can be used in the reservoir, but it should be kept no more than 2/3 full to allow for expansion if the water freezes. In colder climates, use windshield washer system antifreeze, available at any auto parts store, to lower the freezing point of the fluid. Mix the antifreeze with water in accordance with the manufacturer's directions.

Caution: *Don't use cooling system antifreeze - it will damage the vehicle's paint.*

16   To help prevent icing in cold weather, warm the windshield with the defroster before using the washer.

Battery electrolyte

17   These vehicles are equipped with a battery which is permanently sealed (except for vent holes) and has no filler caps. Water doesn't have to be added to these batteries at any time. If a maintenance-type battery is installed, the caps on the top of the battery should be removed periodically to check for a low electrolyte level. If the level is low, add only distilled water until the level is above the plates.

Brake fluid

18 The brake master cylinder is mounted on the upper left of the engine compartment firewall.

19 The translucent plastic reservoir allows the fluid inside to be checked without removing the cap **(see illustration)**. Wipe the top of the reservoir cap with a clean rag to prevent contamination of the brake system before removing it.

20 When adding fluid, pour it carefully into the reservoir to avoid spilling it on surrounding painted surfaces. Be sure the specified fluid is used, since mixing different types of brake fluid can cause damage to the system. See *Recommended lubricants and fluids* in this Chapter's Specifications or your owner's manual.

Warning: *Brake fluid can harm your eyes and damage painted surfaces, so use extreme caution when handling or pouring it. Do not use brake fluid that has been standing open or is more than one year old. Brake fluid absorbs moisture from the air. Moisture in the system can cause a dangerous loss of brake performance.*

21 At this time, the fluid and master cylinder can be inspected for contamination. The system should be drained and refilled if deposits, dirt particles or water droplets are seen in the fluid.

22 After filling the reservoir to the proper level, make sure the cap is installed tight to prevent fluid leakage.

23 The brake fluid level in the master cylinder will drop slightly as the pads at the front wheels wear down during normal operation. If the master cylinder requires repeated additions to keep it at the proper level, it's an indication of leakage in the brake system, which

4.19 The brake fluid reservoir is located at the left side of the firewall

should be corrected immediately. Check all brake lines and connections (see Section 19 for more information).

24 If, upon checking the master cylinder fluid level, you discover the reservoir empty or nearly empty, the brake system should be bled and thoroughly inspected (see Chapter 9).

5 Tire and tire pressure checks (every 250 miles or weekly)

1 Periodic inspection of the tires may spare you the inconvenience of being stranded with a flat tire. It can also provide you with vital

5.2 Use a tire tread depth indicator to monitor tire wear - they are available at auto parts stores and service stations and cost very little

information regarding possible problems in the steering and suspension systems before major damage occurs.

2 The original tires on this vehicle are equipped with 1/2-inch wide wear bands that will appear when tread depth reaches 1/16-inch, at which point the tires can be considered worn out. Tread wear can be monitored with a simple, inexpensive device known as a tread depth indicator **(see illustration)**.

3 Note any abnormal tread wear **(see illustration)**. Tread pattern irregularities such as cupping, flat spots and more wear on one

UNDERINFLATION

INCORRECT TOE-IN OR EXTREME CAMBER

CUPPING

Cupping may be caused by:
- Underinflation and/or mechanical irregularities such as out-of-balance condition of wheel and/or tire, and bent or damaged wheel.
- Loose or worn steering tie-rod or steering idler arm.
- Loose, damaged or worn front suspension parts.

OVERINFLATION

FEATHERING DUE TO MISALIGNMENT

5.3 This chart will help you determine the condition of the tires and the probable cause(s) of abnormal wear

5.4a If a tire loses air on a steady basis, check the valve stem core first to make sure it's snug (special inexpensive wrenches are commonly available at auto parts stores)

5.4b If the valve stem core is tight, raise the corner of the vehicle with the low tire and spray a soapy water solution onto the tread as the tire is turned slowly - leaks will cause small bubbles to appear

5.8 To extend the life of the tires, check the air pressure at least once a week with an accurate gauge (don't forget the spare!)

6.2 The power steering fluid reservoir is located at the right end of the engine

6.6 The power steering fluid level should be between the MIN and MAX marks on the dipstick

side than the other are indications of front end alignment and/or balance problems. If any of these conditions are noted, take the vehicle to a tire shop or service station to correct the problem.

4   Look closely for cuts, punctures and embedded nails or tacks. Sometimes a tire will hold air pressure for a short time or leak down very slowly after a nail has embedded itself in the tread. If a slow leak persists, check the valve stem core to make sure it's tight (**see illustration**). Examine the tread for an object that may have embedded itself in the tire or for a plug that may have begun to leak (radial tire punctures are repaired with a plug that's installed in a puncture). If a puncture is suspected, it can be easily verified by spraying a solution of soapy water onto the puncture area (**see illustration**). The soapy solution will bubble if there's a leak. Unless the puncture is unusually large, a tire shop or service station can usually repair the tire.

5   Carefully inspect the inner sidewall of each tire for evidence of brake fluid leakage. If

you see any, inspect the brakes immediately.

6   Correct air pressure adds miles to the lifespan of the tires, improves mileage and enhances overall ride quality. Tire pressure cannot be accurately estimated by looking at a tire, especially if it's a radial. A tire pressure gauge is essential. Keep an accurate gauge in the vehicle. The pressure gauges attached to the nozzles of air hoses at gas stations are often inaccurate.

7   Always check tire pressure when the tires are cold. Cold, in this case, means the vehicle has not been driven over a mile in the three hours preceding a tire pressure check. A pressure rise of four to eight pounds is not uncommon once the tires are warm.

8   Unscrew the valve cap protruding from the wheel or hubcap and push the gauge firmly onto the valve stem (**see illustration**). Note the reading on the gauge and compare the figure to the recommended tire pressure shown on the placard on the driver's side door pillar. Reinstall the valve cap to keep dirt

and moisture out of the valve stem mechanism. Check all four tires and, if necessary, add enough air to bring them up to the recommended pressure.

9   Don't forget to keep the spare tire inflated to the specified pressure (see your owner's manual or the tire sidewall).

6   Power steering fluid level check (every 3000 miles or 3 months)

1   The power steering system on these models relies on fluid, which may, over a period of time, require replenishing. The fluid should be checked at normal operating temperature.

2   The fluid reservoir for the power steering pump is located at the right end of the engine, under the engine cover (**see illustration**).

3   For the check, the front wheels should be pointed straight ahead and the engine should be off.

4   Remove the engine cover by removing the oil filler cap, then pull the cover up off of its retainers.

5   Use a clean rag to wipe off the reservoir cap and the area around the cap. This will help prevent any foreign matter from entering the reservoir during the check.

6   Twist off the cap. Wipe off the fluid with a clean rag, reinstall the dipstick, then withdraw it and read the fluid level (**see illustration**). The fluid should be at the proper level, depending on whether it was checked hot or cold.

7   If additional fluid is required, pour the specified type directly into the reservoir, using a funnel to prevent spills.

8   If the reservoir requires frequent fluid additions, all power steering hoses, hose connections, steering gear and the power steering pump should be carefully checked for leaks.

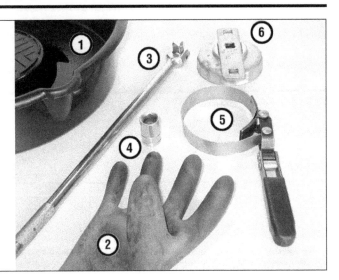

8.3 These tools are required when changing the engine oil and filter

1   **Drain pan** - It should be fairly shallow in depth, but wide to prevent spills
2   **Rubber gloves** - When removing the drain plug and filter, you will get oil on your hands (the gloves will prevent burns)
3   **Breaker bar** - Sometimes the oil drain plug is tight, and a long breaker bar is needed to loosen it
4   **Socket** - To be used with the breaker bar or a ratchet (must be the correct size to fit the drain plug - six-point preferred)
5   **Filter wrench** - This is a metal band-type wrench, which requires clearance around the filter to be effective
6   **Filter wrench** - This type fits on the bottom of the filter and can be turned with a ratchet or breaker bar (different-size wrenches are available for different types of filters)

7   Automatic transaxle fluid level check

1   Checking the automatic transaxle fluid level is not a routine maintenance procedure on these vehicles. The only reason the fluid level would drop would be due to a leak.
2   Checking the fluid level is necessary when the transaxle fluid is drained and refilled. The fluid level checking procedure is included in Section 26.

8   Engine oil and filter change (every 3000 miles or 3 months)

Note: *These vehicles are equipped with an oil life indicator system that illuminates a light on the instrument panel when the system deems it necessary to change the oil. A number of factors are taken into consideration to determine when the oil should be considered worn out. Generally, this system will allow the vehicle to accumulate more miles between oil changes than the traditional 3000 mile interval, but we believe that frequent oil changes are cheap insurance and will prolong engine life. If you do decide not to change your oil every 3000 miles and rely on the oil life indicator instead, make sure you don't exceed 10,000 miles before the oil is changed, regardless of what the oil life indicator shows.*

1   Frequent oil changes are the most important preventive maintenance procedures that can be done by the home mechanic. As engine oil ages, it becomes diluted and contaminated, which leads to premature engine wear.
2   Although some sources recommend oil filter changes every other oil change, we feel that the minimal cost of an oil filter and the relative ease with which it is installed dictate that a new filter be installed every time the oil is changed.
3   Gather together all necessary tools and materials before beginning this procedure (see illustration).

4   You should have plenty of clean rags and newspapers handy to mop up any spills. Access to the underside of the vehicle may be improved if the vehicle can be lifted on a hoist, driven onto ramps or supported by jackstands.
Warning: *Do not work under a vehicle which is supported only by a jack.*
5   If this is your first oil change, familiarize yourself with the locations of the oil drain plug and the oil filter.
6   Warm the engine to normal operating temperature. If the new oil or any tools are needed, use this warm-up time to gather everything necessary for the job. The correct type of oil for your application can be found in *Recommended lubricants and fluids* in this Chapter's Specifications.
7   With the engine oil warm (warm engine oil will drain better and more built-up sludge will be removed with it), raise and support the vehicle. Make sure it's safely supported!
8   Remove the under-vehicle splash shield. Move all necessary tools, rags and newspapers under the vehicle. Set the drain pan under the drain plug. Keep in mind that the oil

will initially flow from the pan with some force; position the pan accordingly.
9   Being careful not to touch any of the hot exhaust components, use a wrench to remove the drain plug near the bottom of the oil pan (see illustration). Depending on how hot the oil is, you may want to wear gloves while unscrewing the plug the final few turns.
10   Allow the oil to drain into the pan. It may be necessary to move the pan as the oil flow slows to a trickle.
11   After all the oil has drained, wipe off the drain plug with a clean rag. Small metal particles may cling to the plug and would immediately contaminate the new oil.
12   Clean the area around the drain plug opening and reinstall the plug. Tighten the plug securely with the wrench. If a torque wrench is available, use it to tighten the plug to the torque listed in this Chapter's Specifications.
13   Move the drain pan into position under the oil filter.
14   Use the oil filter wrench to loosen the oil filter (see illustration).
15   Completely unscrew the old filter. Be

8.9 Use a proper size box-end wrench or socket to remove the oil drain plug and avoid rounding it off

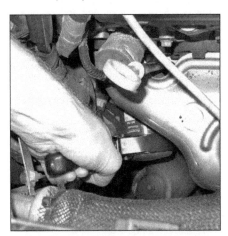

8.14 Since the oil filter is on very tight, you'll need a special wrench for removal - DO NOT use the wrench to tighten the new filter

8.18 Lubricate the oil filter gasket with clean engine oil before installing the filter on the engine

10.4a To release the blade holder, depress the release tab . . .

10.4b . . . and pull the wiper blade in the direction of the windshield to separate it from the arm

careful; it's full of oil. Empty the oil inside the filter into the drain pan.

16   Compare the old filter with the new one to make sure they're the same type.

17   Use a clean rag to remove all oil, dirt and sludge from the area where the oil filter mounts to the engine. Check the old filter to make sure the rubber gasket isn't stuck to the engine. If the gasket is stuck to the engine, remove it.

18   Apply a light coat of clean oil to the rubber gasket on the new oil filter **(see illustration)**.

19   Attach the new filter to the engine, following the tightening directions printed on the filter canister or packing box. Most filter manufacturers recommend against using a filter wrench due to the possibility of overtightening and damage to the seal.

20   Remove all tools, rags, etc. from under the vehicle, being careful not to spill the oil in the drain pan, then lower the vehicle.

21   Move to the engine compartment and locate the oil filler cap.

22   Refer to the engine oil capacity in this Chapter's Specifications and add the proper amount of fresh oil into the engine. Wait a few minutes to allow the oil to drain into the pan, then check the level on the oil dipstick (see Section 4 if necessary). If the oil level is above the upper mark, start the engine and allow the new oil to circulate.

23   Run the engine for only about a minute and then shut it off. Immediately look under the vehicle and check for leaks at the oil pan drain plug and around the oil filter.

24   Wait about five minutes. With the new oil circulated and the filter now completely full, recheck the level on the dipstick and add more oil as necessary.

25   During the first few trips after an oil change, make it a point to check frequently for leaks and proper oil level.

26   The old oil drained from the engine cannot be reused in its present state and should be disposed of. Check with your local auto parts store, disposal facility or environmental

agency to see if they will accept the oil for recycling. After the oil has cooled it can be drained into a container (capped plastic jugs, topped bottles, milk cartons, etc.) for transport to one of these disposal sites. Don't dispose of the oil by pouring it on the ground or down a drain!

Oil Life Monitor

27   The Oil Life Monitor is a function of the PCM that tracks engine operating temperature and rpm. If the PCM determines that your engine's oil has been used long enough, an indicator that shows "Change Engine Oil Soon" will light on the instrument panel.

28   When you change your engine oil and filter, whether you change it at the interval recommended in this Chapter or only when the light comes on, you will have to reset the system to make the indicator go out.

29   Reset the system whenever the engine oil is changed so that the system can calculate the next engine oil change.

30   To reset the system on models that do not have the Driver Information Center (DIC) buttons:

31   Turn the ignition to the ON or RUN position without starting the engine. With the vehicle in the Park position, press the trip odometer reset button until OIL LIFE REMAINING comes on. Press and hold the trip odometer reset button until OIL LIFE REMAINING shows "100%" and wait for three chimes to sound, then the CHANGE ENGINE OIL SOON message will go off. Once the message is off, turn the key to OFF or LOCK position. If the CHANGE ENGINE OIL SOON message comes back on now, or when the engine is started, the engine oil life system has not been reset, and the procedure will need to be repeated.

32   To reset the system on models with the Driver Information Center (DIC) buttons:

33   Turn the ignition to the ON or RUN position without starting the engine. With the vehicle in the Park position, press the vehicle information button until OIL LIFE REMAINING comes on. Press and hold the set/reset button

until "100%" is shown, wait for three chimes to sound, and the CHANGE ENGINE OIL SOON message will go off. Once the message is off, turn the key to the OFF or LOCK position. If the CHANGE ENGINE OIL SOON message comes back on now, or when the engine is started, the engine oil life system has not been reset, and the procedure will need to be repeated.

---

9   Seat belt check (every 6000 miles or 6 months)

---

1   Check seat belts, buckles, latch plates, and guide loops for obvious damage and signs of wear.

2   Where the seat belt receptacle bolts to the floor of the vehicle, check that the bolts are secure.

3   See if the seat belt reminder light comes on when the key is turned to the Run or Start position. A chime should also sound.

---

10   Wiper blade inspection and replacement (every 6000 miles or 6 months)

---

1   The windshield wiper and blade assembly should be inspected periodically for damage, loose components and cracked or worn blade elements.

2   Road film can build up on the wiper blades and affect their efficiency, so they should be washed regularly with a mild detergent solution.

3   If the wiper blade elements are cracked, worn or warped, or no longer clean adequately, they should be replaced with new ones.

4   Lift the arm assembly away from the glass for clearance, depress the release lever, then slide the wiper blade assembly out of the hook in the end of the arm **(see illustrations)**.

5   Attach the new wiper to the arm. Connection can be confirmed by an audible click.

11.1 Tools and materials required for battery maintenance

1   **Face shield/safety goggles** - *When removing corrosion with a brush, the acidic particles can easily fly up into your eyes*

2   **Baking soda** - *A solution of baking soda and water can be used to neutralize corrosion*

3   **Petroleum jelly** - *A layer of this on the battery terminal bolts will help prevent corrosion*

4   **Battery terminal/cable cleaner** - *This wire brush cleaning tool will remove all traces of corrosion from the battery and cable clamps*

5   **Treated felt washers** - *Placing one of these on each terminal, directly under the cable end, will help prevent corrosion (be sure to get the correct type for side terminal batteries)*

6   **Puller** - *Sometimes the cable clamps are very difficult to pull off the posts. This tool pulls the clamp straight up and off the post without damage*

7   **Battery post/cable cleaner** - *Here is another cleaning tool that is a slightly different version of number 4 above, but it does the same thing*

8   **Rubber gloves** - *Another safety item to consider when servicing the battery; remember that's acid inside the battery!*

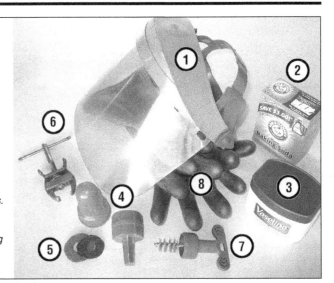

11   Battery check, maintenance and charging (every 6000 miles or 6 months)

Warning: *Certain precautions must be followed when checking and servicing the battery. Hydrogen gas, which is highly flammable, is always present in the battery cells, so keep lighted tobacco and all other open flames and sparks away from the battery. The electrolyte inside the battery is actually dilute sulfuric acid, which will cause injury if splashed on your skin or in your eyes. It will also ruin clothes and painted surfaces. When removing the battery cables, always detach the negative cable first and hook it up last!*

Note: *Some of the covered models have an auxiliary battery in addition to the standard battery. All of the following care and maintenance should be applied to both batteries.*

1   A routine preventive maintenance program for the battery in your vehicle is the only way to ensure quick and reliable starts. But before performing any battery maintenance, make sure that you have the proper equipment necessary to work safely around the battery **(see illustration).**

2   There are also several precautions that should be taken whenever battery maintenance is performed. Before servicing the battery, always turn the engine and all accessories off and disconnect the cable from the negative terminal of the battery.

3   The battery produces hydrogen gas, which is both flammable and explosive. Never create a spark, smoke or light a match around the battery. Always charge the battery in a ventilated area.

4   Electrolyte contains poisonous and corrosive sulfuric acid. Do not allow it to get in your eyes, on your skin or on your clothes. Never ingest it. Wear protective safety glasses when working near the battery. Keep children away from the battery.

5   Note the external condition of the battery. If the positive terminal and cable clamp

11.7a A tool like this one (available at auto parts stores) is used to clean the side-terminal type battery cable contact area

on your vehicle's battery is equipped with a rubber protector, make sure that it's not torn or damaged. It should completely cover the terminal. Look for any corroded or loose connections, cracks in the case or cover or loose hold-down clamps. Also check the entire length of each cable for cracks and frayed conductors.

6   If corrosion, which looks like white, fluffy deposits is evident, particularly around the terminals, the battery should be removed for cleaning. Loosen the cable bolts with a wrench, being careful to remove the ground cable first, and slide them off the terminals. Then remove the hold-down clamp and lift the battery from the engine compartment.

7   Clean the cable ends thoroughly with a battery brush or a terminal cleaner and a solution of warm water and baking soda. Wash the terminals and the side of the battery case with the same solution but make sure that the solution doesn't get into the battery. When cleaning the cables, terminals and battery case, wear safety goggles and rubber gloves to prevent any solution from coming in con-

11.7b Use the brush side of the tool to finish the job

tact with your eyes or hands. Wear old clothes too - even diluted, sulfuric acid splashed onto clothes will burn holes in them. If the terminals have been corroded, clean them up with a terminal cleaner **(see illustrations)**. Thoroughly wash all cleaned areas with plain water.

8   Make sure that the battery tray is in good condition and the hold-down clamp is tight. If the battery is removed from the tray, make sure no parts remain in the bottom of the tray when the battery is reinstalled. When reinstalling the hold-down clamp bolts, do not overtighten them.

9   Any metal parts of the vehicle damaged by corrosion should be covered with a zinc-based primer, then painted.

10   Information on removing and installing the battery can be found in Chapter 5. Information on jump starting can be found at the beginning of this manual. For more detailed battery checking procedures, refer to the *Haynes Automotive Electrical Manual*.

Charging

Warning: *When batteries are being charged, hydrogen gas, which is very explosive and*

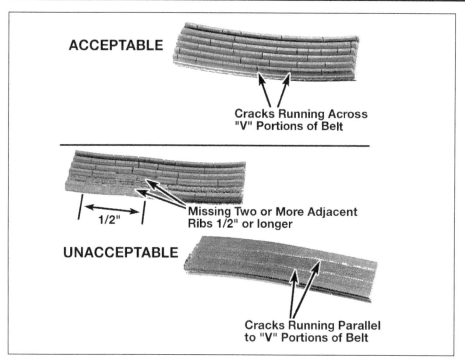

ACCEPTABLE

Cracks Running Across
"V" Portions of Belt

1/2"

Missing Two or More Adjacent
Ribs 1/2" or longer

UNACCEPTABLE

Cracks Running Parallel
to "V" Portions of Belt

**12.2 Check ribbed (serpentine) belts for signs of wear like these - if it looks worn,
replace it**

**12.6 Insert a 3/8-inch drive tool into the
square hole in the tensioner, then turn the
tensioner clockwise for belt removal.**

flammable, is produced. Do not smoke or allow open flames near a charging or a recently charged battery. Wear eye protection when near the battery during charging. Also, make sure the charger is unplugged before connecting or disconnecting the battery from the charger.

Note: *The manufacturer recommends the battery be removed from the vehicle for charging because the gas that escapes during this procedure can damage the paint. Fast charging with the battery cables connected can result in damage to the electrical system.*

11   Slow-rate charging is the best way to restore a battery that's discharged to the point where it will not start the engine. It's also a good way to maintain the battery charge in a vehicle that's only driven a few miles between starts. Maintaining the battery charge is particularly important in the winter when the battery must work harder to start the engine and electrical accessories that drain the battery are in greater use.

12   It's best to use a one or two-amp battery charger (sometimes called a trickle charger). They are the safest and put the least strain on the battery. They are also the least expensive. For a faster charge, you can use a higher amperage charger, but don't use one rated more than 1/10th the amp/hour rating of the battery. Rapid boost charges that claim to restore the power of the battery in one to two hours are hardest on the battery and can damage batteries not in good condition. This type of charging should only be used in emergency situations.

13   The average time necessary to charge a battery should be listed in the instructions that

come with the charger. As a general rule, a trickle charger will charge a battery in 12 to 16 hours.

14   Remove all the cell caps (if equipped) and cover the holes with a clean cloth to prevent spattering electrolyte. Disconnect the negative battery cable and hook the battery charger cable clamps up to the battery posts (positive to positive, negative to negative), then plug in the charger. Make sure it is set at 12-volts if it has a selector switch.

15   If you're using a charger with a rate higher than two amps, check the battery regularly during charging to make sure it doesn't overheat. If you're using a trickle charger, you can safely let the battery charge overnight after you've checked it regularly for the first couple of hours.

16   If the battery has removable cell caps, measure the specific gravity with a hydrometer every hour during the last few hours of the charging cycle. Hydrometers are available inexpensively from auto parts stores - follow the instructions that come with the hydrometer. Consider the battery charged when there's no change in the specific gravity reading for two hours and the electrolyte in the cells is gassing (bubbling) freely. The specific gravity reading from each cell should be very close to the others. If not, the battery probably has a bad cell(s).

17   Some batteries with sealed tops have built-in hydrometers on the top that indicate the state of charge by the color displayed in the hydrometer window. Normally, a bright-colored hydrometer indicates a full charge and a dark hydrometer indicates the battery still needs charging.

18   If the battery has a sealed top and no built-in hydrometer, you can hook up a digital voltmeter across the battery terminals to check the charge. A fully charged battery should read 12.5 volts or higher.

19   Further information on the battery and jump-starting can be found in Chapter 5 and at the beginning of this manual.

---

12   Drivebelt and tensioner check and replacement (every 6000 miles or 6 months)

---

1   A serpentine drivebelt is located at the front of the engine and plays an important role in the overall operation of the engine and its components. Due to its function and material make up, the belt is prone to wear and should be periodically inspected. The serpentine belt drives the alternator, power steering pump, water pump and air conditioning compressor.

2   With the engine off, open the hood and use your fingers (and a flashlight, if necessary), to move along the belt checking for cracks and separation of the belt plies. Also check for fraying and glazing, which gives the belt a shiny appearance **(see illustration)**. Both sides of the belt must be inspected.

3   Check the ribs on the underside of the belt. They should all be the same depth, with none of the surface uneven.

4   The tension of the belt is maintained by a spring-loaded tensioner assembly and isn't adjustable. The belt should be replaced at the mileage specified in Section 1, or if it is damaged or worn.

5   Raise the vehicle and support it securely on jackstands. Remove the passenger's side front tire and inner fender well splash shield (see Chapter 11).

6   Insert a 3/8-inch drive ratchet or breaker bar into the square hole in the tensioner, then rotate the tensioner clockwise to release belt tension **(see illustration)**.

Note: *If your ratchet or breaker bar is too thick to fit in the space between the tensioner and the chassis, special drivebelt tensioner tools are available at most auto parts stores.*

7   Noting how it's routed, remove the belt from the tensioner and auxiliary components, then slowly release the tensioner.

8   After verifying the new belt is the same length as the original belt, route the new belt over the various pulleys, again rotating the tensioner to allow the belt to be installed, then release the belt tensioner.

Tensioner replacement

9   Support the engine with a floor jack and a block of wood, then remove the right-side engine mount and bracket (see Chapter 2A).

10   Remove the drivebelt.

11   Remove the idler pulley center bolt and idler pulley.

12   Remove the tensioner mounting bracket bolts and detach the tensioner/mounting bracket from the engine.

13   Installation is the reverse of removal. Tighten the mounting bolt(s) to the torque listed in this Chapter's Specifications.

---

13   Underhood hose check and replacement (every 6000 miles or 6 months)

Warning: *Replacement of air conditioning hoses must be left to a dealer service department or air conditioning shop that has the equipment to depressurize the system safely and recover the refrigerant. Never remove air conditioning components or hoses until the system has been depressurized.*

1   High temperatures in the engine compartment can cause the deterioration of the rubber and plastic hoses used for engine, accessory and emission systems operation. Periodic inspection should be made for cracks, loose clamps, material hardening and leaks. Information specific to the cooling system hoses can be found in Section 14.

2   Some, but not all, hoses are secured to their fittings with clamps. Where clamps are used, check to be sure they haven't lost their tension, allowing the hose to leak. If clamps aren't used, make sure the hose has not expanded and/or hardened where it slips over the fitting, allowing it to leak.

Vacuum hoses

3   It's quite common for vacuum hoses, especially those in the emissions system, to be color-coded or identified by colored stripes molded into them. Various systems require hoses with different wall thickness, collapse resistance and temperature resistance. When replacing hoses, be sure the new ones are made of the same material.

4   Often the only effective way to check a hose is to remove it completely from the vehicle. If more than one hose is removed, label

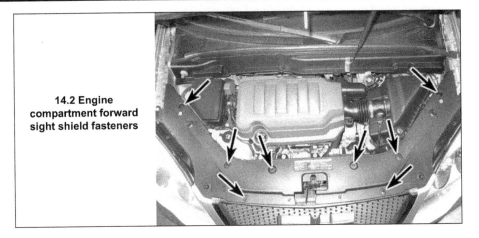

14.2 Engine compartment forward sight shield fasteners

the hoses and fittings to ensure correct installation.

5   When checking vacuum hoses, be sure to include any plastic T-fittings in the check. Inspect the fittings for cracks and the hose where it fits over the fitting for distortion, which could cause leakage.

6   A small piece of vacuum hose (1/4-inch inside diameter) can be used as a stethoscope to detect vacuum leaks. Hold one end of the hose to your ear and probe around vacuum hoses and fittings, listening for the hissing sound characteristic of a vacuum leak.

Warning: *When probing with the vacuum hose stethoscope, be very careful not to come into contact with moving engine components such as the drivebelt, cooling fan, etc.*

Fuel hose

Warning: *Gasoline is extremely flammable, so take extra precautions when you work on any part of the fuel system. Don't smoke or allow open flames or bare light bulbs near the work area, and don't work in a garage where a gas-type appliance (such as a water heater or clothes dryer) is present. Since gasoline is carcinogenic, wear fuel-resistant gloves when there's a possibility of being exposed to fuel, and, if you spill any fuel on your skin, rinse it off immediately with soap and water. Mop up any spills immediately and do not store fuel-soaked rags where they could ignite. When you perform any kind of work on the fuel system, wear safety glasses and have a Class B type fire extinguisher on hand. The fuel system is under pressure, so if any lines must be disconnected, the pressure in the system must be relieved first (see Chapter 4 for more information).*

7   Check all rubber fuel lines for deterioration and chafing. Check especially for cracks in areas where the hose bends and just before fittings, such as where a hose attaches to the fuel filter and fuel injection unit.

8   High quality fuel line, specifically designed for high-pressure fuel injection applications, must be used for fuel line replacement. Never, under any circumstances, use regular fuel line, unreinforced vacuum line,

clear plastic tubing or water hose for fuel lines.

9   Spring-type (pinch) clamps are commonly used on fuel lines. These clamps often lose their tension over a period of time, and can be sprung during removal. Replace all spring-type clamps with screw clamps whenever a hose is replaced.

Metal lines

10   Sections of metal line are routed along the frame, between the fuel tank and the engine. Check carefully to be sure the line has not been bent or crimped and no cracks have started in the line.

11   If a section of metal fuel line must be replaced, only seamless steel tubing should be used, since copper and aluminum tubing don't have the strength necessary to withstand normal engine vibration.

12   Check the metal brake lines where they enter the master cylinder and brake proportioning unit for cracks in the lines or loose fittings. Any sign of brake fluid leakage calls for an immediate and thorough inspection of the brake system.

---

14   Cooling system check (every 6000 miles or 6 months)

Caution: *Never mix green-colored ethylene glycol antifreeze and orange-colored DEX-COOL silicate-free coolant because doing so will destroy the efficiency of the DEX-COOL coolant, which is designed to last for 100,000 miles or five years.*

1   Many major engine failures can be attributed to a faulty cooling system. If the vehicle is equipped with an automatic transaxle, the cooling system also cools the transmission fluid and thus plays an important role in prolonging transmission life. The cooling system should be checked with the engine cold. Do this before the vehicle is driven for the day or after it has been shut off for at least three hours.

2   Remove the engine compartment forward sight shield **(see illustration)**.

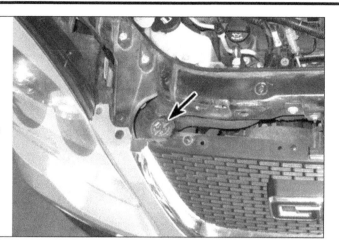

14.3 The radiator cap is located behind the grille, on the right side

Check for a chafed area that could fail prematurely.

Check for a soft area indicating the hose has deteriorated inside.

Overtightening the clamp on a hardened hose will damage the hose and cause a leak.

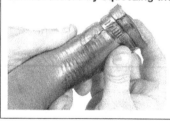

Check each hose for swelling and oil-soaked ends. Cracks and breaks can be located by squeezing the hose.

3   Remove the radiator cap by slowly unscrewing it **(see illustration)**. If you hear any hissing sounds (indicating there is still pressure in the system), wait until it stops. If there is no hissing sound, continue unscrewing it. Thoroughly clean the cap, inside and out, with clean water. Also clean the opening on the radiator filler neck. All traces of corrosion should be removed. The coolant inside the radiator and in the coolant reservoir should be relatively transparent. If it is rust colored, the system should be drained and refilled (see Section 29). If the coolant level is not up to the COLD FILL mark on the reservoir, add additional antifreeze/coolant mixture (see Section 4).
4   Carefully check the large upper and lower radiator hoses along with any smaller diameter heater hoses that run from the engine to the firewall. Inspect each hose along its entire length, replacing any hose that is cracked, swollen or shows signs of deterioration. Cracks may become more apparent if the hose is squeezed **(see illustration)**.
5   Make sure all hose connections are tight. A leak in the cooling system will usually show up as white or rust-colored deposits on the areas adjoining the leak. If wire-type clamps

are used at the ends of the hoses, it may be wise to replace them with more secure, screw-type clamps.
6   Use compressed air or a soft brush to remove bugs, leaves, etc. from the front of the radiator or air conditioning condenser. Be careful not to damage the delicate cooling fins or cut yourself on them.
7   Every other inspection, or at the first indication of cooling system problems, have the cap and system pressure tested. If you don't have a pressure tester, most gas stations and repair shops will do this for a minimal charge.

15   Tire rotation (every 6000 miles or 6 months)

1   The tires should be rotated at the specified intervals and whenever uneven wear is noticed.
2   Tires must be rotated in the recommended pattern **(see illustration)**.
3   Refer to the information in the front of this manual for the proper procedures to follow when raising the vehicle and changing a tire. If the brakes are to be checked, don't apply the parking brake as stated. Make sure the tires are blocked to prevent the vehicle from rolling as it's raised.
4   Preferably, the entire vehicle should be raised at the same time. This can be done on a hoist or by jacking up each corner and then lowering the vehicle onto jackstands placed under the frame rails. Always use four jackstands and make sure the vehicle is safely supported.
5   After rotation, check and adjust the tire pressures as necessary. Tighten the lug nuts to the torque listed in this Chapter's Specifications.

16   Differential lubricant level check (AWD models) (every 6,000 miles or 6 months)

1   Raise the vehicle and support it securely on jackstands.
2   Remove the plug from the filler hole on the left side of the rear axle housing **(see illustration 28.3)**.

14.4 Hoses, like drivebelts, have a habit of failing at the worst possible time - to prevent the inconvenience of a blown radiator or heater hose, inspect them carefully as shown here

3   The lubricant level should be even with the bottom of the filler hole. If not, use a pump or squeeze bottle to add the recommended lubricant.
4   Install the plug and tighten it to the torque listed in this Chapter's Specifications.

17   Driveaxle boot check (every 15,000 miles or 12 months)

1   The driveaxle boots are very important because they prevent dirt, water and foreign material from entering and damaging the constant velocity (CV) joints. Oil and grease can cause the boot material to deteriorate prematurely, so it's a good idea to wash the boots with soap and water. Because it constantly pivots back and forth following the steering action of the front hub, the outer CV boot wears out sooner and should be inspected regularly.

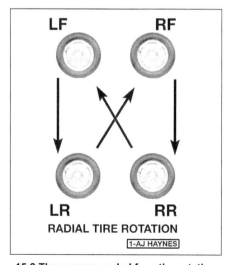

LF   RF

LR   RR

RADIAL TIRE ROTATION

1-AJ HAYNES

15.2 The recommended four-tire rotation pattern for these vehicles

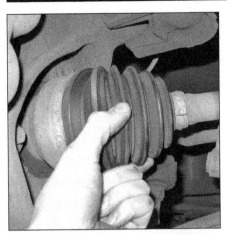

17.2 Inspect the inner and outer driveaxle boots for loose clamps, cracks or signs of leaking lubricant (AWD models have driveaxles at the rear, too)

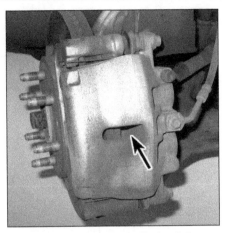

19.7a With the wheel off, check the thickness of the inner pad through the inspection hole in the caliper (front caliper shown, rear caliper similar)

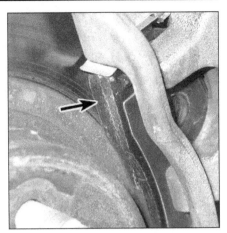

19.7b The outer pad is more easily checked at the edge of the caliper

2   Inspect the boots for tears and cracks as well as loose clamps **(see illustration)**. If there is any evidence of cracks or leaking lubricant, they must be replaced (see Chapter 8).

18   Fuel system check (every 15,000 miles or 12 months)

Warning: *Gasoline is extremely flammable, so take extra precautions when you work on any part of the fuel system. Don't smoke or allow open flames or bare light bulbs near the work area, and don't work in a garage where a gas-type appliance (such as a water heater or clothes dryer) is present. Since gasoline is carcinogenic, wear fuel-resistant gloves when there's a possibility of being exposed to fuel, and, if you spill any fuel on your skin, rinse it off immediately with soap and water. Mop up any spills immediately and do not store fuel-soaked rags where they could ignite. When you perform any kind of work on the fuel system, wear safety glasses and have a Class B type fire extinguisher on hand. The fuel system is under constant pressure, so, before any lines are disconnected, the fuel system pressure must be relieved (see Chapter 4).*

1   If you smell gasoline while driving or after the vehicle has been sitting in the sun, inspect the fuel system immediately.

2   Remove the gas filler cap and inspect it for damage and corrosion. The gasket should have an unbroken sealing imprint. If the gasket is damaged or corroded, install a new cap.

3   Inspect the fuel feed and return lines for cracks. Make sure that the connections between the fuel lines and the fuel injection system are tight.

Warning: *Your vehicle is fuel injected, so you must relieve the fuel system pressure before servicing fuel system components (see Chapter 4).*

4   Since some components of the fuel system - the fuel tank and part of the fuel feed

and return lines, for example - are underneath the vehicle, they can be inspected more easily with the vehicle raised on a hoist. If that's not possible, raise the vehicle and support it on jackstands.

5   With the vehicle raised and safely supported, inspect the gas tank and filler neck for punctures, cracks and other damage. The connection between the filler neck and the tank is particularly critical. Sometimes a rubber filler neck will leak because of loose clamps or deteriorated rubber. Inspect all fuel tank mounting brackets and straps to be sure that the tank is securely attached to the vehicle.

Warning: *Do not, under any circumstances, try to repair a fuel tank (except rubber components). A welding torch or any open flame can easily cause fuel vapors inside the tank to explode.*

6   Carefully check all rubber hoses and metal lines leading away from the fuel tank. Check for loose connections, deteriorated hoses, crimped lines and other damage. Repair or replace damaged sections as necessary (see Chapter 4).

7   The evaporative emissions control system can also be a source of fuel odors. The function of the system is to store fuel vapors from the fuel tank in a charcoal canister until they can be routed to the intake manifold where they mix with incoming air before being burned in the combustion chambers.

8   The most common symptom of a faulty evaporative emissions system is a strong odor of fuel coming from the area of the charcoal canister. If a fuel odor has been detected, and you have already checked the areas described above, check the charcoal canister, located near the fuel tank, and the hoses connected to it (see Chapter 4).

19   Brake system check (every 15,000 miles or 12 months)

Warning: *The dust created by the brake sys-*

tem is harmful to your health. Never blow it out with compressed air and don't inhale any of it. An approved filtering mask should be worn when working on the brakes. Do not, under any circumstances, use petroleum-based solvents to clean brake parts. Use brake system cleaner only!*

Note: *For detailed photographs of the brake system, refer to Chapter 9.*

1   In addition to the specified intervals, the brakes should be inspected every time the wheels are removed or whenever a defect is suspected.

2   Any of the following symptoms could indicate a potential brake system defect: The vehicle pulls to one side when the brake pedal is depressed; the brakes make squealing or dragging noises when applied; brake pedal travel is excessive; the pedal pulsates; or brake fluid leaks, usually onto the inside of the tire or wheel.

3   Loosen the wheel lug nuts.

4   Raise the vehicle and place it securely on jackstands.

5   Remove the wheels (see Section 7, or your owner's manual, if necessary).

6   There are two pads (an outer and an inner) in each caliper. The pads are visible with the wheels removed.

7   Check the pad thickness by looking at each end of the caliper and through the inspection window in the caliper body **(see illustrations)**. If the lining material is less than the thickness listed in this Chapter's Specifications , replace the pads.

Note: *Keep in mind that the lining material is riveted or bonded to a metal backing plate and the metal portion is not included in this measurement.*

8   If it is difficult to determine the exact thickness of the remaining pad material by the above method, or if you are at all concerned about the condition of the pads, remove the caliper(s), then remove the pads from the calipers for further inspection (see Chapter 9).

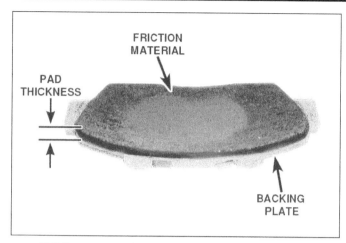

19.9 If a more precise measurement of pad thickness is necessary, remove the pads and measure the remaining friction material

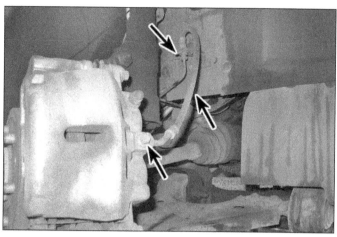

19.11 Check along the brake hoses and at each fitting for deterioration, cracks and leakage

9   Once the pads are removed from the calipers, clean them with brake cleaner and re-measure them with a ruler or a vernier caliper **(see illustration)**.

10   Measure the disc thickness with a micrometer to make sure that it still has service life remaining. If any disc is thinner than the specified minimum thickness, replace it (see Chapter 9). Even if the disc has service life remaining, check its condition. Look for scoring, gouging and burned spots. If these conditions exist, remove the disc and have it resurfaced (see Chapter 9).

11   Before installing the wheels, check all brake lines and hoses for damage, wear, deformation, cracks, corrosion, leakage, bends and twists, particularly in the vicinity of the rubber hoses at the calipers **(see illustration)**. Check the clamps for tightness and the connections for leakage. Make sure that all hoses and lines are clear of sharp edges, moving parts and the exhaust system. If any of the above conditions are noted, repair, reroute or replace the lines and/or fittings as necessary (see Chapter 9).

Brake booster check

12   Sit in the driver's seat and perform the following sequence of tests:

13   With the brake fully depressed, start the engine - the pedal should move down a little when the engine starts.

14   With the engine running, depress the brake pedal several times - the travel distance should not change.

15   Depress the brake, stop the engine and hold the pedal in for about 30 seconds - the pedal should neither sink nor rise.

16   Restart the engine, run it for about a minute and turn it off. Then firmly depress the brake several times - the pedal travel should decrease with each application.

17   If your brakes do not operate as described, the brake booster has failed. Refer to Chapter 9 for the replacement procedure.

Parking brake

18   One method of checking the parking brake is to park the vehicle on a steep hill with the parking brake set and the transaxle in Neutral (be sure to stay in the vehicle for this

check). If the parking brake cannot prevent the vehicle from rolling, it's in need of attention (see Chapter 9).

20   Exhaust system check (every 15,000 miles or 12 months)

1   With the engine cold (at least three hours after the vehicle has been driven), check the complete exhaust system from the manifold to the end of the tailpipe. Be careful around the catalytic converter, which may be hot even after three hours. The inspection should be done with the vehicle on a hoist to permit unrestricted access. If a hoist isn't available, raise the vehicle and support it securely on jackstands.

2   Check the exhaust pipes and connections for signs of leakage and/or corrosion indicating a potential failure. Make sure that all brackets and hangers are in good condition and tight **(see illustrations)**.

3   Inspect the underside of the body for holes, corrosion, open seams, etc. which may allow exhaust gasses to enter the passenger

20.2a Inspect the muffler (A) and all hangers (B) for signs of deterioration

20.2b Inspect all flanged joints for signs of exhaust gas leakage

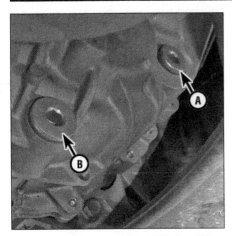

21.2 Transfer case check/fill plug (A) and drain plug (B)

23.6 Remove the cover screws (six total)...

23.7... then lift up the cover and remove the filter element

compartment. Seal all body openings with silicone sealant or body putty.

4　Rattles and other noises can often be traced to the exhaust system, especially the hangers, mounts and heat shields. Try to move the pipes, mufflers and catalytic converter. If the components can come in contact with the body or suspension parts, secure the exhaust system with new brackets and hangers.

21　Transfer case lubricant level check (AWD models) (every 15,000 miles or 12 months)

1　Raise the vehicle and support it securely on jackstands.
2　The transfer case lubricant level is checked by removing the fill plug located at the side of the case **(see illustration)**.
3　The lubricant level should be just at the bottom of the hole. If not, add the appropriate lubricant through the opening.

22　Brake fluid change (every 30,000 miles or 24 months)

Warning: *Brake fluid can harm your eyes and damage painted surfaces, so use extreme caution when handling or pouring it. Do not use brake fluid that has been standing open or is more than one year old. Brake fluid absorbs moisture from the air. Excess moisture can cause a dangerous loss of braking effectiveness.*

1　At the specified intervals, the brake fluid should be drained and replaced. Since the brake fluid may drip or splash when pouring it, place plenty of rags around the master cylinder to protect any surrounding painted surfaces.
2　Before beginning work, purchase the specified brake fluid (see this Chapter's Specifications).
3　Remove the cap from the master cylinder reservoir.
4　Using a hand suction pump or similar device, withdraw the fluid from the master cylinder reservoir.

5　Add new fluid to the master cylinder until it rises to the line indicated on the reservoir.
6　Bleed the brake system at all four brakes until new and uncontaminated fluid is expelled from the bleeder screw (see Chapter 9). Maintain the fluid level in the master cylinder as you perform the bleeding process. If you allow the master cylinder to run dry, air will enter the system.
7　Refill the master cylinder with fluid and check the operation of the brakes. The pedal should feel solid when depressed, with no sponginess.
Warning: *Do not operate the vehicle if you are at all in doubt about the effectiveness of the brake system.*

23　Air filter replacement (every 30,000 miles or 24 months)

1　At the specified intervals, the air filter element should be replaced with a new one.
2　The air filter is housed in a black plastic box mounted on the left side of the engine compartment.
3　Detach the PCV quick-disconnect fitting, if equipped.
4　Disconnect the electrical connector from the Mass Air Flow/Intake Air Temperature (MAF/IAT) sensor (see Chapter 6).
5　Detach the windshield washer hose from the filter housing cover.
6　Loosen the hose clamp and detach the intake duct, then remove the filter housing cover screws **(see illustration)**.
7　Pull up the housing cover, then lift the air filter element out of the housing **(see illustration)**. Wipe out the inside of the air filter housing with a clean rag.
8　Installation is the reverse of removal.

24　Spark plug replacement (see maintenance schedule for service intervals)

1　The spark plugs are threaded into the top of each cylinder head. If you're working on the

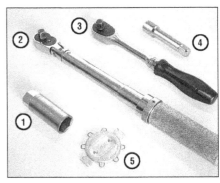

24.2 Tools required for changing spark plugs

1　Spark plug socket - This will have special padding inside to protect the spark plug's porcelain insulator
2　Torque wrench - Although not mandatory, using this tool is the best way to ensure the plugs are tightened properly
3　Ratchet - Standard hand tool to fit the spark plug socket
4　Extension - Depending on model and accessories, you may need special extensions and universal joints to reach one or more of the plugs
5　Spark plug gap gauge - This gauge for checking the gap comes in a variety of styles. Make sure the gap for your engine is included

front cylinder bank (BANK 2) on a 2009 or later model, you'll have to remove the intake manifold (see Chapter 2A) for access.
2　In most cases, the tools necessary for spark plug replacement include a spark plug socket which fits onto a ratchet (spark plug sockets are padded inside to prevent damage to the porcelain insulators on the new plugs), various extensions and a gap gauge to check and adjust the gaps on the new plugs **(see illustration)**. A torque wrench should be used to tighten the new plugs.
3　The best approach when replacing the spark plugs is to purchase the new ones in

24.5 Spark plug manufacturers recommend using a wire-type gauge when checking the gap - the wire should slide between the electrodes with a slight drag

24.8 Use a socket and extension to unscrew the spark plugs - various length extensions and perhaps a flex-joint may be required to reach some plugs

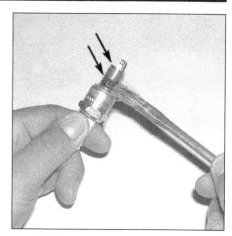

24.9 Apply a thin coat of anti-seize compound to the spark plug threads, being careful not to get any near the lower threads

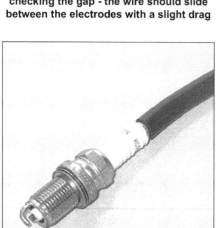

24.10 A length of snug-fitting rubber hose will save time and prevent damaged threads when installing the spark plugs

advance, adjust them to the proper gap and replace them one at a time. When buying the new spark plugs, be sure to obtain the correct plug type for your particular engine. This information can be found in your owner's manual and in this Chapter's Specifications.

4   Allow the engine to cool completely before attempting to remove any of the plugs. While you're waiting for the engine to cool, check the new plugs for defects and adjust the gaps.

5   The gap is checked by inserting the proper-thickness gauge between the electrodes at the tip of the plug **(see illustration)**. The gap between the electrodes should be the same as the one specified in this Chapter's Specifications. The gauge should just slide between the electrodes with a slight amount of drag. If the gap is incorrect, use the adjuster on the gauge body to bend the curved side electrode slightly until the proper gap is obtained. If the side electrode is not exactly over the center electrode, bend it with the adjuster until it is. Check for cracks in the porcelain insulator (if

any are found, the plug should not be used).
Note: *Be careful not to scrape the thin platinum or iridium coating from the electrodes, as this would dramatically shorten the life of the plugs.*

6   Remove the ignition coils (see Chapter 5).

7   If compressed air is available, use it to blow any dirt or foreign material away from around the spark plugs. The idea here is to eliminate the possibility of debris falling into the cylinder as the spark plug is removed.

8   Place the spark plug socket over the plug and remove it from the engine by turning it in a counterclockwise direction **(see illustration)**.

9   Compare each old spark plug with those shown on the inside back cover of this manual to get an indication of the general running condition of the engine. Before installing the new plugs, it is a good idea to apply a thin coat of anti-seize compound to the threads **(see illustration)**.

10   Thread one of the new plugs into the hole until you can no longer turn it with your fingers, then tighten it with a torque wrench (if available) or the ratchet. It's a good idea to slip a short length of rubber hose over the end of the plug to use as a tool to thread it into place **(see illustration)**. The hose will grip the plug well enough to turn it, but will start to slip if the plug begins to cross-thread in the hole - this will prevent damaged threads and the accompanying repair costs.

11   Reinstall the ignition coil.

12   Repeat the procedure for the remaining spark plugs.

---

25   Suspension and steering check (every 30,000 miles or 30 months)

Note: *The steering linkage and suspension components should be checked periodically. Worn or damaged suspension and steering linkage components can result in excessive*

and abnormal tire wear, poor ride quality and vehicle handling and reduced fuel economy. For detailed illustrations of the steering and suspension components, refer to Chapter 10.

Strut and shock absorber check

1   Park the vehicle on level ground, turn the engine off and set the parking brake. Check the tire pressures.

2   Push down at one corner of the vehicle, then release it while noting the movement of the body. It should stop moving and come to rest in a level position within one or two bounces.

3   If the vehicle continues to move up-and-down or if it fails to return to its original position, a worn or weak shock absorber is probably the reason.

4   Repeat the above check at each of the three remaining corners of the vehicle.

5   Raise the vehicle and support it securely on jackstands.

6   Check the struts/shock absorbers for evidence of fluid leakage **(see illustration)**. A light

25.6 Check for signs of fluid leakage at this point on struts/shock absorbers (front strut shown)

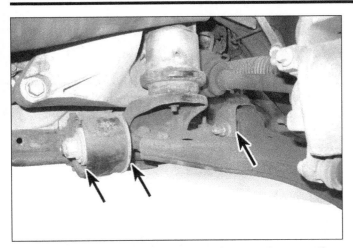

25.9a Examine the mounting points for the control arms on the front suspension

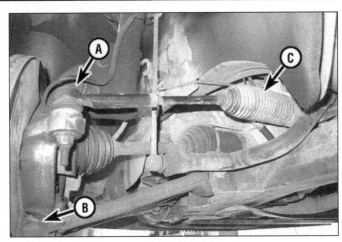

25.9b Inspect the tie-rod ends (A), the balljoints (B) and the steering gear boots (C)

film of fluid is no cause for concern. Make sure that any fluid noted is from the struts/shocks and not from some other source. If leakage is noted, replace the struts/shocks as a set.

7   Check the struts/shocks to be sure that they are securely mounted and undamaged. Check the upper mounts for damage and wear. If damage or wear is noted, replace the struts/shocks as a set (front or rear).

8   If the struts/shocks must be replaced, refer to Chapter 10 for the procedure.

Steering and suspension check

9   Visually inspect the steering and suspension components (front and rear) for damage and distortion. Look for damaged seals, boots and bushings and leaks of any kind. Examine the bushings where the control arms meet the chassis **(see illustrations)**.

10   Clean the lower end of the steering knuckle. Have an assistant grasp the lower edge of the tire and move the wheel in-and-out while you look for movement at the steering knuckle-to-control arm balljoint. If there is

any movement the balljoint must be replaced.

11   Grasp each front tire at the front and rear edges, push in at the front, pull out at the rear and feel for play in the steering system components. If any freeplay is noted, check the tie-rod ends for looseness or wear.

12   Additional steering and suspension system information and illustrations can be found in Chapter 10.

26   Automatic transaxle fluid change (every 30,000 miles or 30 months)

1   At the specified intervals, the transmission fluid should be drained and replaced. Since the fluid will remain hot long after driving, perform this procedure only after the engine has cooled down completely.

2   Before beginning work, purchase the specified transmission fluid (see this Chapter's Specifications).

Note: *On models with a five-speed transaxle, a new sealing washer for the drain plug should be obtained.*

3   Other tools necessary for this job include a floor jack, jackstands to support the vehicle in a raised position, a drain pan capable of holding at least eight quarts, newspapers and clean rags.

4   Raise the vehicle and support it securely on jackstands.

5   Place the drain pan underneath the transaxle drain plug. Remove the drain plug and allow the fluid to drain **(see illustration)**.

6   Install the drain plug and tighten it to the torque listed in this Chapter's Specifications.

7   Lower the vehicle.

8   Remove the transaxle fluid dipstick located on the front side of the transaxle **(see illustration)**.

9   Add approximately 4-1/2 quarts of the specified type of automatic transmission fluid through the dipstick hole. Reinstall the dipstick.

10   Start the engine. With the brake pedal depressed, slowly shift through each gear range, then place the shifter in Park. Let the engine idle for one minute.

11   Using the Driver Information Center, dis-

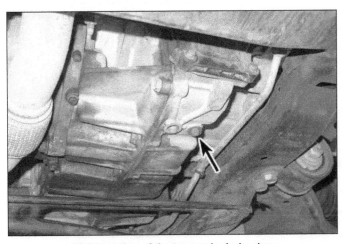

26.5 Location of the transaxle drain plug

26.8 The automatic transaxle fluid dipstick is located towards the front of the engine compartment. The dipstick is mounted to the fluid fill cap

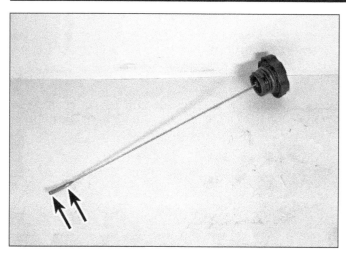

26.13 The transaxle fluid dipstick has COLD and HOT markings, but the final check of the level should always be made when the fluid is hot

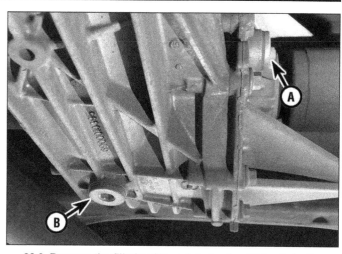

28.3 Remove the fill plug (A) and drain plug (B) to drain the rear differential lubricant

play the transmission fluid temperature (TFT). If the fluid temperature is between 180 and 200-degrees F, proceed to the next Step. If the temperature is not up to 180-degrees F, drive the vehicle until it is. If it's above 200-degrees F, wait until the fluid has cooled back down and falls within the temperature range.

Note: *Some scan tools are capable of displaying transmission fluid temperature.*

12   With the transaxle fluid at the proper temperature, remove the dipstick from the filler tube. Wipe the fluid from the dipstick with a clean rag, then reinsert it.

Note: *It's always a good idea to check the fluid level reading on the dipstick twice to make sure you have a consistent and accurate reading.*

13   Pull the dipstick out again and note the fluid level; it should be in the crosshatched range **(see illustration)**. If additional fluid is required, add fluid a little at a time and keep checking the level until it's correct. It is important to not overfill the transmission.

27   Transfer case lubricant change (AWD models) (every 60,000 miles or 48 months)

1   This procedure should be performed after the vehicle has been driven so the lubricant will be warm and therefore will flow out of the transfer case more easily.

2   Raise the vehicle and support it securely on jackstands.

3   Remove the filler plug from the side of the case.

4   Remove the drain plug from the lower part of the case and allow the lubricant to drain completely **(see illustration 21.2)**.

5   After the case is completely drained, carefully clean and install the drain plug. Tighten the plug to the torque listed in this Chapter's Specifications.

6   Fill the case with the specified lubricant until it is at the proper level (see Section 21).

7   Install the filler plug and tighten it to the torque listed in this Chapter's Specifications.

8   Drive the vehicle for a short distance and recheck the lubricant level. In some instances a small amount of additional lubricant will have to be added.

28   Differential lubricant change (AWD models) (every 60,000 miles or 48 months)

1   This procedure should be performed after the vehicle has been driven, so the lubricant will be warm and therefore will flow out of the differential more easily.

2   Raise the rear of the vehicle and support it securely on jackstands.

3   Remove the check/fill plug, then remove the drain plug and drain the lubricant **(see illustration)**.

Note: *The drain plug is located on the underside of the differential housing.*

4   Reinstall the drain plug and tighten it securely.

5   Add new lubricant until it is up to the proper level (see Section 16). See this Chapter's Specifications for the specified lubricant type.

6   Reinstall the check/fill plug and tighten it to the torque listed in this Chapter's Specifications.

Note: *It may take some driving with sharp right and left turns to distribute the fluid evenly. Recheck the level after about 50 miles of driving.*

29   Cooling system servicing (draining, flushing and refilling) (every 100,000 miles or 60 months)

Warning: *Wait until the engine is completely cool before beginning this procedure.*

Warning: *Do not allow engine coolant (antifreeze) to come in contact with your skin or*

painted surfaces of the vehicle. Rinse off spills immediately with plenty of water. Antifreeze is highly toxic if ingested. Never leave antifreeze lying around in an open container or in puddles on the floor; children and pets are attracted by its sweet smell and may drink it. Check with local authorities about disposing of used antifreeze. Many communities have collection centers which will see that antifreeze is disposed of safely.

Caution: *Never mix green-colored ethylene glycol antifreeze and orange-colored DEX-COOL silicate-free coolant because doing so will destroy the efficiency of the DEX-COOL coolant, which is designed to last for 100,000 miles or five years.*

Note: *Non-toxic coolant is available at local auto parts stores. Although the coolant is non-toxic when fresh, proper disposal is still required.*

1   Periodically, the cooling system should be drained, flushed and refilled to replenish the antifreeze mixture and prevent formation of rust and corrosion, which can impair the performance of the cooling system and cause engine damage. When the cooling system is serviced, all hoses and the expansion tank cap should be checked and replaced if necessary.

Draining

2   Apply the parking brake and block the wheels. If the vehicle has just been driven, wait several hours to allow the engine to cool down before beginning this procedure.

3   Once the engine is completely cool, remove the radiator cap (see Section 14).

4   Move a large container under the lower radiator hose to catch the coolant. Loosen the hose clamp, separate the hose from the radiator, and allow the coolant to drain out.

5   While the coolant is draining, check the condition of the radiator hoses, heater hoses and clamps (see Section 14).

6   Reconnect the hose to the radiator and tighten the clamp securely.

30.2 Squeeze the lock tabs and pull the access door down to open

30.3 Pull the cabin air filter from the housing

Flushing

7   Fill the cooling system with clean water, following the *Refilling* procedure (go to Step 13).

8   Start the engine and allow it to reach normal operating temperature, then rev up the engine a few times.

9   Turn the engine off and allow it to cool completely, then drain the system as described earlier.

10   Repeat Steps 7 through 9 until the water being drained is free of contaminants.

11   In severe cases of contamination or clogging of the radiator, remove the radiator (see Chapter 3) and have a radiator repair facility clean and repair it if necessary.

12   Many deposits can be removed by the chemical action of a cleaner available at auto parts stores. Follow the procedure outlined in the manufacturer's instructions.

Note: *When the coolant is regularly drained and the system refilled with the correct antifreeze/water mixture, there should be no need to use chemical cleaners or descalers.*

Refilling

13   Place the heater temperature control in the maximum heat position.

14   Slowly add new coolant/water mixture to the radiator until it's visible in the filler neck.

15   Install the cap and run the engine in a well-ventilated area until the thermostat opens (coolant will begin flowing through the radiator and the upper radiator hose will become hot; the engine cooling fan will turn on). Allow the engine to run long enough for the cooling fan to cycle on and off twice.

16   Turn the engine off and let it cool. Add more coolant mixture, if necessary to bring the level up into the filler neck. Also add coolant to the coolant reservoir until it is 1/2-inch above the COLD FILL mark.

17   Start the engine, allow it to reach normal operating temperature and check for leaks.

30   Cabin air filter - replacement

1   Remove the glovebox (see Chapter 11).

2   Squeeze the lock tabs to release the cabin air filter access door (see illustration), then pivot the door down.

3   Pull the cabin air filter out of the housing (see illustration).

4   Slide the new cabin air filter into the housing, close the access door and snap it into position.

5   Install the glove box.

Notes

Chapter 2 Part A
3.6L V6 engine

Contents

	Section			Section
Camshafts - removal, inspection and installation	13		Oil pump - removal, inspection and installation	15
Crankshaft front oil seal - removal and installation	11		Rear main oil seal - replacement	17
Crankshaft pulley - removal and installation	10		Repair operations possible with the engine in the vehicle	2
Cylinder heads - removal and installation	9		Rocker arms and hydraulic lash adjusters - removal,	
Driveplate - removal and installation	16		inspection and installation	5
Engine mounts - check and replacement	18		Timing chains and camshaft sprockets - removal,	
Exhaust manifolds - removal and installation	8		inspection and installation	12
General information	1		Top Dead Center (TDC) for number one piston - locating	3
Intake manifold - removal and installation	7		Valve covers - removal and installation	4
Oil pan - removal and installation	14		Valve springs, retainers and seals - replacement	6

Specifications

General

Displacement (3.6L)	217 cubic inches	
Bore and stroke	3.70 x 3.37 inches	
Cylinder numbers (front to rear)		
Left (front) side	2-4-6	
Right (rear) side	1-3-5	
Firing order	1-2-3-4-5-6	

```
 ┌───┐ ┌───┐ ┌───┐      FRONT OF
 │ 1 │ │ 3 │ │ 5 │      VEHICLE
 └───┘ └───┘ └───┘         │
 ┌───┐ ┌───┐ ┌───┐         ▼
 │ 2 │ │ 4 │ │ 6 │
 └───┘ └───┘ └───┘
                    38040-2B-specs HAYNES
```

Cylinder locations

Camshaft

Journal diameters		
Front journal	1.376 inches	34.95 mm
Other journals	1.061 inches	26.95 mm
Camshaft endplay	0.0018 to 0.0085 inch	0.0457 to 0.2159 mm
Valve lift	0.425 inch	10.8 mm
Lobe height		
Intake	1.668 to 1.680 inches	42.367 to 42.672 mm
Exhaust	1.670 to 1.682 inches	42.418 to 42.723 mm

Torque specifications

Ft-lbs (unless otherwise indicated)   **Nm**

Note: *One foot-pound (ft-lb) of torque is equivalent to 12 inch-pounds (in-lbs) of torque. Torque values below approximately 15 foot-pounds are expressed in inch-pounds, because most foot-pound torque wrenches are not accurate at these smaller values*

	Ft-lbs	Nm
Camshaft bearing cap bolts	89 in-lbs	10
Camshaft sprocket bolts (use NEW bolts)	43	58
Camshaft actuator oil valve bolts	89 in-lbs	10
Crankshaft pulley bolt (use NEW bolt)		
Step 1	74	100
Step 2	Tighten an additional 150 degrees	
Cylinder head bolts (in sequence)		
8 mm bolts		
Step 1	132 in-lbs	15
Step 2	Tighten an additional 75 degrees	
11 mm bolts (use NEW bolts)		
Step 1	22	30
Step 2	Tighten an additional 150 degrees	
Driveplate bolts (use NEW bolts)		
Step 1	22	30
Step 2	Tighten an additional 45 degrees	
Engine mount strut bracket-to-engine bolts	66	90
Engine mount strut insulator-to-strut bracket bolts	43	58
Engine mount strut-to-insulator bolt	43	58
Engine mount strut-to-chassis bolt	43	58
Engine mount nuts	37	50
Exhaust manifold bolts	18	24
Exhaust manifold heat shield bolts	89 in-lbs	10
Exhaust manifold-to-catalytic converter nuts	33	45
Intake manifold bolts		
2007 and 2008 models		
Upper intake manifold	17	23
Lower intake manifold	17	23
2009 and later models		
Upper manifold-to-cylinder head	18	25
Upper manifold-to-lower manifold bolts	62 in-lbs	7
Oil pan baffle bolts	89 in-lbs	10
Oil pan-to-block bolts		
All except two rear pan-to-rear main oil seal retainer bolts		
2008 and earlier models	17	23
2009 and later models	18	24
Two rear pan-to-rear main oil seal retainer bolts	89 in-lbs	10
Oil pump mounting bolts	18	24
Rear main oil seal housing bolts	89 in-lbs	10
Timing chain cover bolts		
8 mm		
Step 1	168 in-lbs	19
Step 2	168 in-lbs	19
Step 3	Tighten an additional 60 degrees	
12 mm	48	65
Timing chain guide assembly bolts	18	25
Timing chain tensioner assembly bolts	18	25
Timing chain idler sprocket bolts	43	58
Valve cover bolts	89 in-lbs	10

1   General information

1   This Part of Chapter 2 is devoted to in-vehicle repair procedures for the 3.6L V6 engine. Information concerning engine removal, installation and overhaul can be found in Chapter 2B.

2   Since the repair procedures included in this Part are based on the assumption the engine is still in the vehicle, if they are being used during a complete engine overhaul (with the engine already removed from the vehicle and on a stand) many of the Steps included here will not apply.

3   The engine utilizes an aluminum block with cast-iron sleeves. The six cylinders are arranged in a "V" shape at a 60-degree angle between the two banks. The cylinder heads utilize a twin overhead camshaft arrangement with four valves per cylinder. Variable cam timing is used via adjustable camshaft drive sprockets controlled by the Powertrain Control Module. The engine uses aluminum cylinder heads with powdered metal guides and valve seats. Hydraulic lash adjusters negate the need for valve adjustments and roller rocker arms actuate the valves. The oil pump is mounted at the front of the engine behind the timing chain cover and is driven by the crankshaft.

2   Repair operations possible with the engine in the vehicle

1   Many major repair operations can be accomplished without removing the engine from the vehicle, but due to the vehicle's configuration, there are many jobs that can't. Where this is the case, a reference to Chapter 2B, Section 7 , will be provided.

2   Clean the engine compartment and the exterior of the engine with some type of pressure washer before any work is done. A clean engine will make the job easier and will help keep dirt out of the internal areas of the engine.

3   If oil or coolant leaks develop, indicating a need for gasket or seal replacement, some repairs can be made with the engine in the vehicle. The intake and exhaust manifold gaskets, and the crankshaft oil seals are accessible with the engine in place.

4   Exterior engine components, such as the water pump, the starter motor, the alternator, the power steering pump and the fuel injection components, as well as the intake and exhaust manifolds, can be removed for repair with the engine in place.

5   Since the camshafts can be removed without removing the engine, valve lifter servicing and valve spring/valve stem oil seal replacement can also be accomplished with the engine in the vehicle.

6   Replacement of, repairs to or inspection of the timing chain cover, timing chain, oil pan and oil pump, as well as cylinder head removal and installation, will require engine removal.

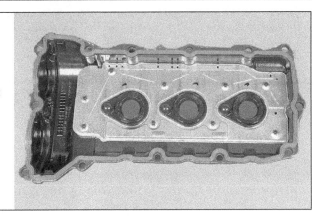

4.9 Make sure the gasket is seated in the valve cover groove

3   Top Dead Center (TDC) for number one piston - locating

1   Top Dead Center (TDC) is the highest point in the cylinder that each piston reaches as it travels up the cylinder bore. Each piston reaches TDC on the compression stroke and again on the exhaust stroke, but TDC generally refers to piston position on the compression stroke.

2   Before beginning this procedure, place the transmission in Neutral and apply the parking brake or block the rear wheels. Remove the spark plugs (see Chapter 1).

3   Disconnect the cable from the negative terminal of the battery (see Chapter 5).

4   In order to bring any piston to TDC, the crankshaft must be turned. When looking at the front of the engine, normal crankshaft rotation is clockwise.

5   Turn the crankshaft with a socket and ratchet attached to the bolt threaded into the front of the crankshaft. Turn the bolt in a clockwise direction.

6   Install a compression gauge in the number one spark plug hole and rotate the crankshaft until pressure registers on the gauge, which indicates the cylinder has started the compression stroke. Once the compression stroke has begun, TDC for the number one cylinder is obtained when the piston reaches the top of the cylinder on the compression stroke.

7   To bring the piston to the top of the cylinder, insert a long wooden dowel into the number one spark plug hole until it touches the top of the piston. Use the dowel (as a feeler gauge) to tell where the top of the piston is located in the cylinder while slowly rotating the crankshaft. As the piston rises, the dowel will be pushed out. The point at which the dowel stops moving outward is TDC.

8   If you go past TDC, rotate the crankshaft counterclockwise until the piston is approximately 1 inch below TDC, then slowly rotate the crankshaft clockwise again until TDC is reached.

9   After the number one piston has been positioned at TDC on the compression stroke, TDC for any of the remaining pistons can be rotating the engine clockwise, 120-degrees at a time, and following the firing order.

4   Valve covers - removal and installation

Removal

1   Disconnect the cable from the negative battery terminal (see Chapter 5).

2   Remove the ignition coils (see Chapter 5).

3   Disconnect the engine wiring harness from the valve cover and position it aside.

4   Remove the upper intake manifold (2008 and earlier models) or intake manifold (2009 and later models) (see Section 7). Detach the power steering fluid reservoir and move it out of the way.

5   Remove the valve cover bolts, then detach the cover from the cylinder head.

Note: *If the cover is stuck to the engine, bump one end with a block of wood and a hammer. If that doesn't work, try to slip a putty knife under it to break the seal. Don't try to pry the cover off as damage to the sealing surface could occur, leading to oil leaks in the future.*

Installation

6   The mating surfaces of each cylinder head and valve cover must be perfectly clean when the covers are reinstalled. Use a gasket scraper to remove all traces of sealant and old gasket material, then clean the mating surfaces with brake system cleaner. If there's sealant or oil on the mating surfaces when the cover is reinstalled, oil leaks may develop.

7   Clean the mounting bolt threads with a die to remove any corrosion and restore damaged threads. Make sure the threaded holes in the cylinder head are clean - run a tap into them to remove corrosion and restore damaged threads.

8   The manufacturer recommends installing tool EN-46101 to the spark plug tube holes to allow the spark plug tube seals to slide over the tubes without tearing. However, applying a small amount of engine oil to the spark plug tube seal inner faces, and using care when installing the seals, the valve cover with the seals attached can slide over the spark plug tubes without the special tools.

9   Position the gasket inside the cover groove **(see illustration)**. If the gasket will not stay in place, apply a thin coat of RTV sealant

4.10 Place some RTV sealant at the joint where the timing chain cover and cylinder head meet

14　Start the engine and check carefully for oil leaks as the engine warms up.

5　Rocker arms and hydraulic lash adjusters - removal, inspection and installation

Removal

1　Remove the valve covers (see Section 4).
2　Remove the camshafts **(see illustration)** (see Section 13).
3　Place markings on each rocker arm to ensure they are returned to the same position, then lift each one from the engine. Once the rocker arms are removed, the hydraulic lash adjusters can be lifted from the cylinder head.
Note: *It is important that any rocker arms or lash adjusters being used again must be returned to their original position on the engine.*

Inspection

4　Check each rocker arm for wear, cracks and other damage, especially where the valve stems contact the rocker arm **(see illustration)**.

to the cover flange so the gasket adheres to the cover.
Note: *If the gasket wasn't leaking, is not damaged and is still pliable, it can be re-used.*
10　Apply an 8 mm wide by 4 mm high bead of RTV sealant to the joints between the timing chain cover and cylinder head mating faces where the valve cover will sit (see illustration).
11　Inspect the valve cover bolt grommets

for damage. If the grommets aren't damaged, they can be reused. Carefully position the valve cover(s) on the cylinder head and install the grommets and bolts.
12　Tighten the bolts in the proper sequence **(see illustrations)**, to the torque listed in this Chapter's Specifications.
13　The remainder of installation is the reverse of removal.

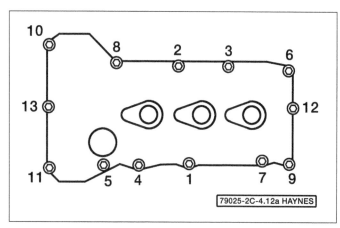

4.12a Valve cover bolt tightening sequence - left side (front)

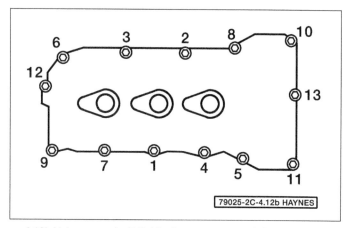

4.12b Valve cover bolt tightening sequence - right side (rear)

5.2 Remove the camshafts (1) to remove the rockers (2). (3) is the lash adjuster socket on the rocker

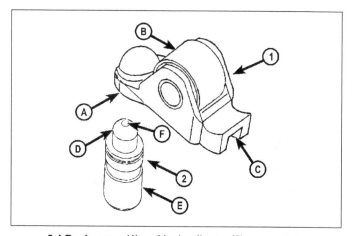

5.4 Rocker arm (1) and lash adjuster (2) wear points

A　Lifter socket
B　Roller
C　Valve stem contact point
D　Lifter-to-rocker contact point
E　Lifter base
F　Lifter oil passage

6.4 Thread the air hose adapter into the spark plug hole - adapters are commonly available at auto parts stores

6.5 After compressing the valve spring, remove the keepers with a magnet or needle-nose pliers

6.7a A screwdriver can be used to pry off the valve seals

5   Check the rollers for binding and roughness. If the bearings are worn or damaged, replacement of the entire rocker arm will be necessary.
Note: *Keep in mind that there is no valve adjustment on these engines, so excessive wear or damage in the valve train can easily result in excessive valve clearance, which in turn will cause valve noise when the engine is running.*
6   Make sure the oil passage hole in each hydraulic lash adjuster is not blocked.

Installation

7   Apply some camshaft/lifter pre-lube to the pivot pockets, roller and the slot that contacts the valve on each rocker arm. Apply clean engine oil into the bores for the hydraulic lash adjusters and install them into the cylinder head. Position the rocker arms onto the lash adjusters.
8   The remainder of installation is the reverse of removal.
9   Before starting and running the engine, change the oil and install a new oil filter (see Chapter 1).

---

6   Valve springs, retainers and seals - replacement

Note: *Broken valve springs and defective valve stem seals can be replaced without removing the cylinder head. Two special tools and a compressed air source are normally required to perform this operation, so read through this Section carefully and rent or buy the tools before beginning the job.*
1   Remove the spark plugs (see Chapter 1).
2   Remove the valve covers (see Section 4), camshafts (see Section 13) and rocker arms (see Section 5).
3   Thread an adapter into the spark plug hole and connect an air hose from a compressed air source to it. Most auto parts stores can supply the air hose adapter.

Note: *Many cylinder compression gauges utilize a screw-in fitting that may work with your air hose quick-disconnect fitting. If a cylinder compression gauge fitting is used, it will be necessary to remove the Schrader valve from the end of the fitting before using it in this procedure.*
4   Apply compressed air to the cylinder. The valves should be held in place by the air pressure **(see illustration)**.
5   Using a socket and a hammer, gently tap on the top of each valve spring retainer several times (this will break the seal between the valve keeper and the spring retainer and allow the keeper to separate from the valve spring retainer as the valve spring is compressed), then use a valve-spring compressor to compress the spring. Remove the keepers with small needle-nose pliers or a magnet **(see illustration)**.
Note: *Several different types of tools are available for compressing the valve springs with the head in place. One type grips the lower spring coils and presses on the retainer as the knob is turned; the lever-type utilizes the rocker arm bolt for leverage. Both types work very well, although the lever type is usually less expensive.*
6   Remove the valve spring and retainer.
Note: *If air pressure fails to retain the valve in the closed position during this operation, the valve face or seat may be damaged. If so, the cylinder head will have to be removed for repair.*
7   Remove the old valve stem seals, noting differences between the intake and exhaust seals **(see illustrations)**.
8   Wrap a rubber band or tape around the top of the valve stem so the valve won't fall into the combustion chamber, then release the air pressure.
9   Inspect the valve stem for damage. Rotate the valve in the guide and check the end for eccentric movement, which would indicate that the valve is bent.
10   Move the valve up-and-down in the guide and make sure it does not bind. If the valve stem binds, either the valve is bent or

6.7b Ensure the seals are installed on the correct valve stems

1   Intake valve seal
2   Exhaust valve seal

the guide is damaged. In either case, the head will have to be removed for repair.
11   Reapply air pressure to the cylinder to retain the valve in the closed position, then remove the tape or rubber band from the valve stem.
12   If you're working on an exhaust valve, fit the new exhaust valve seal on the valve stem and press it down over the valve guide. Don't force the seal against the top of the guide.
13   If you're working on an intake valve, install a new intake valve stem seal over the valve stem and press it down over the valve guide. Don't force the intake valve seal against the top of the guide.
Note: *Do not install an exhaust valve seal on an intake valve, as high oil consumption will result.*
14   Install the spring and retainer in position over the valve.
15   Compress the valve spring assembly only enough to install the keepers in the valve stem.

6.16 Apply a small dab of grease to each keeper as shown here before installation - it'll hold them in place on the valve stem as the spring is released

7.13a Upper intake manifold bolt locations (manifold removed for clarity)

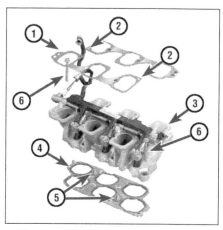

7.13b Typical intake manifold components

1   Upper intake manifold gasket
2   Locating tabs
3   Lower intake manifold
4   Gasket
5   Locating tabs
6   Lower intake manifold-to-cylinder head bolts (2 of 6 shown)

16   Position the keepers in the valve stem groove. Apply a small dab of grease to the inside of each keeper to hold it in place if necessary **(see illustration)**. Remove the pressure from the spring tool and make sure the keepers are seated.

17   Disconnect the air hose and remove the adapter from the spark plug hole.

18   Repeat the above procedure on the remaining cylinders.

19   Install the rocker arm assemblies, camshafts, timing chains and the valve covers (see appropriate Sections).

20   Start the engine, then check for oil leaks and unusual sounds coming from the valve cover area. Allow the engine to idle for at least five minutes before revving the engine.

7   Intake manifold - removal and installation

Warning: *Wait until the engine is completely cool before starting this procedure.*

Removal

1   Disconnect the cable from the negative battery terminal (see Chapter 5).

2008 and earlier models

Upper intake manifold

2   Remove the engine top cover.

3   Remove the air outlet tube from the air filter housing (see Chapter 4).

4   Move the PCV tube from the air filter inlet.

5   Disconnect the electrical connectors to the intake manifold tuning valve, throttle body and Manifold Absolute Pressure (MAP) sensor.

6   Disconnect the brake booster vacuum hose from the intake manifold.

7   Disconnect the wiring from the throttle body.

8   Disconnect the PCV tube from the intake manifold and move it out of the way.

9   Disconnect the quick-connect fitting at the tube from the EVAP canister purge solenoid.

10   Move the EVAP tube aside.

11   Remove the fasteners from the engine wiring harness and position it out of the way.

12   Verify that all interfering wiring and hoses have been disconnected and moved clear of the intake manifold. Use tape and a marker to identify components to avoid confusion later.

13   Completely loosen but do not remove the six mounting bolts from the top of the upper intake manifold, then remove the manifold and upper intake gasket as a unit **(see illustrations)**. Once the assembly has been removed, discard the gasket.

Lower intake manifold

14   Relieve the fuel system pressure (see Chapter 4), disconnect the fuel feed line from the fuel rail, then remove the fuel rail and injectors (see Chapter 4).

15   Remove the lower intake manifold mounting bolts **(see illustration)**, then lift off the manifold.

2009 and later models

Note: *2009 and later models use a single one-piece intake manifold.*

16   Remove the oil filler cap, then lift the engine cover up and off of the ballstuds. Install the oil filler cap.

17   Disconnect the air filter housing outlet duct and remove the air filter housing (see Chapter 4).

18   Disconnect the brake booster hose vacuum and the PCV hose from the manifold.

19   Disconnect the electrical connectors to the EVAP purge solenoid valve and the throttle body (see Chapter 4).

20   Remove the fuel line cover fasteners and cover (see Chapter 4).

21   Remove the engine electrical harness clips and retainers from the intake manifold.

22   Remove the intake manifold sound insulator (if equipped) then remove the intake manifold bolts and intake manifold.

23   Remove and discard the intake manifold gasket.

Installation

Note: *The sealing surfaces of the intake manifold and cylinder heads must be perfectly clean when the manifold is installed.*

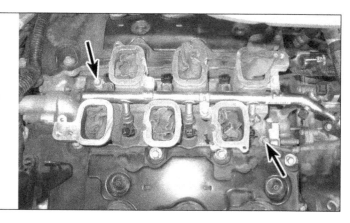

7.15 Lower intake manifold bolts

24 Carefully remove all traces of old gasket material. After the gasket surfaces are cleaned and free of any gasket material, wipe the mating surfaces with a cloth saturated with solvent. If there is old sealant or oil on the mating surfaces when the manifold is reinstalled, oil or vacuum leaks may develop. Use a vacuum cleaner to remove any gasket material that falls into the intake ports in the heads.

25 Use a tap of the correct size to chase the threads in the bolt holes, then use compressed air (if available) to remove the debris from the holes.
Warning: *Wear safety glasses or a face shield to protect your eyes when using compressed air.*

26 Install new gaskets on the cylinder heads, making sure that they're correctly aligned.

2008 and earlier models

27 Install the lower intake manifold. Evenly tighten all the mounting bolts to the torque listed in this Chapter's Specifications.

28 Replace the fuel injectors and the fuel rail (see Chapter 4).

29 Install new gaskets to the top of the lower intake manifold, then replace the upper intake manifold.

30 Install the bolts, then evenly tighten them to the torque listed in this Chapter's Specifications.

31 The remainder of installation is the reverse of removal. Check the coolant level, adding as necessary (see Chapter 1).

2009 and later models

32 Install the intake manifold and tighten the bolts in a criss-cross pattern, to the torque listed in this Chapter's Specifications.

33 The remainder of installation is the reverse of removal. Check the coolant level, adding as necessary (see Chapter 1).

---

8 Exhaust manifolds - removal and installation

Removal

Warning: *Use caution when working around the exhaust manifolds - the sheetmetal heat shields can be sharp on the edges. Also, the engine should be cold when this procedure is followed.*

1 Disconnect the cable from the negative battery terminal (see Chapter 5). Raise the vehicle and support it securely on jackstands.

2 Apply penetrating oil to the studs and nuts that attach the catalytic converters to the exhaust manifolds (they're usually rusty).

3 Remove the catalytic converter bolts and disconnect them from the exhaust manifolds.

4 If you're working on a left (front) manifold, remove the oil dipstick tube.

5 Disconnect the oxygen sensor electrical connectors, then unclip the wiring harness.

6 Remove the exhaust manifold heat shield fasteners and shield.

7 Remove the exhaust manifold mounting bolts **(see illustration)**. The lower bolts for the right (rear) manifold can be removed from under the vehicle.

8 Lift off the exhaust manifolds and discard the gaskets.

Installation

9 Check the manifold for cracks and make sure the bolt threads are clean and undamaged. The manifold and cylinder head mating surfaces must be clean before the manifolds are reinstalled - use a gasket scraper to remove all carbon deposits and gasket material.
Note: *The cylinder heads are made of aluminum, therefore aggressive scraping is not suggested and will damage the sealing surfaces. Also, if any of the manifold bolts are broken, it indicates a warped manifold. Have the manifold machined at a cylinder head reconditioning shop prior to installing the manifold to the vehicle. If this is not done, the manifold may not seal properly and the new manifold bolts will probably break.*

10 Place the manifold on the cylinder head and install the mounting bolts finger tight.

11 When tightening the manifold bolts, work from the center to the ends and be sure to use a torque wrench. Tighten the bolts to the torque listed in this Chapter's Specifications. If required, bend the exposed end of the exhaust manifold gasket back against the cylinder head.

12 Apply an anti-seize compound to the threads of the outer heat shield bolts, then install the bolts and tighten them to the torque listed in this Chapter's Specifications.

13 The remainder of installation is the reverse of removal.

14 Start the engine and check for exhaust leaks.

---

9 Cylinder heads - removal and installation

Warning: *The engine must be completely cool before beginning this procedure.*
Note: *It will be necessary to purchase a new set of 11 mm cylinder head bolts.*

Removal

1 Disconnect the cable from the negative battery terminal (see Chapter 5), then relieve the fuel system pressure (see Chapter 4).

2 Drain the cooling system (see Chapter 1).

3 Remove the engine/transaxle assembly (see Chapter 2B).

4 Remove the intake manifold(s) (see Section 7), the valve covers (see Section 4) and the upper (secondary) timing chains (see Section 12).

5 Disconnect the ground wires from the heads. If you're removing the left (front) cylinder head, remove the dipstick tube and disconnect the electrical connector from the

8.7 Exhaust manifold upper fastener locations (left side shown, right side similar)

Engine Coolant Temperature (ECT) sensor. If you're removing the right (rear) head, detach the power steering hose clip and disconnect the oxygen sensor electrical connector.

6 Disconnect the catalytic converters from the exhaust manifolds (see Section 8). The heads can be removed with the exhaust manifolds attached to them.

7 Remove the camshaft position actuator solenoid valves (see Section 12).

8 Loosen the head bolts in 1/4-turn increments in the reverse order of the tightening sequence **(see illustrations 9.17a and 9.17b)** until they can be removed by hand.
Note: *There will be different lengths and sizes of bolts in different locations. Make notes of where each bolt belongs to aid in assembly.*

9 Carefully lift the heads off the engine block. If you feel resistance, don't pry the head off. To dislodge the head, put a prybar into an intake port and pry with caution. Store the heads on wooden blocks to prevent damage.

Installation

10 The mating surfaces of the cylinder heads and block must be perfectly clean when the heads are reinstalled. Gasket removal solvents are available at auto parts stores and may prove helpful.

11 Use a gasket scraper to remove all traces of carbon and old gasket material, then wipe the mating surfaces with a cloth saturated with brake system cleaner. If there is oil on the mating surfaces when the heads are reinstalled, the gaskets may not seal correctly and leaks may develop. When working on the block, use a vacuum cleaner to remove any debris that falls into the cylinders.
Note: *The cylinder heads are made of aluminum, therefore aggressive scraping is not suggested and will damage the sealing surfaces.*

12 Check the block and head mating surfaces for nicks, deep scratches and other damage. If damage is slight, it can be removed with emery cloth. If it is excessive, machining may be the only alternative.

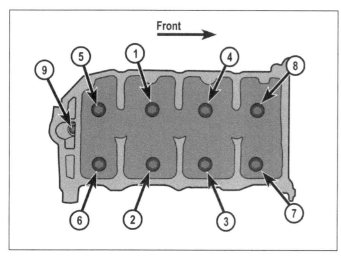

9.17a RH (rear) cylinder head bolt tightening sequence. Bolts 1 to 8 are M11 bolts. Bolt 9 is an M8 bolt

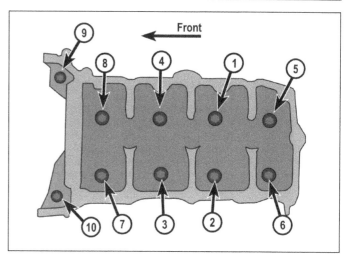

9.17b LH (front) cylinder head bolt tightening sequence. Bolts 1 to 8 are M11 bolts. Bolts 9 and 10 are M8 bolts

10.9 The use of a three jaw puller will be necessary to remove the crankshaft pulley - always place the puller jaws around the hub, not the outer inertia ring

13 Use a tap of the correct size to chase the threads in the head bolt holes in the block. If a tap is not available, spray a liberal amount of brake cleaner into each hole. Use compressed air (if available) to remove the debris from the holes. Corrosion, sealant and damaged threads will affect torque readings, so be sure the threads are clean.
Warning: *Wear safety glasses or a face shield to protect your eyes when using compressed air.*
14 Position the new gaskets over the dowels in the block.
15 Carefully position the heads on the block without disturbing the gaskets.
Warning: *DO NOT reuse 8 mm or 11 mm head bolts - always replace them.*
16 Before installing the *new* 8 mm head bolt(s), coat the threads with a medium-strength thread locking compound. Then install the 8 mm head bolt(s) finger tight (bolt 9 on the

RH cylinder head, bolts 9 and 10 on the LH head).
17 Install *new* 11 mm head bolts (bolts 1 through 8) and tighten them finger tight. Following the recommended sequence **(see illustrations)**, tighten the bolts to the Step 1 torque listed in this Chapter's Specifications. Then tighten them in sequence to the Step 2 angle of rotation listed in this Chapter's Specifications.
18 Tighten the 8 mm bolt(s) to the torque and angle of rotation listed in this Chapter's Specifications.
19 The remainder of installation is the reverse of removal.
20 Reinstall the engine/transaxle assembly (see Chapter 2B).
21 Change the engine oil and filter, and refill the cooling system (see Chapter 1). Start the engine and check for proper operation and coolant or oil leaks.

10   Crankshaft pulley - removal and installation

Removal

Note: *This procedure requires a special pulley installation tool as well as a new crankshaft pulley bolt. Read through the entire procedure and obtain the tools and materials before proceeding.*
1 Disconnect the cable from the negative battery terminal (see Chapter 5).
2 Raise the vehicle and support it securely on jackstands.
3 Remove the drivebelt (see Chapter 1).
4 Install an engine support fixture or an engine hoist to the engine lifting bracket on the right side.
Note: *As an alternative, the engine may be supported from below the oil pan with a floor jack and a block of wood. If this is done, posi-*

tion the wood and the jack to avoid damaging the oil pan.
5 Tighten the engine support chain to remove the weight from the engine mounts.
6 Remove the right side engine mount (see Section 18).
7 Carefully lower the engine about two inches for clearance.
8 Remove the crankshaft pulley bolt. Prevent the crankshaft from turning by removing the starter motor (see Chapter 5) and wedging a large screwdriver or prybar in the driveplate ring gear teeth.
9 Install a large three-jaw puller on the pulley **(see illustration)**, and remove the pulley.
Caution: *The hooks of the puller must contact only the center hub - not the outer (inertia) ring. Also, be sure to use the proper adapter so as not to damage the end of the crankshaft or the threads in the end of the crankshaft.*
Note: *Make note of how far the pulley is pressed onto the crankshaft. It must be installed in the same position. Use the puller to draw the pulley from the crankshaft.*

Installation

10 Lubricate the inside of the pulley bore with clean engine oil.
Note: *Don't let oil contact the outside of the pulley's snout or the oil seal in the timing chain cover. The pulley must only be installed into a dry seal.*
11 Obtain a crankshaft pulley installation tool. They are available at automotive tool dealers and most auto parts stores. Use only this tool to press the pulley into place.
Caution: *Don't try to hammer the pulley on or try to press it into place using a bolt screwed into the crankshaft.*
12 Install the pulley in it's original position. The installation tool should be bottomed on the end of the crankshaft. Remove the installation tool.

10.13a Use a torque wrench to tighten the new crankshaft pulley bolt to the Step 1 torque . . .

10.13b . . . then use a torque angle gauge to finish the tightening sequence

11.2 If you don't have a seal removal tool, thread a self tapping screw partially into the seal, then use pliers as a lever to pull it from the engine

13   Install the NEW crankshaft pulley bolt. Have an assistant wedge the large screwdriver or prybar in the driveplate ring gear teeth, then tighten the bolt to the torque and angle of rotation listed in this Chapter's Specifications **(see illustrations)**.
14   The remainder of installation is the reverse of removal.

11   Crankshaft front oil seal - removal and installation

Removal
1   Remove the crankshaft pulley (see Section 10).
2   Note how the seal is installed - the new one must be installed to the same depth and facing the same way. Carefully pry the oil seal out of the cover **(see illustration)**.
3   If the seal is being replaced with the timing chain cover removed, support the cover on top of two blocks of wood and drive the seal out from the rear with a hammer and punch. **Caution:** *Be careful not to scratch, gouge or distort the area that the seal fits into or a leak will develop.*

Installation
4   Apply clean engine oil or multi-purpose grease to the outer edge of the new seal, then install it in the cover with the lip (spring side) facing IN. Drive the seal into place with a seal driver or a large socket and a hammer. Make sure the seal enters the bore squarely and stop when the front face is at the proper depth. **Note:** *Don't let oil get on the rubber part of the seal or on the part of the crankshaft pulley that touches the seal. The seal must be installed dry.*
5   Check the surface on the pulley hub that the oil seal rides on. If the surface has been

grooved from long-time contact with the seal, the pulley should be replaced with a new one.
6   Lubricate only the inside of the pulley hub with clean engine oil and install the crankshaft pulley (see Section 10).
7   The remainder of installation is the reverse of removal.

12   Timing chains and camshaft sprockets - removal, inspection and installation

Warning: *Wait until the engine is completely cool before beginning this procedure.*
Caution: *The timing system is complex, and severe engine damage will occur if you make any mistakes. Do not attempt this procedure unless you are highly experienced with this type of repair. If you are at all unsure of your abilities, be sure to consult an expert. Double-check all your work and be sure everything is correct before you attempt to start the engine.*
Caution: *Before removing the timing chain(s) from their sprockets(s), check to make sure that all of the sprockets are equipped with timing marks like the ones shown in the illustrations. There are two stages or positions that the engine must be rotated to for alignment of all the timing marks.*
Note: *The engine shown in the illustrations has only one variable camshaft sprocket on each cylinder head. However, procedures for models with dual variable sprockets are identical.*

Removal
Note: *This procedure requires three special tools to lock the camshafts on each cylinder head in position: tool numbers EN 48383-1, 48383-2 and EN 48383-3 are available through SPX Tools. Read through the entire procedure and obtain the tools before proceeding.*

Note: *This procedure can only be performed with the engine removed (see Chapter 2B). In the Steps below, it is assumed that the subframe, engine and transaxle assembly have already been removed.*
1   Disconnect the cable from the negative battery terminal (see Chapter 5).
2   Drain the oil then remove the oil filter (see Chapter 1).
3   Remove the spark plugs (see Chapter 1) to make it easier to rotate the engine.
4   Remove the drivebelt and tensioner (see Chapter 1).
5   Remove the intake manifold (see Section 7).
6   Remove the valve covers (see Section 4).
7   Remove the water pump (see Chapter 3).
8   Remove the power steering reservoir bracket and unbolt the power steering pump and move it out of the way. **Note:** *It is not necessary to disconnect the power steering lines.*
9   Remove the alternator (see Chapter 5).
10   Remove the crankshaft pulley (see Section 10).
11   Remove the Camshaft Position (CMP) sensor fasteners, and remove the sensors from the front cover (see Chapter 6).
12   Disconnect the electrical connectors from the camshaft position actuator valves. Remove the bolts, then remove the four actuator solenoid valves.
13   Remove the engine mount bracket bolts and bracket (see Section 18).
14   Remove the drivebelt idler pulley bolt and pulley from the front of the cover.
15   Remove the thermostat housing (see Chapter 3).
16   Remove all of the bolts from the timing chain cover **(see illustration 12.47** for locations). Screw a 10 mm x 1.5 mm bolt in the threaded hole near the top of the front cover. Tighten this screw while prying at the correct

12.16 A 10 x 1.5 mm bolt can be threaded into the hole near the water pump to press the center portion of the cover away from the engine

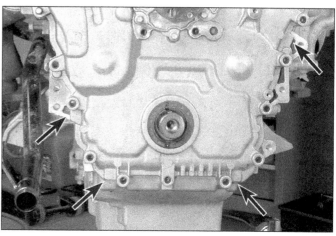

12.17 Use a screwdriver to pry the cover from the engine at the indicated points

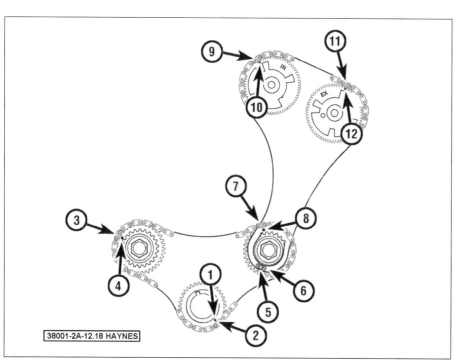

12.18 Stage one timing chain positions

1   Crankshaft sprocket timing mark
2   Primary camshaft drive chain timing plated link for the crankshaft
3   Primary camshaft drive chain timing link for the right (rear bank) primary camshaft intermediate drive sprocket
4   Right (rear bank) primary camshaft intermediate drive chain sprocket timing mark (window) for the right secondary camshaft drive chain
5   Left (front bank) primary camshaft intermediate drive chain sprocket timing window
6   Left (front bank) secondary timing chain timing (plated) link for the left primary camshaft intermediate drive chain sprocket (behind the hole in the sprocket)
7   Primary camshaft drive chain sprocket timing (plated) link for the primary camshaft intermediate drive chain sprocket
8   Left (front bank) primary camshaft intermediate drive chain sprocket timing mark for the primary camshaft drive chain
9   Left (front bank) intake secondary camshaft timing drive chain timing (plated) link
10 Left (front bank) intake camshaft sprocket actuator timing mark (circle)
11 Left (front bank) exhaust camshaft drive chain timing (plated) link
12 Left (front bank) exhaust camshaft sprocket actuator timing mark (circle)

locations to break the cover loose from the engine block **(see illustrations)**.

17   Use a screwdriver in the indicated areas around the perimeter of the cover to pry it from the engine block **(see illustration)**.

18   Reinstall the crankshaft pulley bolt and use it to rotate the crankshaft clockwise until the timing mark on the primary timing chain sprocket aligns with the mark on the oil pump housing in the 5 o'clock position; this is stage one position **(see illustration)**.

Note: *Special tool EN-48589, available from SPX Tools, can also be used to rotate the crankshaft without using the crankshaft pulley bolt.*

Note: *If the RH (rear bank) chain is the only chain being removed, rotate the engine until the timing mark on the primary chain sprocket aligns with the mark on the oil pump housing in the 9 o'clock position.*

19   In this position, the flat segment on the rear of each camshaft on the LH (front) cylinder head should be facing up and parallel with the valve cover mating surface, and the flat segment on the rear of each camshaft. If this is not the case, rotate the crankshaft clockwise one full turn and re-align the crankshaft timing marks. Install the special tool to the camshafts of LH cylinder head **(see illustration)**.

20   Remove the timing chain tensioner bolts for the RH (rear) cylinder head and carefully remove the tensioner and discard the gasket.

Note: *The tensioner is under spring pressure. Use care as the tensioner can fly apart once the bolts are removed.*

21   Remove the RH (rear) cylinder head timing chain guides, then remove the chain **(see illustration)**. Remove the special holding tools from the rear of the camshafts.

22   Repeat the process to remove the lower timing chain or the primary timing chain for the LH (front) cylinder head.

Note: *Again use care as the tensioners are under spring pressure. Do not attempt to remove the timing chain guide from the oil pump housing. At the time of publication, this was only supplied as a complete assembly with the oil pump.*

23   If necessary, after noting their installed

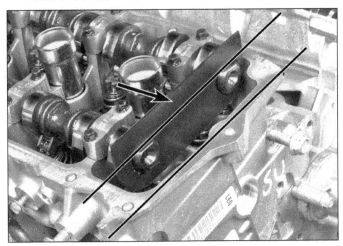

12.19 Typical LH (front) cylinder head with camshaft locking tool EN-48383-1 installed

12.21 RH (rear) cylinder head timing chain tensioner (1), outer guide pivot bolt (2) and inner guide bolts (3)

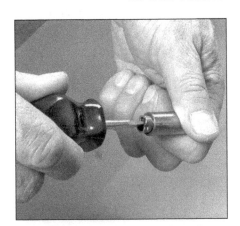

12.29a Remove the plunger from the tensioner and reset it by screwing the bottom of the plunger shaft into the top half of the plunger shaft with a screwdriver

12.29b Slowly push the plunger back into the tensioner body and secure it in place with a paper clip

12.32 Alignment mark between the LH timing chain and idler sprocket

position, remove the bolts retaining the idler sprockets and remove the sprockets.

24   If necessary, mark the relationship between the camshaft sprockets and their respective camshaft. Hold the camshaft steady with an open ended wrench placed on the hex in the camshaft and remove the camshaft sprocket bolt. Remove the sprocket from the camshaft.

Inspection

25   Inspect the timing chain cover for damage. Remove all old sealant from the cover and the engine block.

26   Remove and discard the seal for the coolant passage from the rear of the housing.

27   Clean all components with suitable solvent and inspect the guides and timing chain sprockets for wear or damage. Any gouges or deformities on the guides or tensioners will require replacement parts. Ensure the rear face of the tensioner is clean and there are no marks that would affect the sealing quality once reassembled. Remove any old gasket material from the tensioner mating faces on the engine.

28   Inspect the timing chains for wear, stiff or loose links and binding.

29   Inspect the tensioners for damage or wear. Reset the tensioner by removing the plunger from the tensioner body and screw the rear of the plunger into the plunger shaft using a flat bladed screwdriver. Install the plunger to the body of the tensioner and while maintaining pressure on the plunger, insert a paper clip into the hole in the tensioner body to maintain the tensioner in the retracted position **(see illustrations)**. Repeat this on the other two timing chain tensioners.

Note: *The wire must be inserted into the tensioner to maintain it in the compressed position. If this is not done, the spring pressure will not tension the chain once it is installed to the engine.*

Installation

Caution: *Before starting the engine, carefully rotate the crankshaft by hand through at least two full revolutions (use a socket and breaker bar on the crankshaft pulley center bolt). If you feel any resistance, STOP! There is something wrong - most likely valves are contacting the pistons. You must find the problem before proceeding. Check your work and see if any updated repair information is available.*

30   Align the crankshaft sprocket with the Woodruff key and slide the sprocket onto the crankshaft. Ensure the timing mark on the sprocket is visible and the timing marks are aligned. The crankshaft should be in the stage one position with the crankshaft sprocket mark aligned with the mark on the oil pump housing that is in the 5 o'clock position **(see illustration 12.18)**.

Caution: *If resistance is encountered when installing the sprocket, do not hammer it onto the crankshaft. It may eventually move onto the shaft, but it may be cracked in the process and fail later, causing extensive engine damage.*

31   Install special retaining tool EN-48383-1 onto the LH camshafts **(see illustration 12.19)**, if not already done making sure to fully seat the tool on the ends of the camshaft.

Caution: *Do not rotate the camshaft(s) more than 10-degrees to install the special tool.*

32   Install the left (inner) secondary chain around the sprocket on the left camshaft intermediate drive idler with the timing (plated) link aligned to the alignment access hole in the idler outer sprocket **(see illustration)**.

33   Continue installing the chain over both camshaft actuator sprockets aligning the plated links with the timing marks on the actu-

12.37 Lower timing chain sprocket alignment marks (A). Tensioner (1), upper timing chain guide (2) and lower timing chain guide (3) locations

12.39a Crankshaft sprocket in the correct 9 o'clock position

ators **(see illustration 12.18)** making sure there are 10 links between the two plated links on the secondary chain then check that all of the timing marks are still aligned correctly.

34　Install the outer LH (front) timing chain guide that runs between the idler sprocket and

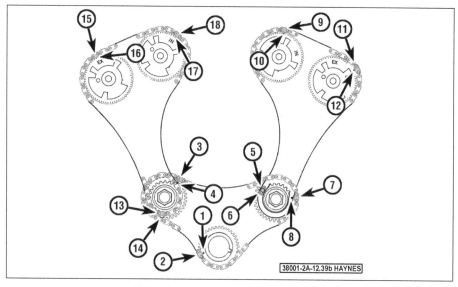

12.39b Stage two timing chain positions

1	Camshaft sprocket timing mark	9	Left intake secondary camshaft timing drive chaintiming (plated) link
2	Primary camshaft drive chain timing plated link for the crankshaft	10	Left intake camshaft sprocket actuator timing mark (circle)
3	Primary camshaft drive chain timing link for the right primary camshaft intermediate drive sprocket	11	Left exhaust camshaft drive chain timing (plated) link
4	Right primary camshaft intermediate drive chain sprocket timing mark (window) for the right secondary camshaft drive chain	12	Left exhaust camshaft sprocket actuator timing mark (circle)
5	Left primary camshaft intermediate drive chain sprocket timing window	13	Right secondary camshaft timing drive chain timing (plated) link for the right primary camshaft intermediate drive chain sprocket
6	Left secondary timing chain timing (plated) link for the left primary camshaft intermediate drive chain sprocket (behind the hole in the sprocket)	14	Right primary camshaft intermediate drive chain sprocket timing (window) mark for the right secondary camshaft timing drive chain
7	Primary camshaft drive chain sprocket timing (plated) link for the primary camshaft intermediate drive sprocket	15	Right exhaust secondary camshaft timing drive chain timing (plated) link
8	Left primary camshaft intermediate drive chain sprocket timing mark for the primary camshaft drive chain	16	Right exhaust camshaft sprocket actuator timing mark (triangle)
		17	Right intake camshaft timing drive chain timing (plated) link
		18	Right intake camshaft sprocket actuator timing mark (triangle)

the exhaust camshaft, then install the inner LH (front) timing chain guide to the engine. Tighten the bolts to the torque listed in this Chapter's Specifications.

Note: *Ensure the inner timing chain guide is not fouling on the mounting pad for the tensioner before tightening the bolts.*

35　Using a new gasket, install the tensioner for the LH (front) timing chain. Ensure the tensioner is aligned with the inner guide and tighten the tensioner bolts to the torque listed in this Chapter's Specifications. Release the pin from the tensioner. Grasp the inner guide and force it into the tensioner. This will allow the tensioner to tension the chain and check that the timing chain alignment marks are all correct.

36　Install the RH (rear) timing chain idler sprocket to the engine ensuring that the small sprocket is facing outward along with the markings "RB Front." Tighten the idler sprocket bolt to the torque listed in this Chapter's Specifications.

37　Ensure the crankshaft sprocket is still aligned with the timing mark on the oil pump housing that is in the 5 o'clock position and install the lower, or primary timing chain. Ensure the shiny chain links align with the arrows on both large idler sprockets and also with the timing mark on the crankshaft sprocket. Install the timing chain guide to the lower chain running between the two idler sprockets, tightening the bolts to the torque listed in this Chapter's Specifications. Using a new gasket, install the tensioner and tighten the tensioner bolts. Remove the pin from the tensioner, force the tensioner guide and push-rod into the tensioner body to unlock it. Allow the tensioner to tension the timing chain. Check that the timing chain alignment marks are all correct. Remove the camshaft special locking tools from the camshafts on both cylinder heads **(see illustration)**.

38　Remove the special retaining tool from the LH camshafts.

39　Rotate the crankshaft 115-degrees, until the timing mark on the crankshaft sprocket is aligned with the mark on the oil pump housing that is in the 9 o'clock position or stage two position **(see illustrations)**.

12.40a RH (rear) cylinder head with camshaft retaining tool installed

12.40b LH (front) cylinder head with camshaft retaining tool installed

40   Install the special locking tool EN-48383-2 onto the LH cylinder head camshafts. Rotate the camshafts for the RH cylinder head and install the other special locking tool to the RH (rear) cylinder head (EN-48383-2). Ensure the tools are correctly seated against the camshafts **(see illustrations)**.
Note: *The camshaft flats for the RH cylinder head must be parallel to the valve cover gasket face on the rear of the cylinder head.*
41   Install the RH cylinder head timing chain, aligning the bright (or marked) chain links with the letter R stamped on the camshaft sprockets. The lower bright link on the chain must align with the hole in the face of the inner idler sprocket. Check that all of the timing marks are still aligned correctly **(see illustration)**.
42   Install the inner RH timing chain guide that runs between the idler sprocket and the intake camshaft, then install the outer RH timing chain guide to the engine. Tighten the bolts to the torque listed in this Chapter's Specifications.
43   Using a new gasket with the tab of the gasket facing towards the center of the engine, install the tensioner for the RH timing chain. Ensure the tensioner is aligned with the outer guide and tighten the tensioner bolts to the torque listed in this Chapter's Specifications. Remove the pin from the tensioner, force the tensioner guide and pushrod into the tensioner body to unlock it. Allow the tensioner to tension the timing chain **(see illustration)**. Check that the timing chain alignment marks are all correct, then remove the camshaft retaining tools and rotate the engine slowly by hand until oil is forced into the three timing chain tensioners.
Caution: *If you encounter any resistance while rotating the engine, STOP! There is something wrong - most likely valves are contacting the pistons. You must find the problem before proceeding. Check your work and see if any updated repair information is available.*
Note: *While doing this, ensure that the timing chains don't jump teeth on any of the timing gears.*

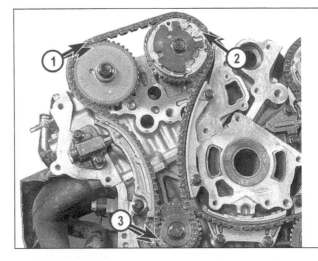

12.41 RH (rear) cylinder head timing chain alignment marks: (1) exhaust camshaft, (2) intake camshaft, (3) idler sprocket

44   Carefully clean all old sealant from the block and the timing chain cover, being careful to avoid scratching the sealing surfaces. Remove all residue with brake system cleaner. Remove the special locking tools from the camshafts.

45   Install a new seal to the inside of the timing cover and apply a bead of RTV sealant to the faces of the cover that will mate with the engine **(see illustration)**. Slide the timing chain cover over the crankshaft and the two

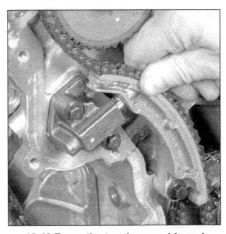

12.43 Force the tensioner guide and chain into the tensioner body to unlock the tensioner

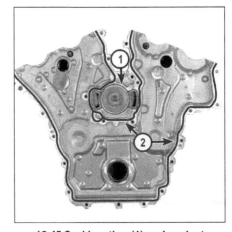

12.45 Seal location (1) and sealant location (2)

12.46 Fabricate two timing chain cover aligning studs from M8 x 1.25 mm bolts with the heads removed and use the plastic sleeves that were between the original timing cover bolts and the timing cover to ensure the cover is in the correct position.

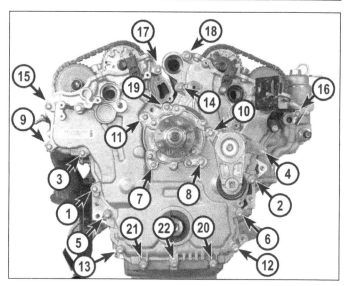

12.47 Timing chain cover bolt tightening sequence

studs, then install the bolts hand tight into the cover.

46 Install two 8 mm studs into the engine block to guide the timing chain cover into the correct position **(see illustration)**.

47 Remove the two 8 mm studs before tightening the timing chain cover bolts in the illustrated order **(see illustration)**.

48 The remainder of installation is the reverse of removal.

13 Camshafts - removal, inspection and installation

Note: *The camshaft bearing caps are marked with a number and letter and a raised arrow. The number signifies the bearing journal (cylinder number) position from the front of the engine. The letter indicates whether it originates from the intake camshaft (I), or the exhaust (E) camshaft and the arrow always points toward the front of the engine.*

Note: *It is necessary to obtain special timing chain holding tools EN 48383-2 and EN 48383-3, available from SPX Tools, to perform this procedure. If you don't use this tool set or a similar fixture, you'll still be able to do the job, however you'll have to remove the timing chain cover to ensure that the chains stay properly attached to their lower sprockets. If you decide to fabricate timing chain holding tools yourself, make sure that they hold the timing chains and upper sprockets securely and tightly in place.*

Removal

1 Disconnect the cable from the negative terminal of the battery (see Chapter 5).

2 Remove the valve cover(s) (see Section 4).

3 Remove the power steering fluid reservoir.

4 On 2008 and earlier models, remove the lower intake manifold (see Section 7).

5 Remove the Camshaft Position (CMP) sensors (see Chapter 6).

6 Remove the camshaft position actuator solenoid valves (see Chapter 6).

7 On 2009 and later models, if you're removing the left side (front cylinder bank) camshafts, remove the high-pressure fuel pump (see Chapter 4).

8 Use a ratchet and socket on the crankshaft pulley bolt to rotate the engine until the flats at the rear of the camshafts you're working on are facing up and are parallel to the surface of the top of the cylinder head. This puts the camshafts in a low-tension position.

9 Check the camshaft endplay by mounting a dial indicator to the front of the engine with the plunger against the end of the camshaft. Lever the camshaft toward the rear of the engine and zero the indicator. Lever it toward the front of the engine and note the reading. Compare this with the endplay listed in this this Chapter's Specifications. Repeat the test on the other camshafts. A camshaft with excessive endplay should be replaced.

Note: *Use care to avoid damaging the camshaft when levering it. It can easily be chipped.*

10 Hold the camshafts with an open-end wrench, then loosen (but don't remove) the camshaft position actuator (sprocket) bolts.

11 Install special tool EN 48313 (or equivalent) to hold the timing chain and sprockets in position. Make sure the tools are engaged securely. Make absolutely sure there's no way an actuator (sprocket) can come loose before proceeding. Make matchmarks on the actuators (sprockets) and the chain in case they get moved.

12 Remove the camshaft actuator (sprocket) bolts, then detach the actuators from the camshafts.

13 Before removing the bearing caps, arrange to store them in a clearly labeled box to ensure that they're installed in their original locations.

14 Remove the camshaft bearing cap bolts in gradual steps in the reverse of the tightening sequence **(see illustrations 13.25 and 13.26)**.

15 Inspect the camshafts for wear and damage. Using a micrometer, measure the journal diameters and lobe heights, comparing your measurements to the values listed in this Chapter's Specifications.

Inspection

16 After the camshaft has been removed, clean it with solvent and dry it, then inspect the bearing journals for uneven wear, pitting and evidence of seizure. If the journals are damaged, the camshaft bearings are probably damaged as well. Both the shaft and bearings will have to be replaced.

17 Measure the bearing journals with a micrometer **(see illustration)** to determine whether they are excessively worn or out-of-round. Measure the camshaft lobes also to check for wear. Measure the camshaft lobes at their highest point, then subtract the measurement of the lobe at its smallest diameter - the difference is the lobe lift **(see illustration)**. Refer to this Chapter's Specifications.

18 Inspect the camshaft lobes for heat discoloration, score marks, chipped areas, pitting and uneven wear. If the lobes are in good condition and if the lobe lift measurements are as specified, you can reuse the camshaft.

19 Check the camshaft bearings in the block

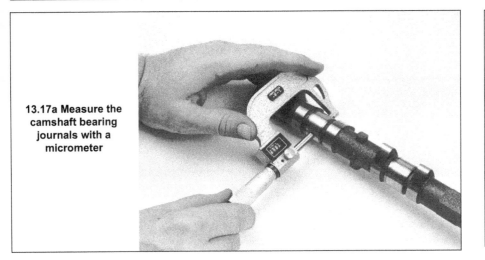

13.17a Measure the camshaft bearing journals with a micrometer

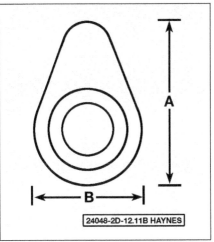

13.17b Measure the camshaft lobe maximum diameter (A) and the minimum diameter (B) - subtract B from A and the difference is the lobe lift

for wear and damage. Look for galling, pitting and discolored areas. Inspect the housing journals for damage and replace them if necessary.

20   The inside diameter of each bearing can be determined with an inside micrometer or a bore gauge and outside micrometer. Subtract the camshaft bearing journal diameter(s) from the corresponding bearing inside diameter(s) to obtain the bearing oil clearance. If it's excessive, new bearings will be required regardless of the condition of the originals. Refer to this Chapter's Specifications.

21   Camshaft bearing replacement requires special tools and expertise that place it outside the scope of the home mechanic. Take the block to an automotive machine shop to ensure the job is done correctly.

Installation

22   Ensure the sealing rings on the front of the camshafts are in place and lubricate the camshaft bearing journals and cam lobes with camshaft installation lube.

23   Working on the RH cylinder head, lubricate the bearing journals and place the cam-

shafts onto the bearing journals. Position the camshafts so the flat portions on the rear of the camshaft are parallel to the valve cover gasket face on the top of the cylinder head. This puts the camshafts in a low-tension position

24   Apply engine oil to the inside of the bearing caps, then install the front bearing cap first, ensuring the thrust face is correctly positioned. Install the remaining bearing caps and hand tighten the bolts.

Note: *Ensure the bearing caps are returned to their original positions and the arrows face the center of the cylinder head.*

25   Tighten the bearing cap bolts for both camshafts in the order illustrated (see illustration) to the torque listed in this Chapter's Specifications. Once the torque procedure is complete, loosen bolts 1, 2 and 3, 4 and retorque them.

26   Repeat Steps 22 through 25 to install the LH cylinder head camshafts (see illustration).

Caution: *If there is any question about the timing chain holding tool retaining the sprockets and chain during the previous Steps, remove*

the timing chain cover to verify that all chains and sprockets are correctly positioned before starting the engine.

27   The remainder of installation is the reverse of removal.

28   Before starting and running the engine, change the oil and install a new oil filter (see Chapter 1).

14   Oil pan - removal and installation

Note: *This procedure can only be done with the engine removed (see Chapter 2B). In the Steps below, it is assumed that the engine has already been removed.*

Removal

1   Remove the engine (see Chapter 2B).

2   Remove the timing chain cover (see Section 12).

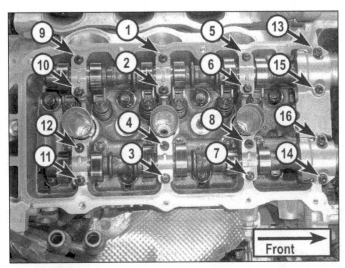

13.25 RH (rear) camshaft bearing cap TIGHTENING sequence

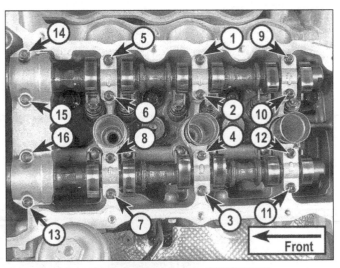

13.26 LH (front) camshaft bearing cap TIGHTENING sequence

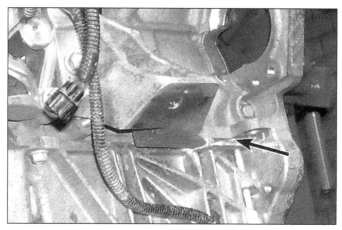

14.3 Arrow indicates one of the pry points to break the seal between the oil pan and cylinder block. There is another one at the front and on the opposite side of the engine

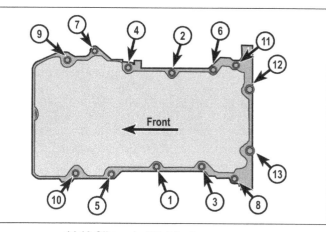

14.10 Oil pan bolt tightening sequence

3   Remove the oil pan mounting bolts, then remove the oil pan by prying only at the proper points **(see illustration)**.

4   The oil pickup can be removed after the bolts in the bottom of the oil pan are removed. The pickup seal must be replaced if removed.

Installation

5   Thoroughly clean the mounting surfaces of the oil pan and engine block of old gasket material and sealer. Wipe the gasket surfaces clean with a rag soaked in brake system cleaner.

6   Check the oil pump pickup for cracks or signs of leakage. Ensure the screen in the bottom of the pickup is not blocked or damaged.

7   Install a new oil seal to the oil pump pickup. Install the oil pickup and tighten the bolts in the bottom of the oil pan. Install the baffle to the inside of the oil pan, and tighten the retaining bolts.

8   Fabricate two aligning studs from M8 x 1.25 mm bolts with the heads removed and screw them into the oil pan bolt holes labeled 1 and 2 to allow for correct oil pan alignment **(see illustration 14.10)**.

9   Apply a 3 mm wide bead of oxygen sensor-safe RTV sealant to the oil pan mating face with the engine block. Also apply some sealant to the corners of the block where the front cover

and the rear cover meet the engine block. Install the oil pan and tighten the bolts finger-tight.

10   Remove the aligning studs, then tighten the oil pan bolts, in sequence **(see illustration)**, to the torque listed in this Chapter's Specifications.

Note: *Check the Specifications for the correct torque. The two bolts holding the oil pan to the rear main oil seal housing are a different torque than the other bolts.*

11   The remainder of installation is the reverse of removal.

12   Install a new oil filter. Add the proper type and quantity of oil (see Chapter 1), start the engine and check for leaks before placing the vehicle back in service.

15   Oil pump - removal, inspection and installation

Removal

1   Remove the timing chains (see Section 12). After noting the installed position, remove the timing chain sprocket from the crankshaft.

2   Remove the oil pump retaining bolts and slide the pump off the end of the crankshaft **(see illustration)**.

Inspection

3   Remove the timing chain guide from the oil pump housing then remove the oil pump cover and withdraw the rotors from the pump body. Remove the snap-ring from the side of the pump housing and withdraw the cap, spring and plunger from the oil pump housing. These components are the oil pressure relief valve assembly. Clean the components with solvent, dry them thoroughly and inspect for any obvious damage. Also check the bolt holes for damaged threads and the splined surfaces on the crankshaft sprocket for any apparent damage. If any of the components are scored, scratched or worn, replace the entire oil pump assembly. There are no serviceable parts currently available.

Installation

4   If re-using the oil pump, assemble the pressure relief valve components and the rotors to the oil pump housing. Prior to installing the cover, prime the pump by pouring clean motor oil between the rotors. Install the cover and tighten the retaining bolts.

5   Position the oil pump over the end of the crankshaft and align the flats on the crankshaft with the flats on the oil pump drive gear. Make sure the pump is fully seated against the block.

6   Install the oil pump mounting bolts and tighten them to the torque listed in this Chapter's Specifications.

7   The remainder of installation is the reverse of removal.

8   Add oil and coolant as necessary. Run the engine and check for oil and coolant leaks. Check the oil pressure (see Chapter 2B).

16   Driveplate - removal and installation

Removal

1   Raise the vehicle and support it securely on jackstands. Remove the transaxle (see Chapter 7A).

15.2 Oil pump retaining bolts

2   Remove the bolts that secure the drive-plate to the crankshaft **(see illustration)**. If the crankshaft turns, insert a screwdriver through one of the holes in the driveplate **(see illustration)**.

Note: *If there is a retaining ring between the bolts and the driveplate, note which side faces the driveplate when removing it.*

3   Remove the driveplate from the crankshaft.

Caution: *When removing a driveplate, wear gloves to protect your fingers - the edges of the ring gear teeth may be sharp.*

4   Clean the driveplate to remove grease and oil. Inspect the surface for cracks, and check for cracked and broken ring gear teeth. Lay the driveplate on a flat surface to check for warpage.

5   Clean and inspect the mating surfaces of the driveplate and the crankshaft. If the crankshaft rear seal is leaking, replace it before reinstalling the driveplate (see Section 17).

Installation

6   Position the driveplate against the crankshaft. Align the marks made during removal. Note that some engines have an alignment dowel or staggered bolt holes to ensure correct installation. Before installing the bolts, apply thread locking compound to the threads and place the retaining ring in position on the driveplate.

7   Wedge a screwdriver through the ring gear teeth to keep the driveplate from turning as you tighten the bolts to the torque listed in this Chapter's Specifications. If the front pump seal/O-ring is leaking, now would be a very good time to replace it.

8   The remainder of installation is the reverse of removal.

---

17   Rear main oil seal - replacement

Note: *At the time of publication, the rear main oil seal is supplied as an assembly consisting*

16.2a Most driveplates have locating dowels - if the one you're working on doesn't have one, make some marks to ensure proper alignment on reassembly

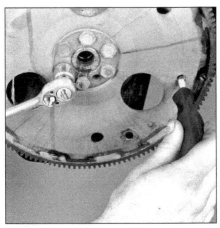

16.2b A large screwdriver wedged in one of the holes in the driveplate can be used to keep the driveplate from turning as the mounting bolts are removed

of the oil seal and housing. The manufacturer recommends using special studs EN 46109, seal alignment tool EN 47839 and handle J 42183 to ensure correct alignment of the seal. These tools are available through SPX Tools.

1   Disconnect the cable from the negative battery terminal (see Chapter 5).

2   Remove the driveplate (see Section 16).

3   Remove the oil pan (see Section 14).

4   Remove the retaining bolts, then remove and discard seal housing from the engine **(see illustration)**. It may be necessary to use a knife to cut the sealant away from the joint between the rear main oil seal housing and the engine block.

Note: *There are specific points provided at the corners for prying the seal housing.*

Caution: *Use care not to scratch the machined faces of the cylinder block, rear main seal housing or oil pan when scraping old gasket material from components.*

5   Thoroughly clean the mounting surfaces of the oil pan, rear main oil seal housing

and engine block of old gasket material and sealer. Wipe the gasket surfaces clean with a rag soaked in brake system cleaner.

6   Install the two dowels into the oil seal housing bolt holes **(see illustration)**.

7   Apply a 1/8-inch wide bead of RTV sealant around the sealing area of the oil seal housing, passing on the inside of the bolt holes.

8   Install the special alignment tool (EN 47839) on the rear of the crankshaft, attaching it with the two supplied screws.

9   Install the new oil seal housing over the alignment tool and studs onto the rear of the engine block. Remove the guide studs but leave the alignment tool in place.

10   Install the housing bolts and tighten them to the torque listed in this Chapter's Specifications.

11   Remove the alignment tool.

12   The remainder of installation is the reverse of removal.

13   Add oil as necessary, run the engine and check for leaks.

17.4 Rear main oil seal housing bolts

17.6 Alignment dowels installed in the rear main oil seal housing bolt holes

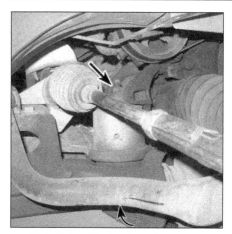

18.9 Right-side (rear) engine mount
fasteners (lower nut not visible)

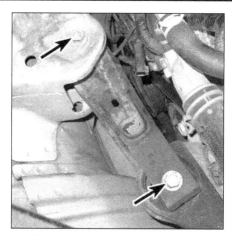

18.16a Engine mount strut fasteners

18.16b Strut mount insulator fasteners

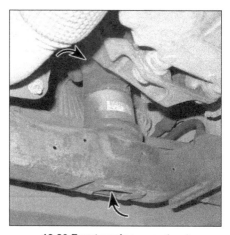

18.20 Front engine mount nut
access points

18   Engine mounts - check and replacement

1   Engine mounts seldom require attention, but broken or deteriorated mounts should be replaced immediately or the added strain placed on the driveline components may cause damage.

Check

2   During the check, the engine must be raised slightly to remove the weight from the mount.

3   Raise the vehicle and support it securely on jackstands, then position the jack under the engine oil pan. Place a large block of wood between the jack head and the oil pan, then carefully raise the engine just enough to take the weight off the mount. Do not use the jack to support the entire weight of the engine.

4   Check the mount to see if the rubber is cracked, hardened or separated from the metal plates. Sometimes the rubber will split right down the center.

5   Check for relative movement between the mount plates and the engine or frame (use a large screwdriver or pry bar to attempt to move the mount). If movement is noted, check the tightness of the mount fasteners first before condemning the mount. Usually when engine mounts are broken, they are very obvious as the engine will easily move away from the mount when pried or under load.

Replacement

Right-side (rear) mount

6   Remove the engine cover.

7   Raise the vehicle and support it securely on jackstands.

8   Support the engine. This can be done by using an engine support fixture that attaches between the fenders and holds the engine from above (the preferred method), and engine hoist, or by using a floor jack with a block of wood placed under the oil pan.

9   Remove the upper and lower nuts from the mount **(see illustration)**, and also remove the upper nut from the front engine mount. Discard the nuts; the manufacturer recommends replacing them with new ones.

10   If equipped with active engine mounts, disconnect the vacuum hose from the mount.

11   Using the support fixture, engine hoist or jack, raise the engine up far enough to remove the mount.

12   Installation is the reverse of removal. Use new fasteners and apply thread-locking compound to the the engine mount studs, then tighten the nuts to the torque listed in this Chapter's Specifications.

Engine mount strut and insulator

Warning: *The engine must be completely cool before beginning this procedure.*

13   If you will be removing the strut mount insulator, drain the cooling system (see Chapter 1).

14   If you will be removing the strut mount insulator, detach the right-side radiator hose from the water outlet tube.

15   Remove the underhood fuse/relay box (see Chapter 12). Reposition the wiring harness as necessary for access to the strut.

16   Unscrew the fasteners and remove the strut from its brackets **(see illustration)**. If necessary, remove the bolts and detach the strut mount insulator from its bracket **(see illustration)**.

17   Installation is the reverse of removal. Refill the cooling system (see Chapter 1).

Front mount

18   Raise the front of the vehicle and support it securely on jackstands.

19   Support the engine. This can be done by using an engine support fixture that attaches between the fenders and holds the engine from above (the preferred method), and engine hoist, or by using a floor jack with a block of wood placed under the oil pan.

20   Remove the upper nut from the right-side (rear) mount, and the upper and lower nuts from the front mount **(see illustration)**. Discard the nuts; the manufacturer recommends replacing them with new ones.

21   Using the support fixture, engine hoist or jack, raise the engine up far enough to remove the mount.

22   Installation is the reverse of removal. Use new fasteners and apply thread-locking compound to the the engine mount studs, then tighten the nuts to the torque listed in this Chapter's Specifications.

Notes

Notes

Chapter 2 Part B
General engine overhaul procedures

Contents

	Section		Section
Crankshaft - removal and installation	10	Engine removal - methods and precautions	6
Cylinder compression check	3	General information - engine overhaul	1
Engine - removal and installation	7	Initial start-up and break-in after overhaul	12
Engine overhaul - disassembly sequence	8	Oil pressure check	2
Engine overhaul - reassembly sequence	11	Pistons and connecting rods - removal and installation	9
Engine rebuilding alternatives	5	Vacuum gauge diagnostic checks	4

Specifications

General

Displacement	
3.6L engine	217 cubic inches
Bore and stroke	3.70 x 3.37
Cylinder compression	Lowest cylinder must be within 75 percent of highest cylinder
Minimum compression pressure	140 psi
Compression ratio	
2008 and earlier 3.6L (LY7)	10.2:1
2009 and later 3.6L (LLT)	11.4:1
Oil pressure (engine at operating temperature)	20 to 45 psi @ 1,850 rpm

Torque specifications

Ft-lbs (unless otherwise indicated)   **Nm**

Note: *One foot-pound (ft-lb) of torque is equivalent to 12 inch-pounds (in-lbs) of torque. Torque values below approximately 15 foot-pounds are expressed in inch-pounds, because most foot-pound torque wrenches are not accurate at these smaller values.*

Driveplate-to-torque converter bolts	44	60
Connecting rod bearing cap bolts*		
Step 1	18	24
Step 2	Tighten an additional 110 degrees	
Main bearing cap bolts*		
Inner		
Step 1	15	20
Step 2	Tighten an additional 80 degrees	
Outer		
Step 1	132 in-lbs	15
Step 2	Tighten an additional 110 degrees	
Side		
Step 1	22	30
Step 2	Tighten an additional 60 degrees	
Subframe mounting bolts	See Chapter 10	

* Bolts must be replaced with NEW bolts

1   General information - engine overhaul

1   Included in this Part of Chapter 2 are general information and diagnostic testing procedures for determining the overall mechanical condition of your engine.

2   The information ranges from advice concerning preparation for an overhaul and the purchase of replacement parts and/or components to detailed, step-by-step procedures covering removal and installation.

3   The following Sections have been written to help you determine whether your engine needs to be overhauled and how to remove and install it once you've determined it needs to be rebuilt. For information concerning in-vehicle engine repair, see Chapter 2A.

4   It's not always easy to determine when, or if, an engine should be completely over-hauled, because a number of factors must be considered.

5   High mileage is not necessarily an indication that an overhaul is needed, while low mileage doesn't preclude the need for an overhaul. Frequency of servicing is probably the most important consideration. An engine that's had regular and frequent oil and filter changes, as well as other required maintenance, will most likely give many thousands of miles of reliable service. Conversely, a neglected engine may require an overhaul very early in its service life.

6   Excessive oil consumption is an indication that piston rings, valve seals and/or valve guides are in need of attention. Make sure that oil leaks aren't responsible before deciding that the rings and/or guides are bad. Perform a cylinder compression check to determine the extent of the work required (see Section 3). Also check the vacuum readings under various conditions (see Section 4).

7   Check the oil pressure with a gauge installed in place of the oil pressure sending unit (see Section 2) and compare it to this Chapter's Specifications. If it's extremely low, the bearings and/or oil pump are probably worn out.

8   Loss of power, rough running, knocking or metallic engine noises, excessive valve train noise and high fuel consumption rates may also point to the need for an overhaul, especially if they're all present at the same time. If a complete tune-up doesn't remedy the situation, major mechanical work is the only solution.

9   An engine overhaul involves restoring the internal parts to the specifications of a new engine. During an overhaul, the piston rings are replaced and the cylinder walls are reconditioned (rebored and/or honed) **(see illustrations)**. If a rebore is done by an auto-

1.9a An engine block being bored. An engine rebuilder will use special machinery to recondition the cylinder bores

1.9b If the cylinders are bored, the machine shop will normally hone the engine on a machine like this

motive machine shop, new oversize pistons will also be installed. The main bearings, connecting rod bearings and camshaft bearings are generally replaced with new ones and, if necessary, the crankshaft may be reground to restore the journals **(see illustration)**. Generally, the valves are serviced as well, since they're usually in less-than-perfect condition at this point. While the engine is being overhauled, other components, such as the distributor, starter and alternator, can be rebuilt as well. The end result should be similar to a new engine that will give many trouble free miles.

Note: *Critical cooling system components such as the hoses, drivebelts, thermostat and water pump should be replaced with new parts when an engine is overhauled. The radiator should be checked carefully to ensure that it isn't clogged or leaking (see Chapter 3). If you purchase a rebuilt engine or short block, some rebuilders will not warranty their engines unless the radiator has been professionally flushed. Also, we don't recommend overhaul-ing the oil pump - always install a new one when an engine is rebuilt.*

10   Overhauling the internal components on today's engines is a difficult and time-consuming task which requires a significant amount of specialty tools and is best left to a professional engine rebuilder **(see illustrations)**. A competent engine rebuilder will handle the inspection of your old parts and offer advice concerning the reconditioning or replacement of the original engine, never purchase parts or have machine work done on other components until the block has been thoroughly inspected by a professional machine shop. As a general rule, time is the primary cost of an overhaul, especially since the vehicle may be tied up for a minimum of two weeks or more. Be aware that some engine builders only have the capability to rebuild the engine you bring them while other rebuilders have a large inventory of rebuilt exchange engines in stock. Also be aware that many machine shops could take as much as two weeks time to completely rebuild your engine depending

on shop workload. Sometimes it makes more sense to simply exchange your engine for another engine that's already rebuilt to save time.

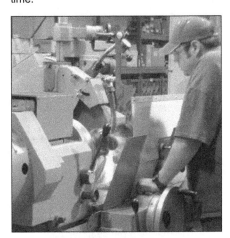

1.9c A crankshaft having a main bearing journal ground

1.10a A machinist checks for a bent connecting rod, using specialized equipment

1.10b A bore gauge being used to check the main bearing bore

1.10c Uneven piston wear like this indicates a bent connecting rod

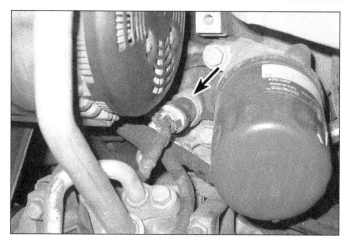

2.2 Oil pressure sending unit location (seen from the front with the radiator removed)

2.3 Remove the oil pressure sending unit and install an oil pressure gauge

3.6 Use a compression gauge with a threaded fitting for the spark plug hole, not the type that requires hand pressure to maintain the seal (typical)

2   Oil pressure check

1   Low engine oil pressure can be a sign of an engine in need of rebuilding. A low oil pressure indicator (often called an "idiot light") is not a test of the oiling system. Such indicators only come on when the oil pressure is dangerously low. Even a factory oil pressure gauge in the instrument panel is only a relative indication, although much better for driver information than a warning light. A better test is with a mechanical (not electrical) oil pressure gauge.

2   Locate the engine oil pressure sending unit on the engine block; it's located on the front of the engine to the lower rear of the alternator, near the oil filter **(see illustration)**.

3   Disconnect the electrical connector, then unscrew and remove the oil pressure sending unit. Screw in the hose for your oil pressure gauge **(see illustration)**. If necessary, install an adapter fitting. Use Teflon tape or thread

sealant on the threads of the adapter and/or the fitting on the end of your gauge's hose.

4   Connect an accurate tachometer to the engine, according to the tachometer manufacturer's instructions.

5   Check the oil pressure with the engine running (normal operating temperature) at the specified engine speed, and compare it to this Chapter's Specifications. If it's extremely low, the bearings and/or oil pump are probably worn out.

3   Cylinder compression check

1   A compression check will tell you what mechanical condition the upper end of your engine (pistons, rings, valves, head gaskets) is in. Specifically, it can tell you if the compression is down due to leakage caused by worn piston rings, defective valves and seats or a blown head gasket.

Note: *The engine must be at normal operating temperature and the battery must be fully charged for this check.*

2   Begin by cleaning the area around the spark plugs before you remove them (compressed air should be used, if available). The idea is to prevent dirt from getting into the cylinders as the compression check is being done.

3   Remove the upper intake manifold (2008 and earlier models) or the intake manifold (2009 and later models) (see Chapter 2A).

4   Remove all of the ignition coils and spark plugs from the engine (see Chapters 1 and 5).

5   Disable the fuel system by removing the fuel pump fuse (see Chapter 12).

6   Install a compression gauge in the number one cylinder spark plug hole **(see illustration)**.

7   Crank the engine over at least seven compression strokes and watch the gauge. The compression should build up quickly in a

healthy engine. Low compression on the first stroke, followed by gradually increasing pressure on successive strokes, indicates worn piston rings. A low compression reading on the first stroke, which doesn't build up during successive strokes, indicates leaking valves or a blown head gasket (a cracked head could also be the cause). Deposits on the undersides of the valve heads can also cause low compression. Record the highest gauge reading obtained.

8   Repeat the procedure for the remaining cylinders and compare the results to this Chapter's Specifications.

9   Add some engine oil (about three squirts from a plunger-type oil can) to each cylinder, through the spark plug hole, and repeat the test.

10   If the compression increases after the oil is added, the piston rings are definitely worn. If the compression doesn't increase significantly, the leakage is occurring at the valves or head gasket. Leakage past the valves may be caused by burned valve seats and/or faces or warped, cracked or bent valves.

11   If two adjacent cylinders have equally low compression, there's a strong possibility that the head gasket between them is blown. The appearance of coolant in the combustion chambers or the crankcase would verify this condition.

12   If one cylinder is slightly lower than the others, and the engine has a slightly rough idle, a worn lobe on the camshaft could be the cause.

13   If the compression is unusually high, the combustion chambers are probably coated with carbon deposits. If that's the case, the cylinder head(s) should be removed and decarbonized.

14   If compression is way down or varies greatly between cylinders, it would be a good idea to have a leak-down test performed by an automotive repair shop. This test will pinpoint exactly where the leakage is occurring and how severe it is.

4   Vacuum gauge diagnostic checks

1   A vacuum gauge provides inexpensive but valuable information about what is going on in the engine. You can check for worn rings or cylinder walls, leaking head or intake manifold gaskets, incorrect carburetor adjustments, restricted exhaust, stuck or burned valves, weak valve springs, improper ignition or valve timing and ignition problems.

2   Unfortunately, vacuum gauge readings are easy to misinterpret, so they should be used in conjunction with other tests to confirm the diagnosis.

3   Both the absolute readings and the rate of needle movement are important for accurate interpretation. Most gauges measure vacuum in inches of mercury (in-Hg). The following references to vacuum assume the diagnosis is being performed at sea level. As elevation increases (or atmospheric pressure decreases), the reading will decrease. For every 1,000 foot increase in elevation above approximately 2,000 feet, the gauge readings will decrease about one inch of mercury.

4   Connect the vacuum gauge directly to the intake manifold vacuum, not to ported (throttle body) vacuum. Be sure no hoses are left disconnected during the test or false readings will result.

5   Before you begin the test, allow the engine to warm up completely. Block the wheels and set the parking brake. With the transaxle in Park, start the engine and allow it to run at normal idle speed.

Warning: *Keep your hands and the vacuum gauge clear of the fans.*

6   Read the vacuum gauge; an average, healthy engine should normally produce about 17 to 22 in-Hg with a fairly steady needle. Refer to the following vacuum gauge readings and what they indicate about the engine's condition **(see illustration)** :

7   A low steady reading usually indicates a leaking gasket between the intake manifold and cylinder head(s) or throttle body, a leaky vacuum hose, late ignition timing or incorrect camshaft timing. Check ignition timing with a timing light and eliminate all other possible causes, utilizing the tests provided in this Chapter before you remove the timing chain cover to check the timing marks.

8   If the reading is three to eight inches below normal and it fluctuates at that low reading, suspect an intake manifold gasket leak at an intake port or a faulty fuel injector.

9   If the needle has regular drops of about two-to-four inches at a steady rate, the valves are probably leaking. Perform a compression check or leak-down test to confirm this.

10   An irregular drop or down-flick of the needle can be caused by a sticking valve or an ignition misfire. Perform a compression check or leak-down test and read the spark plugs.

11   A rapid vibration of about four in-Hg vibration at idle combined with exhaust smoke indicates worn valve guides. Perform a leak-down test to confirm this. If the rapid vibra-

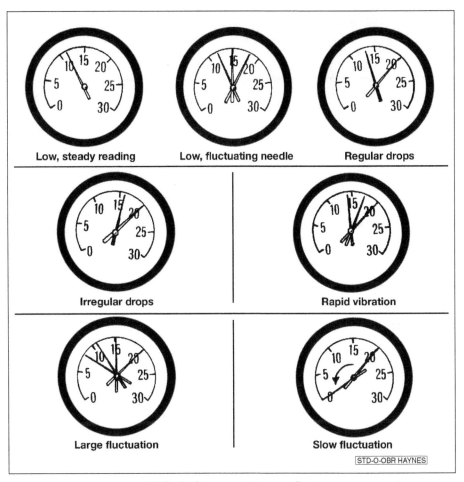

4.6 Typical vacuum gauge readings

tion occurs with an increase in engine speed, check for a leaking intake manifold gasket or head gasket, weak valve springs, burned valves or ignition misfire.

12   A slight fluctuation, say one inch up and down, may mean ignition problems. Check all the usual tune-up items and, if necessary, run the engine on an ignition analyzer.

13   If there is a large fluctuation, perform a compression or leak-down test to look for a weak or dead cylinder or a blown head gasket.

14   If the needle moves slowly through a wide range, check for a clogged PCV system, incorrect idle fuel mixture, throttle body or intake manifold gasket leaks.

15   Check for a slow return after revving the engine by quickly snapping the throttle open until the engine reaches about 2,500 rpm and let it shut. Normally the reading should drop to near zero, rise above normal idle reading (about 5 in-Hg over) and return to the previous idle reading. If the vacuum returns slowly and doesn't peak when the throttle is snapped shut, the rings may be worn. If there is a long delay, look for a restricted exhaust system (often the muffler or catalytic converter). An easy way to check this is to temporarily disconnect the exhaust ahead of the suspected part and redo the test.

5   Engine rebuilding alternatives

1   The do-it-yourselfer is faced with a number of options when purchasing a rebuilt engine. The major considerations are cost, warranty, parts availability and the time required for the rebuilder to complete the project. The decision to replace the engine block, piston/connecting rod assemblies and crankshaft depends on the final inspection results of your engine. Only then can you make a cost effective decision whether to have your engine overhauled or simply purchase an exchange engine for your vehicle.

2   Some of the rebuilding alternatives include:

3   **Individual parts** - If the inspection procedures reveal that the engine block and most engine components are in reusable condition, purchasing individual parts and having a rebuilder rebuild your engine may be the most economical alternative. The block, crankshaft and piston/connecting rod assemblies should all be inspected carefully by a machine shop first.

4   **Short block** - A short block consists of an engine block with a crankshaft and piston/connecting rod assemblies already installed. All new bearings are incorporated and all

6.3a After tightly wrapping water-vulnerable components, use a spray cleaner on everything, with particular concentration on the greasiest areas, usually around the valve cover and lower edges of the block. If one section dries out, apply more cleaner

6.3b Depending on how dirty the engine is, let the cleaner soak in according to the directions and hose off the grime and cleaner. Get the rinse water down into every area you can get at; then dry important components with a hair dryer or paper towels

6.6a Get an engine stand sturdy enough to firmly support the engine while you're working on it. Stay away from three-wheeled models; they have a tendency to tip over more easily, so get a four-wheeled unit

clearances will be correct. The existing camshafts, valve train components, cylinder head and external parts can be bolted to the short block with little or no machine shop work necessary.

5   Long block - A long block consists of a short block plus an oil pump, oil pan, cylinder head, valve cover, camshaft and valve train components, timing sprockets and chain or gears and timing cover. All components are installed with new bearings, seals and gaskets incorporated throughout. The installation of manifolds and external parts is all that's necessary.

6   Low mileage used engines - Some companies now offer low mileage used engines which is a very cost effective way to get your vehicle up and running again. These engines often come from vehicles which have

been in totaled in accidents or come from other countries which have a higher vehicle turn over rate. A low mileage used engine also usually has a similar warranty like the newly remanufactured engines.

7   Give careful thought to which alternative is best for you and discuss the situation with local automotive machine shops, auto parts dealers and experienced rebuilders before ordering or purchasing replacement parts.

6   Engine removal - methods and precautions

1   If you've decided that an engine must be removed for overhaul or major repair work, several preliminary steps should be taken. Read all removal and installation procedures carefully prior to committing to this job.

2   Locating a suitable place to work is extremely important. Adequate work space, along with storage space for the vehicle, will be needed. If a shop or garage isn't available, at the very least a flat, level, clean work surface made of concrete or asphalt is required. A vehicle hoist is necessary for engine removal since the engine and transaxle are removed as an assembly out the bottom of the vehicle.

3   Cleaning the engine compartment and engine before beginning the removal procedure will help keep tools clean and organized **(see illustrations)**.

4   An engine hoist will also be necessary. Make sure the hoist is rated in excess of the combined weight of the engine and transaxle. Safety is of primary importance, considering the potential hazards involved in removing the engine from the vehicle.

5   If you're a novice at engine removal, get at least one helper. One person cannot easily do all the things you need to do to remove a big heavy engine and transaxle assembly from the engine compartment. Also helpful is

to seek advice and assistance from someone who's experienced in engine removal.

6   Plan the operation ahead of time. Arrange for or obtain all of the tools and equipment you'll need prior to beginning the job **(see illustrations)**. Some of the equipment necessary to perform engine removal and installation safely and with relative ease are (in addition to a vehicle hoist and an engine hoist) a heavy duty floor jack (preferably fitted with a transaxle jack head adapter), complete sets of wrenches and sockets as described in the front of this manual, wooden blocks, plenty of rags and cleaning solvent for mopping up spilled oil, coolant and gasoline.

7   Plan for the vehicle to be out of use for quite a while. A machine shop can do the work that is beyond the scope of the home mechanic. Machine shops often have a busy schedule, so before removing the engine, consult the shop for an estimate of how long it will take to rebuild or repair the components that may need work.

6.6b Since many of the fasteners on these engines are tightened using the angle torque method, a torque angle gauge is essential for proper assembly

7   Engine - removal and installation

Warning: *Gasoline is extremely flammable, so take extra precautions when you work on any part of the fuel system. Don't smoke or allow open flames or bare light bulbs near the work area, and don't work in a garage where a gas-type appliance (such as a water heater or clothes dryer) is present. Since gasoline is carcinogenic, wear fuel-resistant gloves when there's a possibility of being exposed to fuel, and, if you spill any fuel on your skin, rinse it off immediately with soap and water. Mop up any spills immediately and do not store fuel-soaked rags where they could ignite. The fuel system is under constant pressure, so, if any fuel lines are to be disconnected, the fuel pressure in the system must be relieved first (see Chapter 4 for more information). When you perform any kind of work on the fuel system, wear safety glasses and have a Class B type fire extinguisher on hand.*

Warning: *The engine must be completely cool before beginning this procedure.*

Note: *Engine removal on these models is a difficult job, especially for the do-it-yourself mechanic working at home. Because of the vehicle's design, the manufacturer states that the engine and transaxle have to be removed as a unit from the bottom of the vehicle, not the top. With a floor jack and jackstands, the vehicle can't be raised high enough and supported safely enough for the engine/transaxle assembly to slide out from underneath. The manufacturer recommends that removal of the engine transaxle assembly only be performed on a frame-contact type vehicle hoist.*

Note: *Read through the entire Section before beginning this procedure. The engine and transaxle are removed as a unit from below, then separated outside the vehicle.*

Removal

Note: *Have the air conditioning system discharged and recovered by an authorized service facility before beginning this procedure.*

1   Park the vehicle on a frame-contact type vehicle hoist, then engage the arms of the hoist with the jacking points of the vehicle. Raise the hoist arms until they contact the vehicle, but not so much that the wheels come off the ground.

2   Position the steering wheel so the front wheels point straight ahead, then disconnect the cable from the negative battery terminal (see Chapter 5).

3   Remove the engine cover.

4   Relieve the fuel system pressure (see Chapter 4).

5   Remove the air filter housing and the air intake duct (see Chapter 4).

6   Disconnect the EVAP hoses and remove the canister purge solenoid valve from the top of the engine (see Chapter 6).

7   Working from under the vehicle, disconnect the main fuel line quick-disconnect connector at the driver's side of the engine compartment, just below the brake booster, and separate the fuel line.

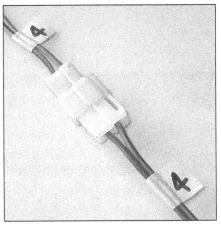

7.15 Label both ends of each wire and hose before disconnecting it

8   Disconnect the fuel line from the fuel rail (see Chapter 4).

9   Drain the cooling system (see Chapter 1) and disconnect the heater hoses at the firewall.

10   Support the radiator/condenser assembly to the body with wire or large plastic tie-wraps, and remove the expansion tank and its hoses.

11   Disconnect and remove the upper and lower radiator hoses.

12   Disconnect the ECM wiring from the underhood fuse center.

13   Disconnect the hoses from the air conditioning compressor and cap the openings at the compressor.

14   Disconnect the shift cable from the transaxle (see Chapter 7A).

15   Clearly label and disconnect all vacuum lines, hoses, wiring harness connectors, negative battery cable extension and fuel lines. Masking tape and/or a touch up paint applicator work well for marking items **(see illustration)**. Take photos or sketch the locations of retainers, clips and brackets. Move the wiring harnesses out of the way.

16   Disconnect the transaxle oil cooler lines from the transaxle. Seal the open ends to prevent contamination.

17   Loosen the wheel lug nuts and the driveaxle/hub nuts, then raise the vehicle on the hoist. Remove the wheels.

18   Unbolt the flexible pipe from the exhaust manifolds and the rear exhaust pipes. Remove the front sections of the exhaust system from the vehicle, then tie up the rear sections using wire.

19   Remove the inner fender splash shields (see Chapter 11).

20   Disconnect the stabilizer bar links, tie-rod ends, and the steering intermediate shaft (see Chapter 10).

Caution: *Don't allow the steering shaft to rotate after the intermediate shaft has been disconnected, as damage to the airbag clockspring could occur. To prevent this, run the seat belt through the steering wheel and click it into its latch.*

21   Disconnect the lower control arms from

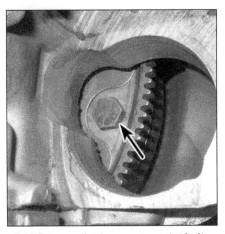

7.29 Remove the torque converter bolts through the starter opening

the steering knuckles (see Chapter 10) and remove the driveaxles (see Chapter 8).

22   On AWD models, remove the rear driveshaft (see Chapter 8).

23   Scribe or make paint marks where the subframe meets the chassis to aid installation.

24   Place four jackstands under the subframe and lower the vehicle until the subframe contacts the jackstands.

25   Remove the subframe bolts and discard them (see Chapter 10).

26   Inspect the engine/transaxle assembly thoroughly once more to make sure that nothing is still attached, then slowly raise the vehicle. Check carefully to make sure nothing is hanging up as this is done.

27   Remove the engine mount fasteners (see Chapter 2A).

28   Connect an engine hoist to the engine, raise the engine/transaxle up a little and support the engine with blocks of wood. Support the transaxle with a floor jack; preferably one with a transmission adapter. Secure the transaxle to the jack with safety chains.

29   Remove the starter (see Chapter 5). Mark the torque converter to the driveplate and remove the driveplate-to-torque converter bolts **(see illustration)**.

30   Remove the engine-to-transaxle brace mounting bolts and brace.

31   On AWD models, remove the transfer case mounting bolts and move the transfer case back.

32   Remove the transaxle-to-engine mounting bolts and separate the engine from the transaxle.

33   Remove the driveplate (see Chapter 2A) and mount the engine on an engine stand.

Installation

34   Installation is the reverse of removal, noting the following points:

a)   Check the engine/transaxle mounts. If they're worn or damaged, replace them.

b)   Replace the subframe mounting bolts with NEW ones.

c)   Attach the transaxle to the engine (see Chapter 7A).

9.1 Before you try to remove the pistons, use a ridge reamer to remove the raised material (ridge) from the top of the cylinders

9.3 Checking the connecting rod endplay (side clearance)

d) When installing the subframe, align the marks made during removal, then tighten the subframe mounting bolts to the torque listed in the Chapter 10 Specifications.

e) Refill the cooling system with the proper mixture of antifreeze. Refill the crankcase with the recommended engine oil (see Chapter 1).

f) Reconnect the battery (see Chapter 5).

g) Run the engine and check for proper operation and leaks. Shut off the engine and recheck fluid levels. Check the transaxle fluid level, adding as necessary (see Chapter 1).

8   Engine overhaul - disassembly sequence

1   It's much easier to remove the external components if it's mounted on an engine stand. A stand can often be rented quite cheaply from an equipment rental yard. Before the engine is mounted on a stand, the driveplate should be removed from the engine.

2   If a stand isn't available, it's possible to remove the external engine components with it blocked up on the floor. Be extra careful not to tip or drop the engine when working without a stand.

3   If you're going to obtain a rebuilt engine, all external components must come off first, to be transferred to the replacement engine. These components include:

   Driveplate
   Intake/exhaust manifolds
   Emissions-related components
   Engine mounts and mount brackets
   Engine rear cover (spacer plate between driveplate and engine block), if equipped
   Fuel injection components
   Oil filter
   Ignition coils and spark plugs
   Thermostat and housing assembly
   Water pump

Note: *When removing the external components from the engine, pay close attention to details that may be helpful or important during installation. Note the installed position of gaskets, seals, spacers, pins, brackets, washers, bolts and other small items.*

4   If you're going to obtain a short block (assembled engine block, crankshaft, pistons and connecting rods), then remove the timing chains, cylinder heads, oil pan, oil pump pick-up tube, oil pump and water pump from your engine so that you can turn in your old short block to the rebuilder as a core. See Section 5 for additional information regarding the different possibilities to be considered.

9   Pistons and connecting rods - removal and installation

Removal

Note: *Prior to removing the piston/connecting rod assemblies, remove the cylinder head and oil pan (see Chapter 2A).*

1   Use your fingernail to feel if a ridge has formed at the upper limit of ring travel (about 1/4-inch down from the top of each cylinder). If carbon deposits or cylinder wear have produced ridges, they must be completely removed with a special tool **(see illustration)**. Follow the manufacturer's instructions provided with the tool. Failure to remove the ridges before attempting to remove the piston/connecting rod assemblies may result in piston breakage.

2   After the cylinder ridges have been removed, turn the engine so the crankshaft is facing up.

3   Before the main bearing caps and connecting rods are removed, check the connecting rod endplay with feeler gauges. Slide them between the first connecting rod and the crankshaft throw until the play is removed **(see illustration)**. Repeat this procedure for each connecting rod. The endplay is equal to the thickness of the feeler gauge(s). Check with an automotive machine shop for the endplay service limit (a typical endplay limit should measure between 0.005 to 0.015 inch [0.127 to 0.381 mm]). If the play exceeds the service limit, new connecting rods will be required. If new rods (or a new crankshaft) are installed, the endplay may fall under the minimum allowable. If it does, the rods will have to be machined to restore it. If necessary, consult an automotive machine shop for advice.

4   Check the connecting rods and caps for identification marks. If they aren't plainly marked, use paint or marker to clearly identify each rod and cap (1, 2, 3, etc., depending on the cylinder they're associated with) **(see illustration)**.

9.4 If the connecting rods and caps are not marked, use paint to mark the caps to the rods by cylinder number (for example, this would be the No. 4 connecting rod)

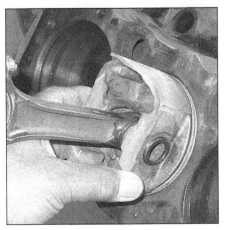

9.13 Install the piston ring into the cylinder then push it down into position using a piston so the ring will be square in the cylinder

9.14 With the ring square in the cylinder, measure the ring end gap with a feeler gauge

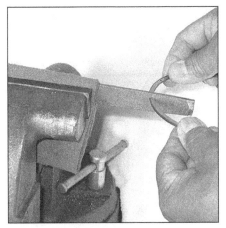

9.15 If the ring end gap is too small, clamp a file in a vise as shown and file the piston ring ends - be sure to remove all raised material

5   Remove the connecting rod cap bolts. **Note:** *New connecting rod cap bolts must be used when reassembling the engine. Save the old bolts for the oil clearance check.*

6   Remove the number one connecting rod cap and bearing insert. Don't drop the bearing insert out of the cap.

7   Remove the bearing insert and push the connecting rod/piston assembly out through the top of the engine. Use a wooden or plastic hammer handle to push on the upper bearing surface in the connecting rod. If resistance is felt, double-check to make sure that all of the ridge was removed from the cylinder.

8   Repeat the procedure for the remaining cylinders.

9   After removal, reassemble the connecting rod caps and bearing inserts in their respective connecting rods and install the cap bolts finger tight. Leaving the old bearing inserts in place until reassembly will help prevent the connecting rod bearing surfaces from being accidentally nicked or gouged.

10   The pistons and connecting rods are now ready for inspection and overhaul at an automotive machine shop.

Piston ring installation

11   Before installing the new piston rings, the ring end gaps must be checked. It's assumed that the piston ring side clearance has been checked and verified correct.

12   Lay out the piston/connecting rod assemblies and the new ring sets so the ring sets will be matched with the same piston and cylinder during the end gap measurement and engine assembly.

13   Insert the top (number one) ring into the first cylinder and square it up with the cylinder walls by pushing it in with the top of the piston **(see illustration)**. The ring should be near the bottom of the cylinder, at the lower limit of ring travel.

14   To measure the end gap, slip feeler gauges between the ends of the ring until a gauge equal to the gap width is found **(see illustra-**

9.19a Installing the spacer/expander in the oil ring groove

tion)**. The feeler gauge should slide between the ring ends with a slight amount of drag. A typical ring gap should fall between 0.010 and 0.020 inch (0.25 to 0.50 mm) for compression rings and up to 0.030 inch (0.76 mm) for the oil ring steel rails. If the gap is larger or smaller than specified, double-check to make sure you have the correct rings before proceeding.

15   If the gap is too small, it must be enlarged or the ring ends may come in contact with each other during engine operation, which can cause serious damage to the engine. If necessary, increase the end gaps by filing the ring ends very carefully with a fine file. Mount the file in a vise equipped with soft jaws, slip the ring over the file with the ends contacting the file face and slowly move the ring to remove material from the ends. When performing this operation, file only by pushing the ring from the outside end of the file towards the vise **(see illustration)**.

16   Excess end gap isn't critical unless it's greater than 0.040 inch (1.01 mm). Again, double-check to make sure you have the correct ring type.

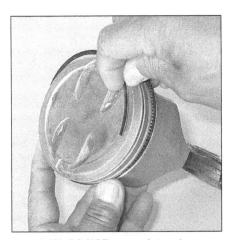

9.19b DO NOT use a piston ring installation tool when installing the oil control side rails

17   Repeat the procedure for each ring that will be installed in the first cylinder and for each ring in the remaining cylinders. Remember to keep rings, pistons and cylinders matched up.

18   Once the ring end gaps have been checked/corrected, the rings can be installed on the pistons.

19   The oil control ring (lowest one on the piston) is usually installed first. It's composed of three separate components. Slip the spacer/expander into the groove **(see illustration)**. If an anti-rotation tang is used, make sure it's inserted into the drilled hole in the ring groove. Next, install the upper side rail in the same manner **(see illustration)**. Don't use a piston ring installation tool on the oil ring side rails, as they may be damaged. Instead, place one end of the side rail into the groove between the spacer/expander and the ring land, hold it firmly in place and slide a finger around the piston while pushing the rail into the groove. Finally, install the lower side rail.

20   After the three oil ring components have been installed, check to make sure that both

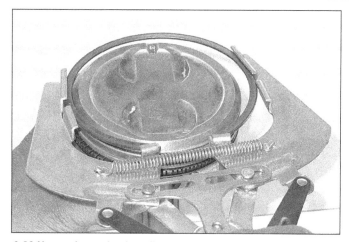

9.22 Use a piston ring installation tool to install the number 2 and the number 1 (top) rings - be sure the directional mark on the piston ring(s) is facing toward the top of the piston

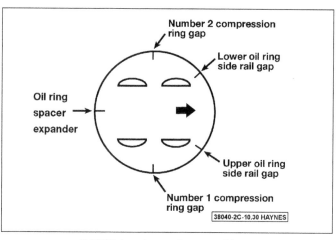

9.30 Piston ring end gap positions

the upper and lower side rails can be rotated smoothly inside the ring grooves.

21 The number two (middle) ring is installed next. It's usually stamped with a mark which must face up, toward the top of the piston. Do not mix up the top and middle rings, as they have different cross-sections.

Note: *Always follow the instructions printed on the ring package or box - different manufacturers may require different approaches.*

22 Use a piston ring installation tool and make sure the identification mark is facing the top of the piston, then slip the ring into the middle groove on the piston **(see illustration)**. Don't expand the ring any more than necessary to slide it over the piston.

23 Install the number one (top) ring in the same manner. Make sure the mark is facing up. Be careful not to confuse the number one and number two rings.

24 Repeat the procedure for the remaining pistons and rings.

Installation

25 Before installing the piston/connecting rod assemblies, the cylinder walls must be perfectly clean, the top edge of each cylinder bore must be chamfered, and the crankshaft must be in place.

26 Remove the cap from the end of the number one connecting rod (refer to the marks made during removal). Remove the original bearing inserts and wipe the bearing surfaces of the connecting rod and cap with a clean, lint-free cloth. They must be kept spotlessly clean.

Connecting rod bearing oil clearance check

27 Clean the back side of the new upper bearing insert, then lay it in place in the connecting rod.

28 Make sure the tab on the bearing fits into the recess in the rod. Don't hammer the bearing insert into place and be very careful not to nick or gouge the bearing face. Don't lubricate the bearing at this time.

29 Clean the back side of the other bearing

insert and install it in the rod cap. Again, make sure the tab on the bearing fits into the recess in the cap, and don't apply any lubricant. It's critically important that the mating surfaces of the bearing and connecting rod are perfectly clean and oil free when they're assembled.

30 Position the piston ring gaps at the specified intervals around the piston as shown **(see illustration)**.

31 Lubricate the piston and rings with clean engine oil and attach a piston ring compressor to the piston. Leave the skirt protruding about 1/4-inch to guide the piston into the cylinder. The rings must be compressed until they're flush with the piston.

32 Rotate the crankshaft until the number one connecting rod journal is at BDC (bottom dead center) and apply a liberal coat of engine oil to the cylinder walls.

33 With the arrow on top of the piston facing the front (timing chain) of the engine, gently insert the piston/connecting rod assembly into the number one cylinder bore and rest the bottom edge of the ring compressor on the engine block.

34 Tap the top edge of the ring compressor to make sure it's contacting the block around its entire circumference.

35 Gently tap on the top of the piston with the end of a wooden or plastic hammer handle **(see illustration)** while guiding the end of the connecting rod into place on the crankshaft journal. The piston rings may try to pop out of the ring compressor just before entering the cylinder bore, so keep some downward force on the ring compressor. Work slowly, and if any resistance is felt as the piston enters the cylinder, stop immediately. Find out what's hanging up and fix it before proceeding. Do not, for any reason, force the piston into the cylinder - you might break a ring and/or the piston.

36 Once the piston/connecting rod assembly is installed, the connecting rod bearing oil clearance must be checked before the rod cap is permanently installed.

37 Cut a piece of the appropriate size Plastigage slightly shorter than the width of the connecting rod bearing and lay it in place on the number one connecting rod journal, parallel with the journal axis **(see illustration)**.

9.35 Use a plastic or wooden hammer handle to push the piston into the cylinder

9.37 Place Plastigage on each connecting rod bearing journal parallel to the crankshaft centerline

9.41 Use the scale on the Plastigage package to determine the bearing oil clearance - be sure to measure the widest part of the Plastigage and use the correct scale; it comes with both standard and metric scales

10.1 Checking crankshaft endplay with a dial indicator

38   Clean the connecting rod cap bearing face and install the rod cap. Make sure the mating mark on the cap is on the same side as the mark on the connecting rod **(see illustration 9.4)**.

39   Install the old rod bolts at this time, and tighten them to the torque listed in this Chapter's Specifications.

Note: *Use a thin-wall socket to avoid erroneous torque readings that can result if the socket is wedged between the rod cap and the bolt. If the socket tends to wedge itself between the fastener and the cap, lift up on it slightly until it no longer contacts the cap. DO NOT rotate the crankshaft at any time during this operation.*

40   Remove the fasteners and detach the rod cap, being very careful not to disturb the Plastigage. Discard the cap bolts at this time as they cannot be reused.

Note: *You MUST use new connecting rod bolts on these engines.*

41   Compare the width of the crushed Plastigage to the scale printed on the Plastigage envelope to obtain the oil clearance **(see illustration)**. The connecting rod oil clearance is usually about 0.001 to 0.002 inch. Consult an automotive machine shop for the clearance specified for the rod bearings on your engine.

42   If the clearance is not as specified, the bearing inserts may be the wrong size (which means different ones will be required). Before deciding that different inserts are needed, make sure that no dirt or oil was between the bearing inserts and the connecting rod or cap when the clearance was measured. Also, recheck the journal diameter. If the Plastigage was wider at one end than the other, the journal may be tapered. If the clearance still exceeds the limit specified, the bearing will have to be replaced with an undersize bearing.

Caution: *When installing a new crankshaft, always use a standard size bearing.*

Final installation

43   Carefully scrape all traces of the Plastigage material off the rod journal and/or bearing face. Be very careful not to scratch the bearing - use your fingernail or the edge of a plastic card.

44   Make sure the bearing faces are perfectly clean, then apply a uniform layer of clean moly-base grease or engine assembly lube to both of them. You'll have to push the piston into the cylinder to expose the face of the bearing insert in the connecting rod.

45   Slide the connecting rod back into place on the journal, install the rod cap, install the *new* bolts and tighten them to the torque listed in this Chapter's Specifications.

Caution: *Install new connecting rod cap bolts on these engines. Do NOT reuse old bolts on these engines - they have stretched and cannot be reused. Also, don't apply any lubricant to the bolts.*

46   Repeat the entire procedure for the remaining pistons/connecting rods.

47   The important points to remember are:

a) *Keep the back sides of the bearing inserts and the insides of the connecting rods and caps perfectly clean when assembling them.*

b) *Make sure you have the correct piston/ rod assembly for each cylinder.*

c) *The arrow on the piston must face the front (timing chain) of the engine.*

d) *Lubricate the cylinder walls liberally with clean oil.*

e) *Lubricate the bearing faces when installing the rod caps after the oil clearance has been checked.*

48   After all the piston/connecting rod assemblies have been correctly installed, rotate the crankshaft a number of times by hand to check for any obvious binding.

49   As a final step, check the connecting rod endplay as described in Step 3. If it was correct before disassembly and the original crankshaft and rods were reinstalled, it should still be correct. If new rods or a new crankshaft were installed, the endplay may be inadequate. If so, the rods will have to be removed and taken to an automotive machine shop for resizing.

10   Crankshaft - removal and installation

Removal

Note: *The crankshaft can be removed only after the engine has been removed from the vehicle. It's assumed that the driveplate, crankshaft pulley, timing chain, oil pan, oil pump body, oil filter and piston/connecting rod assemblies have already been removed. The rear main oil seal retainer must be unbolted and separated from the block before proceeding with crankshaft removal.*

1   Before the crankshaft is removed, measure the endplay. Mount a dial indicator with the indicator in line with the crankshaft and just touching the end of the crankshaft as shown **(see illustration)**.

2   Pry the crankshaft all the way to the rear and zero the dial indicator. Next, pry the crankshaft to the front as far as possible and check the reading on the dial indicator. The distance traveled is the endplay. A typical crankshaft endplay will fall between 0.003 to 0.010 inch (0.076 to 0.254 mm). If it is greater than that, check the crankshaft thrust surfaces for wear after it's removed. If no wear is evident, new main bearings should correct the endplay.

3   If a dial indicator isn't available, feeler gauges can be used. Gently pry the crankshaft all the way to the front of the engine. Slip feeler gauges between the crankshaft and the front face of the thrust bearing or washer to

ENGINE BEARING ANALYSIS

Debris

Babbitt bearing embedded with debris from machinings

Microscopic detail of debris

Microscopic detail of gouges

Overplated copper alloy bearing gouged by cast iron debris

Aluminum bearing embedded with glass beads

Microscopic detail of glass beads

Damaged lining caused by dirt left on the bearing back

Misassembly

Result of a lower half assembled as an upper - blocking the oil flow

Excessive oil clearance is indicated by a short contact arc

Polished and oil-stained backs are a result of a poor fit in the housing bore

Result of a wrong, reversed, or shifted cap

Overloading

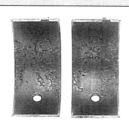

Damage from excessive idling which resulted in an oil film unable to support the load imposed

Damaged upper connecting rod bearings caused by engine lugging; the lower main bearings (not shown) were similarly affected

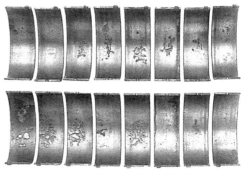

The damage shown in these upper and lower connecting rod bearings was caused by engine operation at a higher-than-rated speed under load

Misalignment

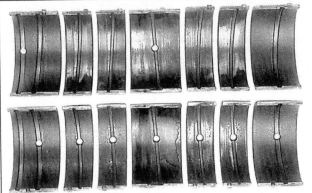

A poorly finished crankshaft caused the equally spaced scoring shown

A tapered housing bore caused the damage along one edge of this pair

A bent connecting rod led to the damage in the "V" pattern

A warped crankshaft caused this pattern of severe wear in the center, diminishing toward the ends

Lubrication

Result of dry start: The bearings on the left, farthest from the oil pump, show more damage

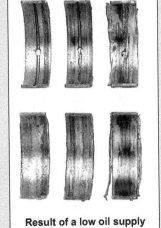

Result of a low oil supply or oil starvation

Severe wear as a result of inadequate oil clearance

Corrosion

Microscopic detail of corrosion

Corrosion is an acid attack on the bearing lining generally caused by inadequate maintenance, extremely hot or cold operation, or inferior oils or fuels

Microscopic detail of cavitation

Example of cavitation - a surface erosion caused by pressure changes in the oil film

Damage from excessive thrust or insufficient axial clearance

Bearing affected by oil dilution caused by excessive blow-by or a rich mixture

10.3 Checking the crankshaft endplay with feeler gauges at the thrust bearing journal

10.18 Place the Plastigage onto the crankshaft bearing journal as shown

determine the clearance **(see illustration)**.

4   Remove the side bearing cap bolts followed by the outermost bearing cap bolts and finally the innermost cap bolts.

5   Remove the main bearing caps. Pull each main bearing cap straight up and off the cylinder block. Gently tap the main bearing cap with a soft-face hammer, if necessary. Try not to drop the bearing inserts if they come out with the caps.

6   Carefully lift the crankshaft out of the engine. It may be a good idea to have an assistant available, since the crankshaft is quite heavy and awkward to handle. With the bearing inserts in place inside the engine block and main bearing caps, reinstall the main bearing caps onto the engine block and tighten the bolts finger tight. Make sure the caps are in the exact order they were removed with the arrow pointing toward the front (timing chain and front cover) of the engine.

Installation

7   Crankshaft installation is the first step in engine reassembly. It's assumed at this point that the engine block and crankshaft have been cleaned, inspected and repaired or reconditioned.

8   Position the engine block with the bottom facing up.

9   Remove the old bolts and lift off the main bearing caps.

10   If they're still in place, remove the original bearing inserts from the block and from the main bearing caps. Wipe the bearing surfaces of the block and main bearing caps with a clean, lint-free cloth. They must be kept spotlessly clean. This is critical for determining the correct bearing oil clearance.

Main bearing oil clearance check

11   Without mixing them up, clean the back sides of the new upper main bearing inserts (with grooves and oil holes) and lay one in each main bearing saddle in the engine block. Each upper bearing (engine block) has an oil

groove and oil hole in it.

Caution: *The oil holes in the block must line up with the oil holes in the engine block inserts.*

12   The thrust washer or thrust bearing insert must be installed in the correct location.

Note: *Clean the back sides of the lower main bearing inserts and lay them in the corresponding location in the main bearing caps. Make sure the tab on the bearing insert fits into the recess in the block or main bearing caps.*

Caution: *Do not hammer the bearing insert into place and don't nick or gouge the bearing faces. DO NOT apply any lubrication at this time.*

13   Clean the faces of the bearing inserts in the block and the crankshaft main bearing journals with a clean, lint-free cloth.

14   Check or clean the oil holes in the crankshaft, as any dirt here can go only one way - straight through the new bearings.

15   Once you're certain the crankshaft is clean, carefully lay it in position in the cylinder block.

16   Before the crankshaft can be permanently installed, the main bearing oil clearance must be checked.

17   Cut several strips of the appropriate size of Plastigage. They must be slightly shorter than the width of the main bearing journal.

18   Place one piece on each crankshaft main bearing journal, parallel with the journal axis as shown **(see illustration)**.

19   Clean the faces of the bearing inserts in the main bearing caps or lower crankcase. Install the caps without disturbing the Plastigage.

20   Apply clean engine oil to all bolt threads prior to installation, install all the old bolts finger-tight, then tighten them to the torque listed in this Chapter's Specifications. DO NOT rotate the crankshaft at any time during this operation.

21   Remove the bolts and carefully lift the main bearing caps straight up and off the block. Do not disturb the Plastigage or rotate the crankshaft.

22   Compare the width of the crushed Plastigage on each journal to the scale printed on the Plastigage envelope to determine the main bearing oil clearance **(see illustration)**. Check with an automotive machine shop for the crankshaft bearing oil clearance for your engine.

23   If the clearance is not as specified, the bearing inserts may be the wrong size (which means different ones will be required). Before deciding if different inserts are needed, make sure that no dirt or oil was between the bearing inserts and the caps or block when the clearance was measured. If the Plastigage was wider at one end than the other, the crankshaft journal may be tapered. If the clearance still exceeds the limit specified, the bearing insert(s) will have to be replaced with an undersize bearing insert(s).

Caution: *When installing a new crankshaft, always install a standard bearing insert set.*

24   Carefully scrape all traces of the Plastigage material off the main bearing journals

10.22 Use the scale on the Plastigage package to determine the bearing oil clearance - be sure to measure the widest part of the Plastigage and use the correct scale; it comes with both standard and metric scales

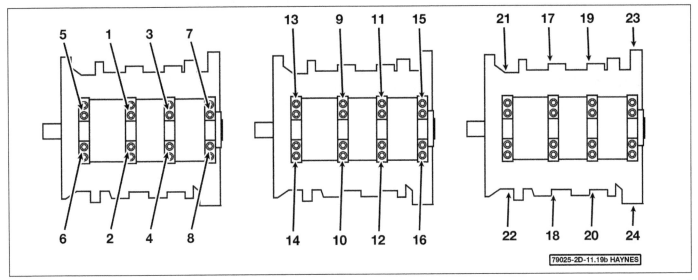

10.30 Crankshaft main bearing cap TIGHTENING sequence; all bolts must be replaced with new ones whenever they have been loosened

and/or the bearing insert faces. Be sure to remove all residue from the oil holes. Use your fingernail or the edge of a plastic card - don't nick or scratch the bearing faces.

Final installation

25 Carefully lift the crankshaft out of the cylinder block.

26 Clean the bearing insert faces in the cylinder block, then apply a thin, uniform layer of moly-base grease or engine assembly lube to each of the bearing surfaces. Be sure to coat the thrust faces as well as the journal face of the thrust bearing.

27 Make sure the crankshaft journals are clean, then lay the crankshaft back in place in the cylinder block.

28 Clean the bearing insert faces and apply the same lubricant to their faces only. Clean the engine block and the bearing cap mating surfaces thoroughly. The surfaces must be free of oil residue.

29 Prior to installation, apply clean engine oil to the NEW bolt threads, wiping off any excess, then install all bolts finger-tight.

Caution: *Install NEW main bearing cap bolts. Do NOT reuse old bolts on these engines - they have stretched insert and cannot be reused.*

30 Tighten the new main bearing cap bolts in sequence to the torque listed in this Chapter's Specifications **(see illustration)**.

31 Install new side bolts and tighten them, in sequence **(see illustration 10.30)**, to the torque listed in this Chapter's Specifications. **Note:** *The side bolts have a special seal. If they are not replaced with new ones, oil leaks can develop. The four corner bolts are longer than the four inner ones.*

32 Recheck the crankshaft endplay with a feeler gauge or a dial indicator. The endplay should be correct if the crankshaft thrust faces aren't worn or damaged and if new bearings have been installed.

33 Rotate the crankshaft a number of times by hand to check for any obvious binding. It

should rotate with a running torque of 50 in-lbs or less. If the running torque is too high, correct the problem at this time.

34 Install the new rear main oil seal (see Chapter 2A).

---

11   Engine overhaul - reassembly sequence

1 Before beginning engine reassembly, make sure you have all the necessary new parts, gaskets and seals as well as the following items on hand:

Common hand tools
A 1/2-inch drive torque wrench
New engine oil
Gasket sealant
Thread locking compound

2 If you obtained a short block it will be necessary to install the cylinder head, the oil pump and pick-up tube, the oil pan, the water pump, the timing belt and timing cover, and the valve cover (see Chapter   2A). In order to save time and avoid problems, the external components must be installed in the following general order:

Thermostat and housing cover
Water pump
Intake and exhaust manifolds
Fuel injection components
Emission control components
Spark plugs
Ignition coils
Oil filter
Engine mounts and mount brackets
Driveplate

---

12   Initial start-up and break-in after overhaul

Warning: *Have a fire extinguisher handy when starting the engine for the first time.*

1 Once the engine has been installed in the vehicle, double-check the engine oil and coolant levels.

2 With the spark plugs out of the engine, and the ignition system and fuel pump disabled (see Chapter 4), crank the engine until oil pressure registers on the gauge or the light goes out.

3 Install the spark plugs, hook up the plug wires and restore the ignition system and fuel pump functions.

4 Start the engine. It may take a few moments for the fuel system to build up pressure, but the engine should start without a great deal of effort.

5 After the engine starts, it should be allowed to warm up to normal operating temperature. While the engine is warming up, make a thorough check for fuel, oil and coolant leaks.

6 Shut the engine off and recheck the engine oil and coolant levels.

7 Drive the vehicle to an area with minimum traffic, accelerate from 30 to 50 mph, then allow the vehicle to slow to 30 mph with the throttle closed. Repeat the procedure 10 or 12 times. This will load the piston rings and cause them to seat properly against the cylinder walls. Check again for oil and coolant leaks.

8 Drive the vehicle gently for the first 500 miles (no sustained high speeds) and keep a constant check on the oil level. It is not unusual for an engine to use oil during the break-in period.

9 At approximately 500 to 600 miles, change the oil and filter.

10 For the next few hundred miles, drive the vehicle normally. Do not pamper it or abuse it.

11 After 2,000 miles, change the oil and filter again and consider the engine broken in.

COMMON ENGINE OVERHAUL TERMS

B

Backlash - The amount of play between two parts. Usually refers to how much one gear can be moved back and forth without moving the gear with which it's meshed.

Bearing Caps - The caps held in place by nuts or bolts which, in turn, hold the bearing surface. This space is for lubricating oil to enter.

Bearing clearance - The amount of space left between shaft and bearing surface. This space is for lubricating oil to enter.

Bearing crush - The additional height which is purposely manufactured into each bearing half to ensure complete contact of the bearing back with the housing bore when the engine is assembled.

Bearing knock - The noise created by movement of a part in a loose or worn bearing.

Blueprinting - Dismantling an engine and reassembling it to EXACT specifications.

Bore - An engine cylinder, or any cylindrical hole; also used to describe the process of enlarging or accurately refinishing a hole with a cutting tool, as to bore an engine cylinder. The bore size is the diameter of the hole.

Boring - Renewing the cylinders by cutting them out to a specified size. A boring bar is used to make the cut.

Bottom end - A term which refers collectively to the engine block, crankshaft, main bearings and the big ends of the connecting rods.

Break-in - The period of operation between installation of new or rebuilt parts and time in which parts are worn to the correct fit. Driving at reduced and varying speed for a specified mileage to permit parts to wear to the correct fit.

Bushing - A one-piece sleeve placed in a bore to serve as a bearing surface for shaft, piston pin, etc. Usually replaceable.

C

Camshaft - The shaft in the engine, on which a series of lobes are located for operating the valve mechanisms. The camshaft is driven by gears or sprockets and a timing chain. Usually referred to simply as the cam.

Carbon - Hard, or soft, black deposits found in combustion chamber, on plugs, under rings, on and under valve heads.

Cast iron - An alloy of iron and more than two percent carbon, used for engine blocks and heads because it's relatively inexpensive and easy to mold into complex shapes.

Chamfer - To bevel across (or a bevel on) the sharp edge of an object.

Chase - To repair damaged threads with a tap or die.

Combustion chamber - The space between the piston and the cylinder head, with the piston at top dead center, in which air-fuel mixture is burned.

Compression ratio - The relationship between cylinder volume (clearance volume) when the piston is at top dead center and cylinder volume when the piston is at bottom dead center.

Connecting rod - The rod that connects the crank on the crankshaft with the piston. Sometimes called a con rod.

Connecting rod cap - The part of the connecting rod assembly that attaches the rod to the crankpin.

Core plug - Soft metal plug used to plug the casting holes for the coolant passages in the block.

Crankcase - The lower part of the engine in which the crankshaft rotates; includes the lower section of the cylinder block and the oil pan.

Crank kit - A reground or reconditioned crankshaft and new main and connecting rod bearings.

Crankpin - The part of a crankshaft to which a connecting rod is attached.

Crankshaft - The main rotating member, or shaft, running the length of the crankcase, with offset throws to which the connecting rods are attached; changes the reciprocating motion of the pistons into rotating motion.

Cylinder sleeve - A replaceable sleeve, or liner, pressed into the cylinder block to form the cylinder bore.

D

Deburring - Removing the burrs (rough edges or areas) from a bearing.

Deglazer - A tool, rotated by an electric motor, used to remove glaze from cylinder walls so a new set of rings will seat.

E

Endplay - The amount of lengthwise movement between two parts. As applied to a crankshaft, the distance that the crankshaft can move forward and back in the cylinder block.

F

Face - A machinist's term that refers to removing metal from the end of a shaft or the face of a larger part, such as a flywheel.

Fatigue - A breakdown of material through a large number of loading and unloading cycles. The first signs are cracks followed shortly by breaks.

Feeler gauge - A thin strip of hardened steel, ground to an exact thickness, used to check clearances between parts.

Free height - The unloaded length or height of a spring.

Freeplay - The looseness in a linkage, or an assembly of parts, between the initial application of force and actual movement. Usually perceived as slop or slight delay.

Freeze plug - See Core plug.

G

Gallery - A large passage in the block that forms a reservoir for engine oil pressure.

Glaze - The very smooth, glassy finish that develops on cylinder walls while an engine is in service.

H

Heli-Coil - A rethreading device used when threads are worn or damaged. The device is installed in a retapped hole to reduce the thread size to the original size.

I

Installed height - The spring's measured length or height, as installed on the cylinder head. Installed height is measured from the spring seat to the underside of the spring retainer.

J

Journal - The surface of a rotating shaft which turns in a bearing.

K

Keeper - The split lock that holds the valve spring retainer in position on the valve stem.

Key - A small piece of metal inserted into matching grooves machined into two parts fitted together - such as a gear pressed onto a shaft - which prevents slippage between the two parts.

Knock - The heavy metallic engine sound, produced in the combustion chamber as a result of abnormal combustion - usually detonation. Knock is usually caused by a loose or worn bearing. Also referred to as detonation, pinging and spark knock. Connecting rod or main bearing knocks are created by too much oil clearance or insufficient lubrication.

L

Lands - The portions of metal between the piston ring grooves.

Lapping the valves - Grinding a valve face and its seat together with lapping compound.

Lash - The amount of free motion in a gear train, between gears, or in a mechanical assembly, that occurs before movement can begin. Usually refers to the lash in a valve train.

Lifter - The part that rides against the cam to transfer motion to the rest of the valve train.

M

Machining - The process of using a machine to remove metal from a metal part.

Main bearings - The plain, or babbit, bearings that support the crankshaft.

Main bearing caps - The cast iron caps, bolted to the bottom of the block, that support the main bearings.

O

O.D. - Outside diameter.

Oil gallery - A pipe or drilled passageway in the engine used to carry engine oil from one area to another.

Oil ring - The lower ring, or rings, of a piston; designed to prevent excessive amounts of oil from working up the cylinder walls and into the combustion chamber. Also called an oil-control ring.

Oil seal - A seal which keeps oil from leaking out of a compartment. Usually refers to a dynamic seal around a rotating shaft or other moving part.

O-ring - A type of sealing ring made of a special rubberlike material; in use, the O-ring is compressed into a groove to provide the sealing action.

Overhaul - To completely disassemble a unit, clean and inspect all parts, reassemble it with the original or new parts and make all adjustments necessary for proper operation.

P

Pilot bearing - A small bearing installed in the center of the flywheel (or the rear end of the crankshaft) to support the front end of the input shaft of the transmission.

Pip mark - A little dot or indentation which indicates the top side of a compression ring.

Piston - The cylindrical part, attached to the connecting rod, that moves up and down in the cylinder as the crankshaft rotates. When the fuel charge is fired, the piston transfers the force of the explosion to the connecting rod, then to the crankshaft.

Piston pin (or wrist pin) - The cylindrical and usually hollow steel pin that passes through the piston. The piston pin fastens the piston to the upper end of the connecting rod.

Piston ring - The split ring fitted to the groove in a piston. The ring contacts the sides of the ring groove and also rubs against the cylinder wall, thus sealing space between piston and wall. There are two types of rings: Compression rings seal the compression pressure in the combustion chamber; oil rings scrape excessive oil off the cylinder wall.

Piston ring groove - The slots or grooves cut in piston heads to hold piston rings in position.

Piston skirt - The portion of the piston below the rings and the piston pin hole.

Plastigage - A thin strip of plastic thread, available in different sizes, used for measuring clearances. For example, a strip of plastigage is laid across a bearing journal and mashed as parts are assembled. Then parts are disassembled and the width of the strip is measured to determine clearance between journal and bearing. Commonly used to measure crankshaft main-bearing and connecting rod bearing clearances.

Press-fit - A tight fit between two parts that requires pressure to force the parts together. Also referred to as drive, or force, fit.

Prussian blue - A blue pigment; in solution, useful in determining the area of contact between two surfaces. Prussian blue is commonly used to determine the width and location of the contact area between the valve face and the valve seat.

R

Race (bearing) - The inner or outer ring that provides a contact surface for balls or rollers in bearing.

Ream - To size, enlarge or smooth a hole by using a round cutting tool with fluted edges.

Ring job - The process of reconditioning the cylinders and installing new rings.

Runout - Wobble. The amount a shaft rotates out-of-true.

S

Saddle - The upper main bearing seat.

Scored - Scratched or grooved, as a cylinder wall may be scored by abrasive particles moved up and down by the piston rings.

Scuffing - A type of wear in which there's a transfer of material between parts moving against each other; shows up as pits or grooves in the mating surfaces.

Seat - The surface upon which another part rests or seats. For example, the valve seat is the matched surface upon which the valve face rests. Also used to refer to wearing into a good fit; for example, piston rings seat after a few miles of driving.

Short block - An engine block complete with crankshaft and piston and, usually, camshaft assemblies.

Static balance - The balance of an object while it's stationary.

Step - The wear on the lower portion of a ring land caused by excessive side and back-clearance. The height of the step indicates the ring's extra side clearance and the length of the step projecting from the back wall of the groove represents the ring's back clearance.

Stroke - The distance the piston moves when traveling from top dead center to bottom dead center, or from bottom dead center to top dead center.

Stud - A metal rod with threads on both ends.

T

Tang - A lip on the end of a plain bearing used to align the bearing during assembly.

Tap - To cut threads in a hole. Also refers to the fluted tool used to cut threads.

Taper - A gradual reduction in the width of a shaft or hole; in an engine cylinder, taper usually takes the form of uneven wear, more pronounced at the top than at the bottom.

Throws - The offset portions of the crankshaft to which the connecting rods are affixed.

Thrust bearing - The main bearing that has thrust faces to prevent excessive endplay, or forward and backward movement of the crankshaft.

Thrust washer - A bronze or hardened steel washer placed between two moving parts. The washer prevents longitudinal movement and provides a bearing surface for thrust surfaces of parts.

Tolerance - The amount of variation permitted from an exact size of measurement. Actual amount from smallest acceptable dimension to largest acceptable dimension.

U

Umbrella - An oil deflector placed near the valve tip to throw oil from the valve stem area.

Undercut - A machined groove below the normal surface.

Undersize bearings - Smaller diameter bearings used with re-ground crankshaft journals.

V

Valve grinding - Refacing a valve in a valve-refacing machine.

Valve train - The valve-operating mechanism of an engine; includes all components from the camshaft to the valve.

Vibration damper - A cylindrical weight attached to the front of the crankshaft to minimize torsional vibration (the twist-untwist actions of the crankshaft caused by the cylinder firing impulses). Also called a harmonic balancer.

W

Water jacket - The spaces around the cylinders, between the inner and outer shells of the cylinder block or head, through which coolant circulates.

Web - A supporting structure across a cavity.

Woodruff key - A key with a radiused back-side (viewed from the side).

Notes

Chapter 3
Cooling, heating and air conditioning systems

Contents

	Section
Air conditioning compressor - removal and installation	12
Air conditioning condenser - removal and installation	13
Air conditioning receiver-drier - removal and installation	14
Air conditioning refrigerant pressure sensor - replacement	15
Air conditioning Thermostatic Expansion Valve (TXV) - general information	16
Blower motor resistor/module and blower motor - replacement	9
Coolant reservoir - removal and installation	6
Engine cooling fans - replacement	5

	Section
General information	1
Heater/air conditioning control assembly - removal and installation	10
Heater core - replacement	11
Radiator - removal and installation	7
Thermostat - replacement	4
Troubleshooting	2
Water pump - replacement	8

Specifications

General

Radiator cap pressure rating	15 psi (103 kPa)
Cooling system capacity	See Chapter 1
Refrigerant type	R-134a
Refrigerant capacity	Refer to HVAC specification tag

Torque specifications

Note: *One foot-pound (ft-lb) of torque is equivalent to 12 inch-pounds (in-lbs) of torque. Torque values below approximately 15 foot-pounds are expressed in inch-pounds, because most foot-pound torque wrenches are not accurate at these smaller values.*

	Ft-lbs (unless otherwise indicated)	Nm
Compressor mounting bolts		
2007 through 2012 models	18	25
2013 and later models	43	58
Refrigerant line fitting nuts	16	22
Refrigerant line fitting bolts	80 in-lbs	9
Engine strut mount/bracket bolts	See Chapter 2A	
Thermostat housing bolts	89 in-lbs	10
Radiator hose tube bracket-to-engine	37	50
Thermostatic expansion valve mounting bolts	44 in-lbs	5
Radiator bracket mounting bolts	89 in-lbs	10
Water pump fasteners*		
Step 1	89 in-lbs	10
Step 2	89 in-lbs	10
Step 3	Tighten an additional 45 degrees	
Water pump pulley bolts	89 in-lbs	10

* Use new bolts

1   General information

Warning: *Do not allow antifreeze to come in contact with your skin or painted surfaces of the vehicle. Rinse off spills immediately with plenty of water. Antifreeze is highly toxic if ingested. Never leave antifreeze lying around in an open container or in puddles on the floor; children and pets are attracted by it's sweet smell and may drink it. Check with local authorities about disposing of used antifreeze. Many communities have collection centers which will see that antifreeze is disposed of safely. Never dump used antifreeze on the ground or pour it into drains.*

Engine cooling system

1   All modern vehicles employ a pressurized engine cooling system with thermostatically controlled coolant circulation. The cooling system consists of a radiator, an expansion tank or coolant reservoir, a pressure cap (located on the expansion tank or radiator), a thermostat, a cooling fan, and a water pump.

2   The water pump circulates coolant through the engine. The coolant flows around each cylinder and around the intake and exhaust ports, near the spark plug areas and in close proximity to the exhaust valve guides.

3   A thermostat controls engine coolant temperature. During warm up, the closed thermostat prevents coolant from circulating through the radiator. As the engine nears normal operating temperature, the thermostat opens and allows hot coolant to travel through the radiator, where it's cooled before returning to the engine.

Heating system

4   The heating system consists of a blower fan and heater core located in a housing under the dash, the hoses connecting the heater core to the engine cooling system and the heater/air conditioning control head on the dashboard. Hot engine coolant is circulated through the heater core. When the heater mode is activated, a flap door in the housing opens to expose the heater core to the passenger compartment through air ducts. A fan switch on the control head activates the blower motor, which forces air through the core, heating the air.

Air conditioning system

5   The air conditioning system consists of a condenser mounted in front of the radiator, an evaporator mounted adjacent to the heater core, a compressor mounted on the engine, a receiver-drier or accumulator and the plumbing connecting all of the above components.

6   A blower fan forces the warmer air of the passenger compartment through the evaporator core (sort of a radiator-in-reverse), transferring the heat from the air to the refrigerant. The liquid refrigerant boils off into low pressure vapor, taking the heat with it when it leaves the evaporator.

2.2 The cooling system pressure tester is connected in place of the pressure cap, then pumped up to pressurize the system

2   Troubleshooting

Coolant leaks

1   A coolant leak can develop anywhere in the cooling system, but the most common causes are:

    A loose or weak hose clamp
    A defective hose
    A faulty pressure cap
    A damaged radiator
    A bad heater core
    A faulty water pump
    A leaking gasket at any joint that
      carries coolant

2   Coolant leaks aren't always easy to find. Sometimes they can only be detected when the cooling system is under pressure. Here's where a cooling system pressure tester comes in handy. After the engine has cooled completely, the tester is attached in place of the pressure cap, then pumped up to the pressure value equal to that of the pressure cap rating **(see illustration)**. Now, leaks that only exist when the engine is fully warmed up will become apparent. The tester can be left connected to locate a nagging slow leak.

Coolant level drops, but no external leaks

3   If you find it necessary to keep adding coolant, but there are no external leaks, the probable causes include:

   a)   *A blown head gasket*
   b)   *A leaking intake manifold gasket (only on engines that have coolant passages in the manifold)*
   b)   *A cracked cylinder head or cylinder block*

4   Any of the above problems will also usually result in contamination of the engine oil, which will cause it to take on a milkshake-like appearance. A bad head gasket or cracked head or block can also result in engine oil contaminating the cooling system.

5   Combustion leak detectors (also known as block testers) are available at most auto parts stores. These work by detecting exhaust gases in the cooling system, which indicates a compression leak from a cylinder into the coolant. The tester consists of a large bulb-type syringe and bottle of test fluid **(see illustration)**. A measured amount of the fluid is added to the syringe. The syringe is placed over the cooling system filler neck and, with the engine running, the bulb is squeezed and a sample of the gases present in the cooling system are drawn up through the test fluid **(see illustration)**. If any combustion gases are present in the sample taken, the test fluid will change color.

6   If the test indicates combustion gas is present in the cooling system, you can be sure that the engine has a blown head gasket or a crack in the cylinder head or block, and will require disassembly to repair.

Pressure cap

Warning: *Wait until the engine is completely cool before beginning this check.*

7   The cooling system is sealed by a spring-loaded cap, which raises the boiling point of the coolant. If the cap's seal or spring are worn out, the coolant can boil and escape past the cap. With the engine completely cool, remove the cap and check the seal; if it's cracked, hardened or deteriorated in any way, replace it with a new one.

8   Even if the seal is good, the spring might not be; this can be checked with a cooling system pressure tester **(see illustration)**. If the cap can't hold a pressure within approximately 1-1/2 lbs of its rated pressure (which is marked on the cap), replace it with a new one.

9   The cap is also equipped with a vacuum relief spring. When the engine cools off, a vacuum is created in the cooling system. The vacuum relief spring allows air back into the system, which will equalize the pressure and prevent damage to the radiator (the radiator tanks could collapse if the vacuum is great enough). If, after turning the engine off and allowing it to cool down you notice any of the cooling system hoses collapsing, replace the pressure cap with a new one.

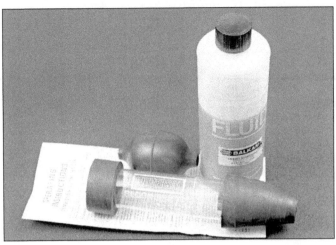

2.5a The combustion leak detector consists of a bulb, syringe and test fluid

2.5b Place the tester over the cooling system filler neck and use the bulb to draw a sample into the tester

Thermostat

10 Before assuming the thermostat **(see illustration)** is responsible for a cooling system problem, check the coolant level (see Chapter 1), drivebelt tension (see Chapter 1) and temperature gauge (or light) operation.

11 If the engine takes a long time to warm up (as indicated by the temperature gauge or heater operation), the thermostat is probably stuck open. Replace the thermostat with a new one.

12 If the engine runs hot or overheats, a thorough test of the thermostat should be performed.

13 Definitive testing of the thermostat can only be made when it is removed from the vehicle. If the thermostat is stuck in the open position at room temperature, it is faulty and must be replaced.

Caution: *Do not drive the vehicle without a thermostat. The computer may stay in open loop and emissions and fuel economy will suffer.*

14 To test a thermostat, suspend the (closed) thermostat on a length of string or wire in a pot of cold water.

15 Heat the water on a stove while observing thermostat. The thermostat should fully open before the water boils.

16 If the thermostat doesn't open and close as specified, or sticks in any position, replace it.

Cooling fan

Electric cooling fan

17 If the engine is overheating and the cooling fan is not coming on when the engine temperature rises to an excessive level, unplug the fan motor electrical connector(s) and connect the motor directly to the battery with fused jumper wires. If the fan motor doesn't come on, replace the motor.

18 If the fan motor is okay, but it isn't coming on when the engine gets hot, the cooling fan control module or the Powertrain Control Module (PCM) might be defective.

19 These control circuits are fairly complex, and checking them should be left to a qualified automotive technician.

20 Check all wiring and connections to the fan motor. Refer to the wiring diagrams at the end of Chapter 12.

21 If no obvious problems are found, the problem could be the Cylinder Head Temperature (CHT) sensor or the Powertrain Control Module (PCM). Have the cooling fan system and circuit diagnosed by a dealer service department or repair shop with the proper diagnostic equipment.

Belt-driven cooling fan

22 Disconnect the cable from the negative terminal of the battery and rock the fan back and forth by hand to check for excessive bearing play.

23 With the engine cold (and not running), turn the fan blades by hand. The fan should turn freely.

24 Visually inspect for substantial fluid leakage from the clutch assembly. If problems are noted, replace the clutch assembly.

25 With the engine completely warmed up, turn off the ignition switch and disconnect the negative battery cable from the battery. Turn the fan by hand. Some drag should be evident. If the fan turns easily, replace the fan clutch.

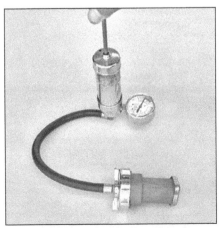

2.8 Checking the cooling system pressure cap with a cooling system pressure tester

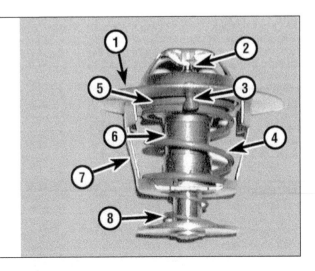

2.10 Typical thermostat:

1 Flange
2 Piston
3 Jiggle valve
4 Main coil spring
5 Valve seat
6 Valve
7 Frame
8 Secondary coil spring

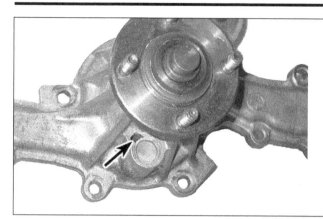

2.28 The water pump weep hole is generally located on the underside of the pump

Water pump

26   A failure in the water pump can cause serious engine damage due to overheating.

Drivebelt-driven water pump

27   There are two ways to check the operation of the water pump while it's installed on the engine. If the pump is found to be defective, it should be replaced with a new or rebuilt unit.

28   Water pumps are equipped with weep (or vent) holes **(see illustration)**. If a failure occurs in the pump seal, coolant will leak from the hole.

29   If the water pump shaft bearings fail, there may be a howling sound at the pump while it's running. Shaft wear can be felt with the drivebelt removed if the water pump pulley is rocked up and down (with the engine off). Don't mistake drivebelt slippage, which causes a squealing sound, for water pump bearing failure.

Timing chain or timing belt-driven water pump

30   Water pumps driven by the timing chain or timing belt are located underneath the timing chain or timing belt cover.

31   Checking the water pump is limited because of where it is located. However, some basic checks can be made before deciding to remove the water pump. If the pump is found to be defective, it should be replaced with a new or rebuilt unit.

32   One sign that the water pump may be failing is that the heater (climate control) may not work well. Warm the engine to normal operating temperature, confirm that the coolant level is correct, then run the heater and check for hot air coming from the ducts.

33   Check for noises coming from the water pump area. If the water pump impeller shaft or bearings are failing, there may be a howling sound at the pump while the engine is running. **Note:** *Be careful not to mistake drivebelt noise (squealing) for water pump bearing or shaft failure.*

34   It you suspect water pump failure due to noise, wear can be confirmed by feeling for play at the pump shaft. This can be done by rocking the drive sprocket on the pump shaft up and down. To do this you will need to

remove the tension on the timing chain or belt as well as access the water pump.

All water pumps

35   In rare cases or on high-mileage vehicles, another sign of water pump failure may be the presence of coolant in the engine oil. This condition will adversely affect the engine in varying degrees.
Note: *Finding coolant in the engine oil could indicate other serious issues besides a failed water pump, such as a blown head gasket or a cracked cylinder head or block.*

36   Even a pump that exhibits no outward signs of a problem, such as noise or leakage, can still be due for replacement. Removal for close examination is the only sure way to tell. Sometimes the fins on the back of the impeller can corrode to the point that cooling efficiency is diminished significantly.

Heater system

37   Little can go wrong with a heater. If the fan motor will run at all speeds, the electrical part of the system is okay. The three basic heater problems fall into the following general categories:

a) Not enough heat
b) Heat all the time
c) No heat

38   If there's not enough heat, the control valve or door is stuck in a partially open position, the coolant coming from the engine isn't hot enough, or the heater core is restricted. If the coolant isn't hot enough, the thermostat in the engine cooling system is stuck open, allowing coolant to pass through the engine so rapidly that it doesn't heat up quickly enough. If the vehicle is equipped with a temperature gauge instead of a warning light, watch to see if the engine temperature rises to the normal operating range after driving for a reasonable distance.

39   If there's heat all the time, the control valve or the door is stuck wide open.

40   If there's no heat, coolant is probably not reaching the heater core, or the heater core is plugged. The likely cause is a collapsed or plugged hose, core, or a frozen heater control valve. If the heater is the type that flows coolant all the time, the cause is a stuck door or a broken or kinked control cable.

Air conditioning system

41   If the cool air output is inadequate:

a) *Inspect the condenser coils and fins to make sure they're clear*
b) *Check the compressor clutch for slippage.*
c) *Check the blower motor for proper operation.*
d) *Inspect the blower discharge passage for obstructions.*
e) *Check the system air intake filter for clogging.*

42   If the system provides intermittent cooling air:

a) *Check the circuit breaker, blower switch and blower motor for a malfunction.*
b) *Make sure the compressor clutch isn't slipping.*
c) *Inspect the plenum door to make sure it's operating properly.*
d) *Inspect the evaporator to make sure it isn't clogged.*
e) *If the unit is icing up, it may be caused by excessive moisture in the system, incorrect super heat switch adjustment or low thermostat adjustment.*

43   If the system provides no cooling air:

a) *Inspect the compressor drivebelt. Make sure it's not loose or broken.*
b) *Make sure the compressor clutch engages. If it doesn't, check for a blown fuse.*
c) *Inspect the wire harness for broken or disconnected wires.*
d) *If the compressor clutch doesn't engage, bridge the terminals of the A/C pressure switch(es) with a jumper wire; if the clutch now engages, and the system is properly charged, the pressure switch is bad.*
e) *Make sure the blower motor is not disconnected or burned out.*
f) *Make sure the compressor isn't partially or completely seized.*
g) *Inspect the refrigerant lines for leaks.*
h) *Check the components for leaks.*
i) *Inspect the receiver-drier/accumulator or expansion valve/tube for clogged screens.*

44   If the system is noisy:

a) *Look for loose panels in the passenger compartment.*
b) *Inspect the compressor drivebelt. It may be loose or worn.*
c) *Check the compressor mounting bolts. They should be tight.*
d) *Listen carefully to the compressor. It may be worn out.*
e) *Listen to the idler pulley and bearing and the clutch. Either may be defective.*
f) *The winding in the compressor clutch coil or solenoid may be defective.*
g) *The compressor oil level may be low.*
h) *The blower motor fan bushing or the motor itself may be worn out.*
i) *If there is an excessive charge in the system, you'll hear a rumbling noise in the high pressure line, a thumping noise in the compressor, or see bubbles or cloudiness in the sight glass.*

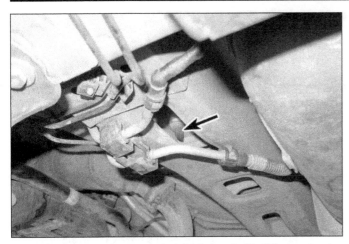

3.1a The front evaporator drain tube is located in this opening in the forward structural crossmember of the chassis

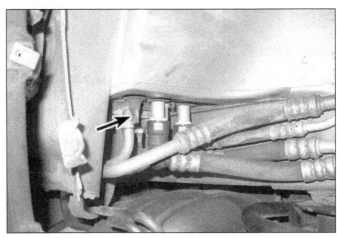

3.1b The rear auxiliary evaporator drain tube is located behind the right-rear wheel and inner fender splash shield

j) If there's a low charge in the system, you might hear hissing in the evaporator case at the expansion valve, or see bubbles or cloudiness in the sight glass.

---

3   Air conditioning and heating system - check and maintenance

Air conditioning system

Warning: *The air conditioning system is under high pressure. Do not loosen any hose fittings or remove any components until after the system has been discharged. Air conditioning refrigerant should be properly discharged into an EPA-approved recovery/recycling unit at a dealer service department or an automotive air conditioning repair facility. Always wear eye protection when disconnecting air conditioning system fittings.*

Caution: *All models covered by this manual use environmentally friendly R-134a. This refrigerant (and its appropriate refrigerant oils) are not compatible with R-12 refrigerant system components and must never be mixed or the components will be damaged.*

Caution: *When replacing entire components, additional refrigerant oil should be added equal to the amount that is removed with the component being replaced. Read the can before adding any oil to the system, to make sure it is compatible with the R-134a system.*

1   The following maintenance checks should be performed on a regular basis to ensure that the air conditioning continues to operate at peak efficiency.

a) Inspect the condition of the compressor drivebelt. If it is worn or deteriorated, replace it (see Chapter 1).
b) Check the drivebelt tension (see Chapter 1).
c) Inspect the system hoses. Look for cracks, bubbles, hardening and deterioration. Inspect the hoses and all fittings for oil bubbles or seepage. If there is any evidence of wear, damage or leakage, replace the hose(s).

d) Inspect the condenser fins for leaves, bugs and any other foreign material that may have embedded itself in the fins. Use a fin comb or compressed air to remove debris from the condenser.
e) Make sure the system has the correct refrigerant charge.
f) Check the evaporator housing drain tube(s) for blockage (see illustration).

2   It's a good idea to operate the system for about ten minutes at least once a month. This is particularly important during the winter months because long term non-use can cause hardening, and subsequent failure, of the seals. Note that using the Defrost function operates the compressor.

3   If the air conditioning system is not working properly, proceed to Step 6 and perform the general checks outlined below.

4   Because of the complexity of the air conditioning system and the special equipment necessary to service it, in-depth troubleshooting and repairs beyond checking the refrigerant charge and the compressor clutch operation are not included in this manual. However, simple checks and component replacement procedures are provided in this Chapter. For more complete information on the air conditioning system, refer to the *Haynes Automotive Heating and Air Conditioning Manual.*

5   The most common cause of poor cooling is simply a low system refrigerant charge. If a noticeable drop in system cooling ability occurs, one of the following quick checks will help you determine if the refrigerant level is low.

Checking the refrigerant charge

6   Warm the engine up to normal operating temperature.

7   Place the air conditioning temperature selector at the coldest setting and put the blower at the highest setting.

8   After the system reaches operating temperature, feel the larger pipe exiting the evaporator at the firewall. The outlet pipe should be cold (the tubing that leads back to the com-

3.9 Insert a thermometer in the center vent, turn on the air conditioning system and wait for it to cool down; depending on the humidity, the output air should be 35 to 40 degrees cooler than the ambient air temperature

pressor). If the evaporator outlet pipe is warm, the system probably needs a charge.

9   Insert a thermometer in the center air distribution duct **(see illustration)** while operating the air conditioning system at its maximum setting - the temperature of the output air should be 35 to 40 degrees F below the ambient air temperature (down to approximately 40 degrees F). If the ambient (outside) air temperature is very high, say 110 degrees F, the duct air temperature may be as high as 60 degrees F, but generally the air conditioning is 35 to 40 degrees F cooler than the ambient air.

10   Further inspection or testing of the system requires special tools and techniques and is beyond the scope of the home mechanic.

Adding refrigerant

Caution: *Make sure any refrigerant, refrigerant oil or replacement component you purchase is designated as compatible with R-134a systems.*

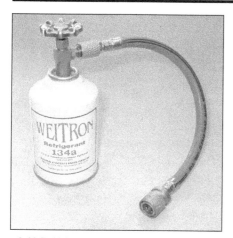

3.11 R-134a automotive air conditioning charging kit

3.13 Location of the low-side charging port

3.24 Insert the nozzle of the disinfectant can into the return-air intake behind the glove box

11   Purchase an R-134a automotive charging kit at an auto parts store **(see illustration)**. A charging kit includes a can of refrigerant, a tap valve and a short section of hose that can be attached between the tap valve and the system low side service valve.
Caution: *Never add more than one can of refrigerant to the system. If more refrigerant than that is required, the system should be evacuated and leak tested.*
12   Back off the valve handle on the charging kit and screw the kit onto the refrigerant can, making sure first that the O-ring or rubber seal inside the threaded portion of the kit is in place.
Warning: *Wear protective eyewear when dealing with pressurized refrigerant cans.*
13   Remove the dust cap from the low-side charging port and attach the hose's quick-connect fitting to the port **(see illustration)**.
Warning: *DO NOT hook the charging kit hose to the system high side! The fittings on the charging kit are designed to fit only on the low side of the system.*
14   Warm up the engine and turn On the air conditioning. Keep the charging kit hose away from the fan and other moving parts.
Note: *The charging process requires the compressor to be running. If the clutch cycles off, you can put the air conditioning switch on High and leave the car doors open to keep the clutch on and compressor working. The compressor can be kept on during the charging by removing the connector from the pressure switch and bridging it with a paper clip or jumper wire during the procedure.*
15   Turn the valve handle on the kit until the stem pierces the can, then back the handle out to release the refrigerant. You should be able to hear the rush of gas. Keep the can upright at all times, but shake it occasionally. Allow stabilization time between each addition.
Note: *The charging process will go faster if you wrap the can with a hot-water-soaked rag to keep the can from freezing up.*
16   If you have an accurate thermometer, you can place it in the center air condition-

ing duct inside the vehicle and keep track of the output air temperature. A charged system that is working properly should cool down to approximately 40 degrees F. If the ambient (outside) air temperature is very high, say 110 degrees F, the duct air temperature may be as high as 60 degrees F, but generally the air conditioning is 35 to 40 degrees F cooler than the ambient air.
17   When the can is empty, turn the valve handle to the closed position and release the connection from the low-side port. Reinstall the dust cap.
18   Remove the charging kit from the can and store the kit for future use with the piercing valve in the UP position, to prevent inadvertently piercing the can on the next use.

Heating systems

19   If the carpet under the heater core is damp, or if antifreeze vapor or steam is coming through the vents, the heater core is leaking. Remove it (see Section 11) and install a new unit (most radiator shops will not repair a leaking heater core).
20   If the air coming out of the heater vents isn't hot, the problem could stem from any of the following causes:

a) *The thermostat is stuck open, preventing the engine coolant from warming up enough to carry heat to the heater core. Replace the thermostat (see Section 4).*
b) *There is a blockage in the system, preventing the flow of coolant through the heater core. Feel both heater hoses at the firewall. They should be hot. If one of them is cold, there is an obstruction in one of the hoses or in the heater core, or the heater control valve is shut. Detach the hoses and back flush the heater core with a water hose. If the heater core is clear but circulation is impeded, remove the two hoses and flush them out with a water hose.*
c) *If flushing fails to remove the blockage from the heater core, the core must be replaced (see Section 11).*

Eliminating air conditioning odors

21   Unpleasant odors that often develop in air conditioning systems are caused by the growth of a fungus, usually on the surface of the evaporator core. The warm, humid environment there is a perfect breeding ground for mildew to develop.
22   The evaporator core on most vehicles is difficult to access, and factory dealerships have a lengthy, expensive process for eliminating the fungus by opening up the evaporator case and using a powerful disinfectant and rinse on the core until the fungus is gone. You can service your own system at home, but it takes something much stronger than basic household germ-killers or deodorizers.
23   Aerosol disinfectants for automotive air conditioning systems are available in most auto parts stores, but remember when shopping for them that the most effective treatments are also the most expensive. The basic procedure for using these sprays is to start by running the system in the RECIRC mode for ten minutes with the blower on its highest speed. Use the highest heat mode to dry out the system and keep the compressor from engaging by disconnecting the wiring connector at the compressor.
24   The disinfectant can usually comes with a long spray hose. Insert the nozzle into an intake port inside the cabin, and spray according to the manufacturer's recommendations **(see illustration)**. Try to cover the whole surface of the evaporator core, by aiming the spray up, down and sideways. Follow the manufacturer's recommendations for the length of spray and waiting time between applications.

Automatic heating and air conditioning systems

25   Some vehicles are equipped with an optional automatic climate control system. This system has its own computer that receives inputs from various sensors in the heating and air conditioning system. This

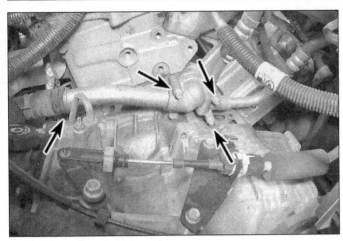

4.8 Remove the radiator hose tube support bolt and the three bolts attaching the thermostat housing to the engine

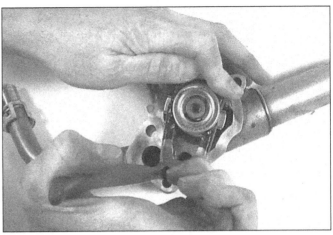

4.9 Push down on the ears of the thermostat and turn it to remove it from the housing

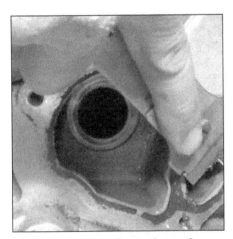

4.10a Clean the engine mating surface...

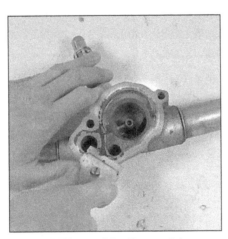

4.10b... and the thermostat housing surface

4.11 When installing the new thermostat, use a new seal and position the thermostat with the jiggle pin at the 12 o'clock position

computer, like the PCM, has self-diagnostic capabilities to help pinpoint problems or faults within the system. Vehicles equipped with automatic heating and air conditioning systems are very complex and considered beyond the scope of the home mechanic. Vehicles equipped with automatic heating and air conditioning systems should be taken to dealer service department or other qualified facility for repair.

4   Thermostat - replacement

Warning: *The engine must be completely cool before beginning this procedure.*

1 Disconnect the cable from the negative battery terminal (see Chapter 5).

2 Drain the cooling system (see Chapter 1). If the coolant is relatively new or in good condition, save it and reuse it. Read the **Warning** in Section 1.

3 The thermostat housing assembly is mounted to the engine, on the opposite end

from the drivebelt, under the throttle body.

4 Remove the oil filler cap and remove the engine cover by pulling firmly upwards to disengage the attachment points.

5 Remove the intake air duct from between the air filter housing and the throttle body.

6 If equipped, remove the single attaching bolt and the fuel pipe shield from below the throttle body.

7 Disconnect the radiator hose and both heater inlet and outlet hoses from the thermostat housing assembly.

8 Remove the three bolts attaching the thermostat housing assembly to the engine, and the one bolt for the radiator hose tube support bracket, then remove the thermostat housing **(see illustration)**.

Note: *Make note of the installed position of the original thermostat before removing to aid in installation.*

9 Remove the thermostat from the housing **(see illustration)**.

10 Clean the housing and engine mating surfaces **(see illustrations)**.

11 Install a new thermostat and seal **(see illustration)**. Install the thermostat housing assembly with the bolts finger tight at first, then tighten the bolts to the torque listed in this Chapter's Specifications.

12 The remainder of installation is the reverse of removal.

13 Refill the cooling system (see Chapter 1).

14 Start the engine and allow it to reach normal operating temperature, then check for leaks and proper thermostat operation.

5   Engine cooling fans - replacement

Warning: *To avoid possible injury or damage, DO NOT operate the engine with a damaged fan. Do not attempt to repair fan blades - replace a damaged fan with a new one.*

Warning: *Wait until the engine is completely cool before beginning this procedure.*

1 Remove the front bumper cover and hood latch (see Chapter 11).

5.2 Remove front bumper impact absorber bolts

5.3 Mounting brackets for the radiator

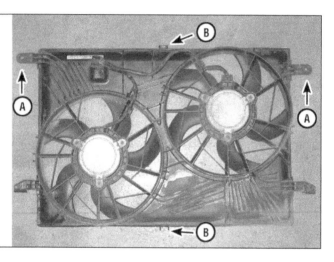

5.5a Fan shroud-to-radiator bolt (A) and fastener (B) locations

remove the mounting screws.

Note: *On some designs, the fan motor is riveted to the shroud and requires drilling the rivets out to remove the fan motor. When working with this type, tape the openings of the fan motors to protect them from debris, if necessary. Use locking nuts and bolts to attach the motor to the fan shroud.*

7   To detach the fan blade from the motor, remove the motor shaft clip or bolt, depending on design.

Note: *On clip-type models only, mark the relationship of the fan to the fan motor shaft if the fan is going to be reinstalled. On nut retained fans, the fan cannot be reused. Heat the new fan's hub with HOT tap water (120-degrees for 1 minute) before installation or the fan could crack and cause damage. Also, a special nut may be used to retain the fan.*

8   Installation is the reverse of removal.

9   Run the engine to normal operating temperature and operate the air conditioning to ensure the fans operate properly.

---

6   Coolant reservoir - removal and installation

Warning: *Wait until the engine is completely cool before beginning this procedure.*

1   Remove as much coolant as possible from the reservoir using a suction device. If the coolant is relatively new or in good condition, save it and reuse it.

Note: *If necessary, remove the upper radiator cover to gain access to the reservoir mounting bolts. The cover is secured by push-in retainers (see Chapter 11 for fastener types).*

2   Remove the two bolts securing the reservoir to the vehicle **(see illustration)**.

3   Lift the reservoir up enough to disconnect the overflow hose.

4   If the reservoir is to be reused, clean it with soapy water and a brush to remove any deposits inside. Inspect the reservoir carefully for cracks. If you find a crack, replace the reservoir.

5   Installation is the reverse of removal.

6   Fill the reservoir with the proper mixture of antifreeze and water (see Chapter 1).

5.5b Remove the plastic fasteners at the top and bottom of the fan shroud

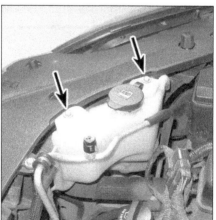

6.2 Remove the reservoir bolts

2   Detach any electrical connectors from the bumper support and remove the front bumper impact absorber **(see illustration)**.

3   Remove the two radiator upper mounting brackets **(see illustration)**.

4   Tilt the radiator and condenser forward for additional access and disconnect the cooling fan electrical connectors.

5   Remove the two fan shroud-to-radiator bolts and fasteners, and lift the cooling fan assembly from the lower radiator mounts and out of the engine compartment **(see illustrations)**.

Caution: *Use care not to damage the radiator fins when removing the cooling fan assembly from the engine compartment.*

6   To detach the fan motor from the shroud,

7.4a Air guide retainer locations (passenger's side shown)

7.4b Remove the air guide retainers by prying them off

7.5 Remove the spring clips to disconnect the transaxle fluid cooler lines

7.6 Pinch the upper condenser-to-radiator mounting clip and tab

8.4 Remove the water pump pulley bolts (fourth bolt not visible)

7 Radiator - removal and installation

Warning: *Wait until the engine is completely cool before beginning this procedure.*

Removal

1   Drain the cooling system (see Chapter 1). If the coolant is relatively new or in good condition, save it and reuse it.
2   Remove the engine cooling fan assembly (see Section 5).
3   Disconnect the upper and lower radiator hoses and coolant reservoir overflow hose from the radiator.
4   Detach the air guides from the sides of the radiator **(see illustrations)**.
5   Remove the air filter housing air inlet duct and disconnect the transaxle fluid cooler lines from the radiator **(see illustration)**. Remove the air filter housing if additional clearance is required.
6   Disconnect the condenser from the radiator by pinching the fastening tabs at the top of the condenser **(see illustration)**.
7   Carefully lift the condenser to release the lower retaining tabs, separate the condenser from the radiator and position it forward.
Caution: *Be careful not to damage the fins on the radiator or condenser when removing the radiator from the vehicle.*

8   Life the radiator from behind the condenser to remove it from the vehicle.
9   Inspect the radiator for leaks and damage. If it needs repair, have a radiator shop or dealer service department perform the work, as special techniques are required.
10   Bugs and dirt can be removed from the radiator by spraying it with a garden hose nozzle from the back side. The radiator should be flushed out with a garden hose before installation.
11   Check the rubber mounts on the bottom of the radiator for wear or deterioration and replace them if necessary.

Installation

12   If a new radiator is being installed, transfer any components from the old radiator to the new radiator, specifically the side air deflectors from the tanks.
13   Installation is the reverse of the removal procedure noting the following points:

a) *Carefully guide the radiator into position and make certain that the rubber mounts on the bottom are properly seated into the support.*
b) *Carefully insert the bottom of the condenser into the lower mounts on the radiator and clip the top of the condenser into the radiator.*

14   Fill the cooling system with the proper mixture of antifreeze and water (see Chapter 1).
15   Start the engine and check for leaks. Allow the engine to reach normal operating temperature, then recheck the coolant level and add more if required.
16   Check the transaxle fluid and add more as needed (see Chapter 1).

8 Water pump - replacement

Warning: *Wait until the engine is completely cool before beginning this procedure.*
1   Disconnect the cable from the negative battery terminal (see Chapter 5).
2   Drain the cooling system (see Chapter 1). If the coolant is relatively new or in good condition, save it and reuse it. Read the **Warning** in Section 1.
3   Remove the drivebelt (see Chapter 1). Also remove the right-side engine strut mount and mount bracket (see Chapter 2A).
4   Hold the water pump pulley with a strap wrench or pin spanner, then remove the four pulley mounting bolts and pulley **(see illustration)**.

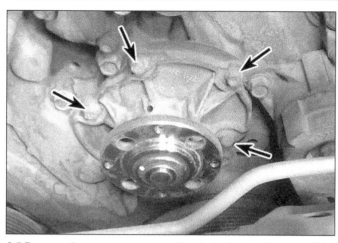

8.5 Remove the water pump mounting bolts (two bolts not visible)

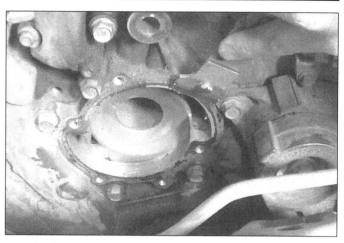

8.7 Clean the water pump mating surface on the timing chain cover

5   Remove the water pump mounting bolts **(see illustration)**.

6   Carefully remove the pump from the engine. If the water pump is stuck, gently tap it with a soft-faced hammer to break the seal.

7   Thoroughly clean the mating surface of the engine (and the water pump if it's going to be reinstalled) **(see illustration)**.

8   Compare the replacement pump with the old one to make sure that they're identical.

9   Place the gasket and water pump into position. Install the *new* mounting bolts until they are all finger tight.

Caution: *When installing the water pump, new bolts are required to achieve the proper torque values and sealing strength.*

10   Tighten the water pump mounting bolts a little at a time, in a criss-cross pattern, to the torque listed in this Chapter's Specifications.

11   The remainder of installation is the reverse of removal. Tighten the water pump pulley bolts to the torque listed in this Chapter's Specifications. Refill the cooling system with the proper concentration of antifreeze (see Chapter 1). Start the engine and allow it to reach normal operating temperature while inspecting the system for leaks.

9   Blower motor resistor/module and blower motor - replacement

Warning: *The models covered by this manual are equipped with Supplemental Restraint Systems (SRS), more commonly known as airbags. Always disable the airbag system before working in the vicinity of any airbag system component to avoid the possibility of accidental deployment of the airbag, which could cause personal injury (see Chapter 12).*

Front blower motor resistor/ module

1   Working in the passenger compartment under the glove box, remove the passenger's side insulator panel **(see illustration)**.

2   Remove the resistor/module retainers, and remove the resistor/module from the blower housing to the left of the blower motor **(see illustration)**.

3   Disconnect the electrical connectors from the blower motor resistor/module.

4   Installation is the reverse of removal.

Rear auxiliary blower motor resistor/module

Warning: *The models covered by this manual are equipped with Supplemental Restraint Systems (SRS), more commonly known as airbags. Always disable the airbag system before working in the vicinity of any airbag system component to avoid the possibility of accidental deployment of the airbag, which could cause personal injury (see Chapter 12).*

5   Remove the passenger side rear quarter trim panel (see Chapter 11).

6   Disconnect the electrical connectors from the blower motor resistor/module.

7   Remove the two screws and remove the rear auxiliary resistor/module from the blower housing below the rear auxiliary blower motor **(see illustration)**.

8   Installation is the reverse of removal.

Front blower motor

9   Working in the passenger compartment under the glove box, remove the passenger's side insulator panel **(see illustration 9.2)**.

10   Disconnect the blower motor electrical connector, remove the mounting screw

9.1 Remove the two retainers and pull the front of the panel down and towards you

9.2 Blower motor resistor/ module retainers

9.7 Rear auxiliary blower motor resistor/ module fasteners

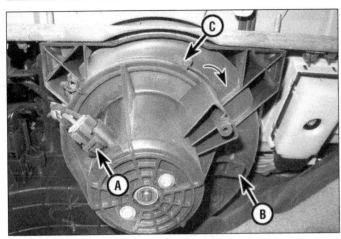

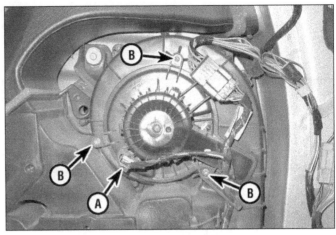

9.10 Disconnect the blower motor electrical connector (A), remove the mounting screw (B), pull down on the tab (C) and rotate the blower motor clockwise to remove it

9.14 Disconnect the rear auxiliary blower motor electrical connector (A) and remove the three mounting screws (B)

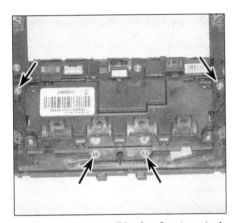

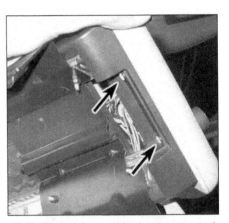

10.3 Heater/air conditioning front control assembly mounting fasteners

10.7a Heater/air conditioning rear control assembly - upper mounting screws

10.7b Heater/air conditioning rear control assembly - lower mounting screws

and remove the blower motor assembly **(see illustration)**.

11   Remove the blower motor fan retaining clip and remove the blower fan from the motor.

12   Installation is the reverse of removal.

Rear auxiliary blower motor

Warning: *The models covered by this manual are equipped with Supplemental Restraint Systems (SRS), more commonly known as airbags. Always disable the airbag system before working in the vicinity of any airbag system component to avoid the possibility of accidental deployment of the airbag, which could cause personal injury (see Chapter 12).*

13   Remove the third row seat and passenger's side rear quarter trim panel (see Chapter 11).

14   Disconnect the rear auxiliary blower motor electrical connector, remove the three blower motor mounting screws and remove the blower motor assembly **(see illustration)**.

15   Remove the blower motor fan retain-

ing clip and remove the blower fan from the motor.

16   Installation is the reverse of removal.

---

10   Heater/air conditioning control assembly - removal and installation

---

Warning: *The models covered by this manual are equipped with Supplemental Restraint Systems (SRS), more commonly known as airbags. Always disable the airbag system before working in the vicinity of any airbag system component to avoid the possibility of accidental deployment of the airbag, which could cause personal injury (see Chapter 12).*

Front control assembly

1   Remove the instrument panel center bezel (see Chapter 11).

2   Disconnect the electrical connectors from the rear of the front control assembly.

3   Working at the back of the trim, remove the four mounting screws for the front control assembly and separate the assembly from the bezel **(see illustration)**.

4   Installation is the reverse of removal. If the front control assembly has been replaced with a new component, it will require programming by a dealer or qualified repair facility.

Rear auxiliary control assembly

5   Open the center console armrest and remove the five screws attaching the hinge to the center console.

6   Remove the mounting screws and the center console armrest storage compartment.

7   Remove the screws from the top and bottom of the rear auxiliary control bezel **(see illustrations)**. Disconnect the electrical connector from the control assembly and remove the bezel from the center console.

8   Remove the screws securing the rear auxiliary control assembly to the bezel.

9   Installation is the reverse of removal.

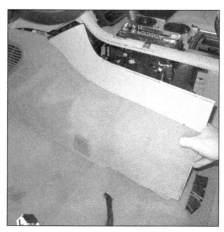

11.4 Pull the driver's side floor console extension panel away from the floor console to release the clips

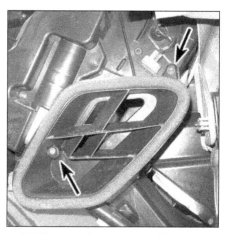

11.5 Remove the screws and the floor outlet duct

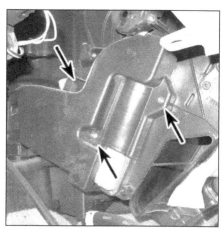

11.6 Remove the heater core cover screws and remove the cover

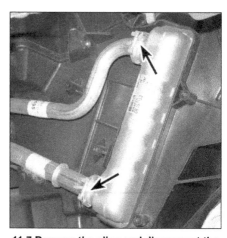

11.7 Remove the clips and disconnect the heater core tubes from the heater core

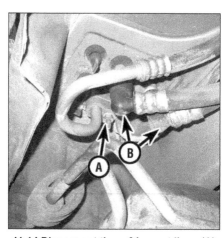

11.14 Disconnect the refrigerant lines (A) and the heater hoses (B), and remove the mounting nut from the unit

---

11 Heater core - replacement

Warning: *The models covered by this manual are equipped with Supplemental Restraint Systems (SRS), more commonly known as airbags. Always disable the airbag system before working in the vicinity of any airbag system component to avoid the possibility of accidental deployment of the airbag, which could cause personal injury (see Chapter 12).*

1 Disconnect the cable from the negative battery terminal (see Chapter 5).
2 Drain the cooling system (see Chapter 1). If the coolant is relatively new or in good condition, save it and reuse it.

Front heater core

Note: *The heater hoses remain connected to the heater core tubes during the procedure.*
3 Remove the driver's seat (see Chapter 11).
Caution: *Failure to remove the driver's seat will result in damage to the driver's side floor console extension panel.*

4 Working in the passenger's compartment under the steering column, remove the driver's side floor console extension panel **(see illustration)**.
5 Remove the two floor outlet duct screws and remove the floor outlet duct **(see illustration)**.
6 Remove the three heater core cover screws and remove the heater core cover **(see illustration)**.
Caution: *Cap or plug heater core and tube openings to prevent spillage of coolant into the passenger compartment during replacement.*
7 Remove and discard the clips connecting the heater core tubes to the heater core, then disconnect the tubes from the heater core **(see illustration)**.
Caution: *Place towels or plastic sheeting on the floor under the heater core tubes to catch any coolant that might spill out.*
Caution: *New clips must be used when installing the heater core tubes on the new heater core.*
8 Carefully pull the heater core from the heater case.

Caution: *Ensure the foam seal for the new heater core is properly installed on the heater core. Use care when inserting the new heater core into the heater case to prevent damage to the foam seal.*
9 Installation is the reverse of removal.
Caution: *Use NEW clips when connecting the heater core tubes to the heater core.*
10 Reconnect the cable to the negative terminal of the battery (see Chapter 5).
11 Refill the cooling system with the proper concentration of antifreeze (see Chapter 1). Start the engine and allow it to reach normal operating temperature while inspecting the system for leaks.

Rear auxiliary heater core

Note: *This is a time-consuming and complex procedure, as the entire rear auxiliary HVAC module requires removal. Make sure that you have sufficient time as well as the proper equipment and parts, and read through the entire procedure before you begin.*
12 Have the air conditioning system recovered by a dealer service department or by an automotive air conditioning shop prior to performing any work on the rear auxiliary heater core.
13 Loosen the right-rear wheel lug nuts. Raise the rear of the vehicle and support it securely on jackstands. Remove the right rear wheel and inner fender splash shield (see Chapter 11).
14 From below the rear of the vehicle, disconnect the heater hoses and refrigerant lines from the rear auxiliary HVAC module fittings and remove the single HVAC module mounting nut **(see illustration)**.
15 Working from inside the vehicle, remove the rear storage compartment located behind the third row seat (if equipped). Remove the third row seat and passenger's side rear quarter trim panel (see Chapter 11).
16 Disconnect the electrical connector for the rear auxiliary HVAC unit.
17 Remove the air conditioning air outlet duct mounting bolt and squeeze both ends of

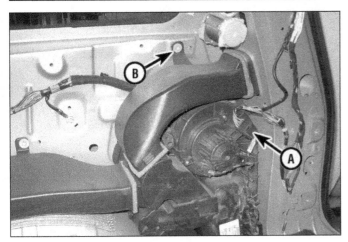

11.17 Disconnect the HVAC connector (A) and remove the bolt attaching the outlet duct (B)

11.18 Remove the heater air outlet duct bolts

the duct to remove it from the HVAC unit **(see illustration)**.

18   Remove the two heater air outlet duct mounting bolts and squeeze both ends of the duct to remove it from the HVAC unit **(see illustration)**.

19   Remove the three module mounting bolts and remove the module from the vehicle.

20   Place the module on its back on a flat working surface, with the blower motor facing upwards.

21   Working from the bottom of the module, remove the rear auxiliary evaporator tube clips. Remove the three HVAC module lower cover bolts (and any clips), and remove the cover and gasket.

22   Working from the front of the module, remove the HVAC module case screws (and any clips) and separate the case halves to access the heater core.

23   Carefully pull the heater core from the HVAC case.

Caution: *Ensure the foam seal for the new heater core is properly installed on the heater core. Use care when inserting the new heater core into the heater case to prevent damage to the foam seal.*

24   Installation is the reverse of removal.

25   Reconnect the cable to the negative terminal of the battery (see Chapter 5).

26   Refill the cooling system with the proper concentration of antifreeze (see Chapter 1). Start the engine and allow it to reach normal operating temperature while inspecting the system for leaks.

27   Have the system evacuated, recharged and leak tested by the same shop that discharged the system.

---

12   Air conditioning compressor - removal and installation

Warning: *The air conditioning system is under high pressure. Do not loosen any hose fittings or remove any components until after the*

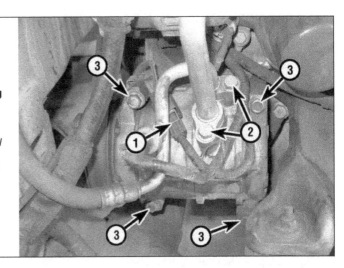

12.6 Air conditioning compressor mounting details:

1   *Clutch electrical connector*
2   *Refrigerant line fitting bolts*
3   *Mounting bolts*

system has been discharged. Air conditioning refrigerant must be properly discharged into an EPA-approved recovery/recycling unit at a dealer service department or an automotive air conditioning repair facility. Always wear eye protection when disconnecting air conditioning system fittings.

Caution: *When replacing entire components, additional refrigerant oil must be added to them. Be sure to read the label on the container before adding any oil to the system; make sure it is compatible with the R-134a system.*

Note: *The receiver-drier should be replaced whenever the compressor is replaced.*

Removal

1   Have the air conditioning system recovered by a dealer service department or by an automotive air conditioning shop before proceeding (see **Warning** above).

2   Disconnect the cable from the negative battery terminal (see Chapter 5).

3   Remove the drivebelt (see Chapter 1).

4   Set the parking brake, block the rear wheels and raise the front of the vehicle, supporting it securely on jackstands.

5   Remove the right front wheel and inner fender splash shield (see Chapter 11).

6   Disconnect the compressor clutch electrical connector **(see illustration)**.

7   Remove the bolts for the refrigerant line fittings, then pull the fittings from the compressor and discard the seals. Seal all open connections to avoid contamination.

8   Remove the compressor mounting bolts.

9   Remove the compressor from the vehicle. Use care not to tilt compressor during removal to prevent spilling the refrigerant oil.

Installation

10   The clutch may have to be transferred from the old compressor to the new unit.

11   If a new or rebuilt compressor is being installed, follow the directions supplied with the compressor to properly adjust the oil level before installing it. If no specifications are provided, drain the oil from the new compressor and add 2.5 oz (75 ml) of fresh refrigerant oil.

12   Installation is the reverse of removal. Use new seals where the hose fitting attaches to the compressor. Lubricate any O-rings with clean refrigerant oil of the correct type.

13.5 Refrigerant line fitting-to-condenser bolt

14.3a Remove the desiccant cartridge plug cap...

13   Reconnect the battery (see Chapter 5).
14   Have the system evacuated, recharged and leak-tested by the shop that discharged it.

13   Air conditioning condenser - removal and installation

Warning: *The air conditioning system is under high pressure. Do not loosen any hose fittings or remove any components until after the system has been discharged. Air conditioning refrigerant must be properly discharged into an EPA-approved recovery/recycling unit at a dealer service department or an automotive air conditioning repair facility. Always wear eye protection when disconnecting air conditioning system fittings.*
Caution: *When replacing entire components, additional refrigerant oil must be added to them. Be sure to read the label on the container before adding any oil to the system; make sure it is compatible with the R-134a system.*
Note: *The receiver-drier is part of the condenser and is replaced whenever the condenser is replaced.*
1   Have the air conditioning system recovered by a dealer service department or by an automotive air conditioning shop before proceeding (see **Warning** above).
2   Disconnect the cable from the negative battery terminal (see Chapter 5).
3   Remove the front bumper cover, the front bumper cover support and the hood latch (see Chapter 11).
4   Remove the front bumper impact absorber **(see illustration 5.3)**.
5   Disconnect the refrigerant lines from the condenser **(see illustration)**. Cap all fittings on the condenser and lines to prevent entry of dirt or moisture.
6   On 2013 and later models, detach the air conditioning refrigerant heat exchanger from the condenser.
7   Disconnect the condenser from the radiator by pinching the fastening tabs at the top of the condenser **(see illustration 7.6)**.
8   Carefully lift the condenser to release the lower retaining tabs and separate it from the

14.3b . . . then tap upwards on the plug with a hammer...

radiator. Be careful not to damage the fins on either unit while removing it from the engine compartment.
Note: *The receiver-drier is part of the condenser. Ensure the desiccant cartridge is installed in the condenser (see Section 14).*
9   On 2013 and later models, connect the air conditioning refrigerant heat exchanger to the condenser, using new O-rings.
10   Installation is the reverse of removal. Tighten the line fitting mounting bolts to the torque listed in this Chapter's Specifications. If a new condenser is being installed, add 0.6 oz (20 ml) of fresh refrigerant oil. Assemble all connections with new O-rings, lightly lubricated with R-134a refrigerant oil.
11   Have the system evacuated, charged and leak tested by the shop that discharged it.

14   Air conditioning receiver-drier - removal and installation

Warning: *The air conditioning system is under high pressure. Do not loosen any hose fittings or remove any components until after the*

14.3c . . . and remove the snap ring

system has been discharged. Air conditioning refrigerant must be properly discharged into an EPA-approved recovery/recycling unit at a dealer service department or an automotive air conditioning repair facility. Always wear eye protection when disconnecting air conditioning system fittings.
Caution: *When replacing entire components, additional refrigerant oil must be added to them. Be sure to read the label on the container before adding any oil to the system; make sure it is compatible with the R-134a system.*
1   Have the air conditioning system recovered by a dealer service department or by an automotive air conditioning shop before proceeding (see **Warning** above).
2   Remove the front bumper cover (see Chapter).
3   Working at the lower left side of the condenser, remove the cap covering the snap-ring and desiccant cartridge plug. Lightly tap upwards on the desiccant cartridge plug and remove the snap-ring **(see illustrations)**.
4   Remove the desiccant cartridge plug (a 5mm screw can be inserted into the cartridge plug to assist with removal) and remove the

14.4a Remove the cartridge plug...

14.4b... and remove the desiccant cartridge

15.2 Disconnect the pressure cycling switch (pressure transducer) electrical connector

desiccant cartridge (use pliers if necessary) **(see illustrations)**.

5 Installation is the reverse of removal. Apply a coat of refrigerant oil to the desiccant plug before installing. Add 0.3 oz (10 ml) of fresh R-134a refrigerant oil.

6 Have the system evacuated, charged and leak tested by the shop that discharged it.

15 Air conditioning refrigerant pressure sensor - replacement

Note: *The AC system does not need to be discharged to replace the refrigerant pressure sensor.*

1 The refrigerant pressure sensor is located on top of the hard line at the AC compressor. The refrigerant pressure sensor detects low refrigerant pressure, and switches the AC system off. If the refrigerant pressure sensor fails and AC pressure increases too high, the pressure cut-off switch (located in the high-pressure side of the system) shuts the system off.

2 Unplug the electrical connector from the pressure cycling switch **(see illustration)**.

3 Unscrew the pressure cycling switch.

4 Lubricate the switch O-ring with clean refrigerant oil of the correct type.

5 Screw the new switch into place until hand tight, then tighten it securely.

6 Reconnect the electrical connector.

16 Air conditioning Thermostatic Expansion Valve (TXV) - general information

Warning: *The air conditioning system is under high pressure. DO NOT loosen any hose fittings or remove any components until the system has been discharged. Air conditioning*

refrigerant must be properly discharged into an EPA-approved recovery/recycling unit by a dealer service department or an automotive air conditioning repair facility. Always wear eye protection when disconnecting air conditioning system fittings.

1 There are several ways that air conditioning systems convert the high-pressure liquid refrigerant from the compressor to lower-pressure vapor. The conversion takes place at the air conditioning evaporator; the evaporator is chilled as the refrigerant passes through, cooling the airflow through the evaporator for delivery to the vents. The conversion is usually accomplished by a sudden change in the tubing size. Many vehicles have a removable controlled orifice in one of the AC pipes at the firewall.

2 The models covered by this manual use a thermostatic expansion valve (TXV) that accomplishes the same purpose as a controlled orifice.

Removal

3 Have the air conditioning system recovered by a dealer service department or by an automotive air conditioning shop before proceeding (see **Warning** above).

Front expansion valve

4 Remove the air intake duct (see Chapter 4) and if equipped, remove the windshield washer fluid heater.

5 Remove the single nut and disconnect the refrigerant lines from the TXV **(see illustration)**.

6 If needed for access, disconnect the power brake booster vacuum line and move it out of the way.

7 Remove the two bolts securing the TXV to the firewall, then remove the valve.

Rear auxiliary expansion valve

8 Remove the passenger rear wheel and inner fender splash shield (see Chapter 11).

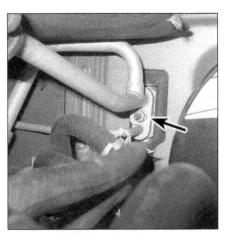

16.5 Remove the nut securing the refrigerant lines to the TXV and disconnect the lines

9 From below the rear of the vehicle, remove the single nut and disconnect the refrigerant lines from the rear auxiliary HVAC module fittings **(see illustration 11.14)**.

10 Remove the two bolts securing the TXV to the vehicle, then remove the valve.

Installation

11 Installation is the reverse of removal. Assemble all connections with new O-rings or seals as necessary, lightly lubricated with R-134a refrigerant oil.

Note: *When installing the TXV to the vehicle, insert a small Phillips head screwdriver into one of the TXV mounting holes to align the valve and insert a bolt into the other mounting hole. Remove the screwdriver and insert the second bolt, then torque both bolts to this Chapter's Specifications.*

12 Have the system evacuated, charged and leak tested by the shop that discharged it.

Notes

Chapter 4
Fuel and exhaust systems

Contents

	Section		Section
Air filter housing - removal and installation...................................	12	Fuel pump module - removal and installation...............................	7
Exhaust system servicing - general information...........................	6	Fuel rail and injectors - removal and installation	14
Fuel lines and fittings - general information		Fuel tank - removal and installation..	11
and disconnection	5	General information	1
Fuel pressure - check..	4	High pressure fuel pump - removal and installation	10
Fuel pressure relief procedure...	3	Throttle body - removal and installation.......................................	13
Fuel pressure sensor - replacement..	8	Troubleshooting...	2
Fuel Pump Flow Control Module (FPFCM) - removal			
and installation	9		

Specifications

Fuel system pressure (key on, engine off except where indicated)		
2009 and earlier models...	56 to 62 psi	384 to 425 kPa
2010 models..	50 to 60 psi	345 to 414 kPa
2011 and later models		
Key on, engine off..	50 to 100 psi	345 to 690 kPa
Idle...	43 to 58 psi	300 to 400 kPa

Torque specifications

	Ft-lbs (unless otherwise indicated)	Nm

Note: *One foot-pound (ft-lb) of torque is equivalent to 12 inch-pounds (in-lbs) of torque. Torque values below approximately 15 foot-pounds are expressed in inch-pounds, because most foot-pound torque wrenches are not accurate at these smaller values.*

Throttle body..	89 in-lbs	10
Fuel pressure sensor (2009 and later models)		
Fuel rail fuel pressure sensor...	25	34
Fuel feed pipe pressure sensor..	Not available	

Torque specifications

Ft-lbs (unless otherwise indicated)　　**Nm**

Note: One foot-pound (ft-lb) of torque is equivalent to 12 inch-pounds (in-lbs) of torque. Torque values below approximately 15 foot-pounds are expressed in inch-pounds, because most foot-pound torque wrenches are not accurate at these smaller values.

	Ft-lbs	Nm
High-pressure fuel pump bolts	132 in-lbs	15
High-pressure fuel line fittings		
Step 1	144 in-lbs	16
Step 2	24	32
Fuel rail bolts		
2008 and earlier models	89 in-lbs	10
2009 and later models		
Step 1	106 in-lbs	12
Step 2	17	23

1　General information

Fuel system warnings

1　*Gasoline is extremely flammable and repairing fuel system components can be dangerous. Consider your automotive repair knowledge and experience before attempting repairs which may be better suited for a professional mechanic.*

a) *Don't smoke or allow open flames or bare light bulbs near the work area*

b) *Don't work in a garage with a gas-type appliance (water heater, clothes dryer)*

c) *Use fuel-resistant gloves. If any fuel spills on your skin, wash it off immediately with soap and water*

d) *Clean up spills immediately*

e) *Do not store fuel-soaked rags where they could ignite*

f) *Prior to disconnecting any fuel line, you must relieve the fuel pressure (see Section 3)*

g) *Wear safety glasses*

h) *Have a proper fire extinguisher on hand*

Fuel system

2　The fuel system consists of the fuel tank, electric fuel pump/fuel level sending unit (located in the fuel tank) fuel rail and fuel injectors. The fuel injection system is a multi-port system; the system uses timed impulses to inject the fuel directly into the intake port of each cylinder. The Powertrain Control Module (PCM) controls the injectors and monitors various engine parameters, and delivers the exact amount of fuel required into the intake ports.

3　Fuel is circulated from the fuel pump to the fuel rail through fuel lines running along the underside of the vehicle. Various sections of the fuel line are either rigid metal or nylon, or flexible fuel hose. The various sections of the fuel hose are connected either by quick-connect fittings or threaded metal fittings.

Exhaust system

4　The exhaust system consists of the exhaust manifold(s), catalytic converter(s), muffler(s), tailpipe and all connecting pipes, flanges and clamps. The catalytic converters are an emission control device added to the exhaust system to reduce pollutants.

2　Troubleshooting

Fuel pump

1　The low pressure fuel pump is located inside the fuel tank. Sit inside the vehicle with the windows closed, turn the ignition key to ON (not START) and listen for the sound of the fuel pump as it's briefly activated. You will only hear the sound for a second or two, but that sound tells you that the pump is working. Alternatively, have an assistant listen at the fuel filler cap.

2　If the pump does not come on, check the fuel pump fuse and relay **(see illustration)**. If the fuse and relay are okay, check the wiring back to the fuel pump. If the fuse, relay and wiring are okay, the fuel pump is probably defective. If the pump runs continuously with the ignition key in the ON position, the Powertrain Control Module (PCM) is probably defective. Have the PCM checked by a professional mechanic.

Fuel injection system

Note: *The following procedure is based on the assumption that the fuel pump is working and the fuel pressure is adequate (see Section 4).*

3   Check all electrical connectors that are related to the system. Check the ground wire connections for tightness.

4   Verify that the battery is fully charged (see Chapter 1).

5   Inspect the air filter element (see Chapter 1).

6   Check all fuses related to the fuel system (see Chapter 12).

7   Check the air induction system between the throttle body and the intake manifold for air leaks. Also inspect the condition of all vacuum hoses connected to the intake manifold and to the throttle body.

8   Remove the air intake duct from the throttle body and look for dirt, carbon, varnish, or other residue in the throttle body, particularly around the throttle plate. If it's dirty, clean it with carb cleaner, a toothbrush and a clean shop towel.

9   Check to see if any trouble codes are stored in the PCM (see Chapter 6).

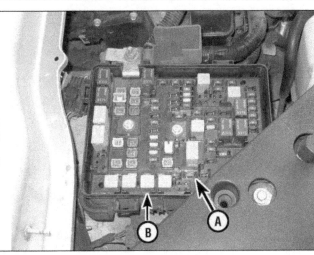

2.2 The fuel pump fuse (A) is located in the engine compartment fuse box; (B) is the fuel pump relay (this is on a 2007 model; check the guide on your fuse box cover for the locations on your model)

3   Fuel pressure relief procedure

Warning: *Gasoline is extremely flammable. See **Fuel system warnings** in Section 1.*

Warning: *The fuel delivery system on 2009 and later models is made up of a low-pressure system and a high-pressure system. Once the pressure on the low-pressure side of the system has been relieved, wait at least two hours before loosening any fuel line fittings in the engine compartment.*

1   Disconnect the cable from the negative terminal of the battery (see Chapter 5).

2   Remove the fuel filler cap to relieve any pressure built-up in the fuel tank.

3   Remove the engine cover, then unscrew

the cap on the fuel pressure test port **(see illustration 4.1b)**.

4   Surround and cover the test port with shop rags, then depress the Schrader valve inside the test port with a small screwdriver until the pressure in the fuel system is relieved. Properly dispose of the rags.

5   It's a good idea to cover any fuel connection to be disassembled with rags to absorb the residual fuel that may leak out. Properly dispose of the rags.

4   Fuel pressure - check

Warning: *Gasoline is extremely flammable. See **Fuel system warnings** in Section 1.*

Note: The following procedure assumes that the fuel pump is receiving voltage and runs.

1   On 2008 and earlier models, locate the fuel pressure test port on the fuel rail, unscrew the cap and connect a fuel pressure gauge **(see illustrations)**.

2   On 2009 and later models, connect the pressure gauge to the Schrader valve on the fuel feed pipe.

3   On 2010 and earlier models, turn the key to the On position.

4   On 2011 and later models, start the engine and allow it to idle.

5   Note the gauge reading as soon as the pressure stabilizes, and compare it with the pressure listed in this Chapter's Specifications.

6   If the fuel pressure is not within specifications, check the following:

a) *Check for a restriction in the fuel system (kinked fuel line, plugged fuel pump inlet strainer or clogged fuel filter). If no restrictions are found, replace the fuel pump module (see Chapter 4).*

b) *If the fuel pressure is higher than specified, replace the fuel pump module (see Chapter 4).*

7   Turn off the engine. Fuel pressure should not fall more than 8 psi over five minutes. If it does, the problem could be a leaky fuel injector, fuel line leak, or faulty fuel pump module.

8   Disconnect the fuel pressure gauge. Wipe up any spilled gasoline.

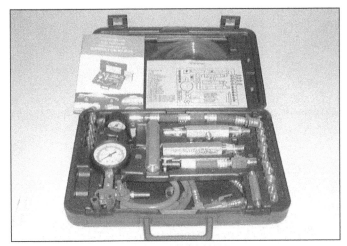

4.1a This fuel pressure testing kit contains all the necessary fittings and adapters, along with the fuel pressure gauge, to test most automotive systems

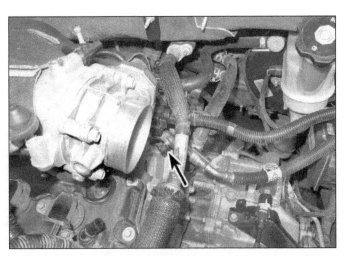

4.1b The fuel pressure test port is located on the fuel rail (2008 and earlier models)

Disconnecting Fuel Line Fittings

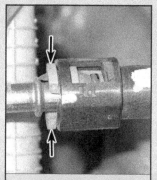

Two-tab type fitting; depress both tabs with your fingers, then pull the fuel line and the fitting apart

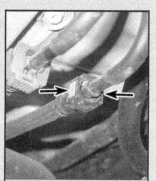

On this type of fitting, depress the two buttons on opposite sides of the fitting, then pull it off the fuel line

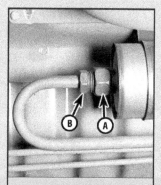

Threaded fuel line fitting; hold the stationary portion of the line or component (A) while loosening the tube nut (B) with a flare-nut wrench

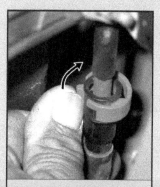

Plastic collar-type fitting; rotate the outer part of the fitting

Metal collar quick-connect fitting; pull the end of the retainer off the fuel line and disengage the other end from the female side of the fitting . . .

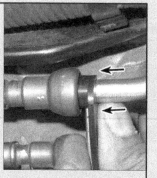

. . . insert a fuel line separator tool into the female side of the fitting, push it into the fitting and pull the fuel line off the pipe

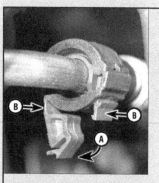

Some fittings are secured by lock tabs. Release the lock tab (A) and rotate it to the fully-opened position, squeeze the two smaller lock tabs (B) . . .

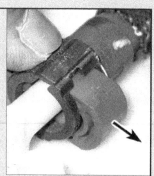

. . . then push the retainer out and pull the fuel line off the pipe

Spring-lock coupling; remove the safety cover, install a coupling release tool and close the tool around the coupling . . .

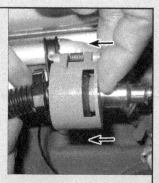

. . . push the tool into the fitting, then pull the two lines apart

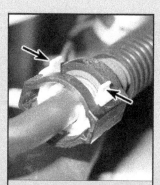

Hairpin clip type fitting: push the legs of the retainer clip together, then push the clip down all the way until it stops and pull the fuel line off the pipe

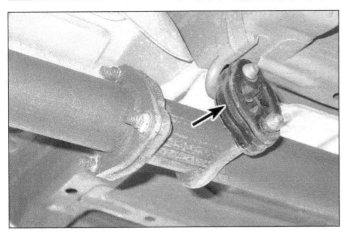

6.1 A typical exhaust system hanger. Inspect regularly and replace at the first sign of damage or deterioration

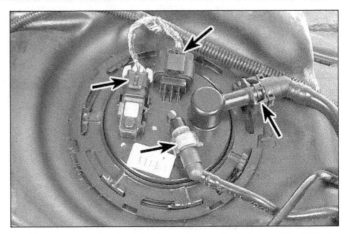

7.4 Disconnect the electrical connectors, fuel line and the EVAP line from the fuel pump module

5 Fuel lines and fittings - general information and disconnection

Warning: *Gasoline is extremely flammable. See* **Fuel system warnings** *in Section 1.*

1   Relieve the fuel pressure before servicing fuel lines or fittings (see Section 3), then disconnect the cable from the negative battery terminal (see Chapter 5) before proceeding.

2   The fuel supply line connects the fuel pump in the fuel tank to the fuel rail on the engine. The Evaporative Emission (EVAP) system lines connect the fuel tank to the EVAP canister and connect the canister to the intake manifold.

3   Whenever you're working under the vehicle, inspect all fuel and evaporative emission lines for leaks, kinks, dents and other damage. Always replace a damaged fuel or EVAP line immediately.

4   If you find signs of dirt in the lines during disassembly, disconnect all lines and blow them out with compressed air. Inspect the fuel strainer on the fuel pump pick-up unit for damage and deterioration.

Steel tubing

5   It is critical that the fuel lines be replaced with lines of equivalent type and specification.

6   Some steel fuel lines have threaded fittings. When loosening these fittings, hold the stationary fitting with a wrench while turning the tube nut.

Plastic tubing

7   When replacing fuel system plastic tubing, use only original equipment replacement plastic tubing.

Caution: *When removing or installing plastic fuel line tubing, be careful not to bend or twist it too much, which can damage it. Plastic fuel tubing is NOT heat resistant, so keep it away from excessive heat.*

Flexible hoses

8   When replacing fuel system flexible hoses, use only original equipment replacements.

9   Don't route fuel hoses (or metal lines) within four inches of the exhaust system or within ten inches of the catalytic converter. Make sure that no rubber hoses are installed directly against the vehicle, particularly in places where there is any vibration. If allowed to touch some vibrating part of the vehicle, a hose can easily become chafed and it might start leaking. A good rule of thumb is to maintain a minimum of 1/4-inch clearance around a hose (or metal line) to prevent contact with the vehicle underbody.

6 Exhaust system servicing - general information

Warning: *Allow exhaust system components to cool before inspection or repair. Also, when working under the vehicle, make sure it is securely supported on jackstands.*

1   The exhaust system consists of the exhaust manifolds, catalytic converter, muffler, tailpipe and all connecting pipes, flanges and clamps. The exhaust system is isolated from the vehicle body and from chassis components by a series of rubber hangers (**see illustration**). Periodically inspect these hangers for cracks or other signs of deterioration, replacing them as necessary.

2   Conduct regular inspections of the exhaust system to keep it safe and quiet. Look for any damaged or bent parts, open seams, holes, loose connections, excessive corrosion or other defects which could allow exhaust fumes to enter the vehicle. Do not repair deteriorated exhaust system components; replace them with new parts.

3   If the exhaust system components are extremely corroded, or rusted together, a cutting torch is the most convenient tool for removal. Consult a properly-equipped repair shop. If a cutting torch is not available, you can use a hacksaw, or if you have compressed air, there are special pneumatic cutting chisels that can also be used. Wear safety goggles to protect your eyes from metal chips and wear work gloves to protect your hands.

4   Here are some simple guidelines to follow when repairing the exhaust system:

a) *Work from the back to the front when removing exhaust system components.*
b) *Apply penetrating oil to the exhaust system component fasteners to make them easier to remove.*
c) *Use new gaskets, hangers and clamps.*
d) *Apply anti-seize compound to the threads of all exhaust system fasteners during reassembly.*
e) *Be sure to allow sufficient clearance between newly installed parts and all points on the underbody to avoid overheating the floor pan and possibly damaging the interior carpet and insulation. Pay particularly close attention to the catalytic converter and heat shield.*

7 Fuel pump module - removal and installation

Warning: *Gasoline is extremely flammable. See* **Fuel system warnings** *in Section 1.*

1   This vehicle is equipped with a single fuel pump module. The module includes the fuel pump, a fuel level sending unit, a fuel filter, and, on mechanical returnless pumps, a fuel pressure regulator. Starting in 2009, a Direct Injection system is used, which incorporated an additional high-pressure fuel pump that is mounted on the engine.

2   Disconnect the cable from the negative battery terminal (see Chapter 5). Relieve the fuel system pressure (see Section 3).

3   Remove the fuel tank (see Section 11).

4   Disconnect the electrical and the fuel supply line quick-connect fitting from the fuel pump module (**see illustration**).

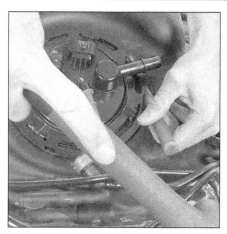

7.5 Rotate the lock ring counterclockwise using a hammer and brass punch

7.6 Carefully pull the fuel pump module from the tank; angle it as necessary to protect the fuel level sensor float arm

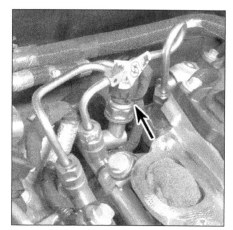

8.5 Location of the fuel rail fuel pressure sensor (2009 and later models)

10.4 High-pressure fuel line connections

10.6 Remove the high-pressure fuel pump mounting bolts (one of two shown) and discard them (obtain new ones for reassembly)

5   Using a hammer and brass punch, rotate the lock ring counterclockwise to remove it **(see illustration)**.

6   Carefully pull the fuel pump module out of the tank. Angle it as necessary to protect the fuel level sensor float arm and remove it from the tank **(see illustration)**.

7   To remove the fuel level sensor, push down on the retainer to release it.

8   Disengage the wiring terminals and replace the fuel level sensor.

9   Inspect the O-ring and replace it if it shows any sign of deterioration.

10   Installation is the reverse of removal.

8   Fuel pressure sensor - replacement

Warning: *Gasoline is extremely flammable. See* **Fuel system warnings** *in Section 1.*
Warning: *The manufacturer states that the fuel pressure sensor must be replaced with a new one whenever it is removed.*

1   Disconnect the cable from the negative battery terminal (see Chapter 5).

2   Remove the engine cover.

3   Relieve the fuel system pressure (see Section 3). Wait at least two hours before proceeding.

4   Remove the intake manifold (see Chapter 2A) and remove the foam insulator.

5   Disconnect the electrical connector from the sensor **(see illustration)**.

6   Unscrew the sensor from the fuel rail.
Note: *Have rags ready to catch any fuel that is remaining in the fuel rail.*

7   Installation is the reverse of removal. Tighten the sensor to the torque listed in this Chapter's Specifications.

9   Fuel Pump Flow Control Module (FPFCM) - removal and installation

Note: *The FPFCM is located on the left side of the rear suspension subframe.*

1   Raise the rear of the vehicle and support it securely on jackstands.

2   Disconnect the electrical connector from the FPFCM.

3   Loosen the bolts that retain the FPFCM, then slide the module mounting bracket to disengage it from the bolt heads.
Note: *Before installing the FPFCM (either the same one or a new one), make sure the area around the seam of the module is clean, then apply a bead of RTV sealant to the seam. This will prevent the intrusion of moisture which can cause module failure.*

4   Installation is the reverse of removal.

10   High pressure fuel pump - removal and installation

Warning: *Gasoline is extremely flammable. See* **Fuel system warnings** *in Section 1.*
Warning: *The fuel delivery system is made up of a low-pressure system and a high-pressure system. Once the pressure of the low-pressure side of the system has been relieved, wait at least two hours before loosening any fuel fittings in the engine compartment.*
Note: *This procedure applies to 2009 and later models.*

1   Disconnect the cable from the negative battery terminal (see Chapter 5).

2   Relieve the fuel system pressure (see Section 3).

3   Disconnect the low-pressure feed line to the pump.

4   Using a flare nut wrench, unscrew the high-pressure fuel line fittings from the pump and the fuel rail **(see illustration)**.
Caution: *The manufacturer states that the pipe must be replaced if it has been removed.*

5   Disconnect the electrical connector from the high-pressure fuel pump.

6   Remove the high-pressure fuel pump bolts **(see illustration)** and remove the pump. Always replace the bolts and O-ring.

7   With the pump removed, rotate the engine by hand and make sure the camshaft lobe is at its base circle before trying to install the pump.

8   Lubricate the pump roller with engine oil and install the gasket and bolts.

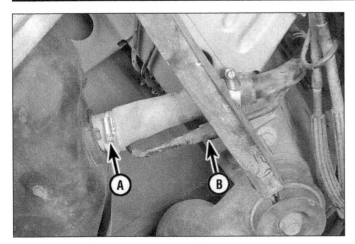

11.6a Fuel tank filler hose clamp (A) and vent hose quick-connect fitting (B)

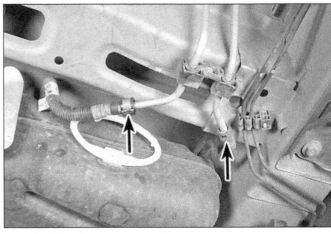

11.6b Disconnect the fuel and EVAP lines at the front of the tank

9 Set the pump into the cylinder head and tighten the bolts evenly.
Note: *As the pump bolts are tightened, it will get harder to tighten the bolts until the spring in the pump is compressed.*
10 Tighten the bolts to the torque listed in this Chapter's Specifications.
11 The remainder of installation is the reverse of removal. Clean the high-pressure fuel line fittings on the fuel rails, then apply a little clean engine oil to the threads. Install the new high-pressure fuel line, tightening the fittings to the torque listed in this Chapter's Specifications.
12 Reconnect the cable to the negative battery terminal (see Chapter 5), then start the engine and check for fuel leaks.

11 Fuel tank - removal and installation

Warning: *Gasoline is extremely flammable. See **Fuel system warnings** in Section 1.*
Warning: *The following procedure is much easier to perform if the fuel tank is empty.*
1 Disconnect the cable from the negative battery terminal (see Chapter 5).
2 Remove the fuel tank filler cap to relieve fuel tank pressure.
3 Relieve the fuel system pressure (see Section 3).
4 Use a siphoning kit or a suitable hose to remove as much gasoline as possible from the fuel tank.
Warning: *Never start a siphon by mouth! Siphoning kits are available at most auto parts stores.*
5 Raise the rear of the vehicle and support it securely on jackstands.
6 Disconnect the filler and vent hoses **(see illustration)**. Also detach the fuel hose(s) at the fitting(s) near the tank, the EVAP hoses from the canister, and the fuel pump module electrical connector **(see illustrations)**.
Warning: *Be sure to catch fuel spillage with rags. Dispose of the fuel-soaked rags in an approved container.*
7 Remove the left underbody side rail.

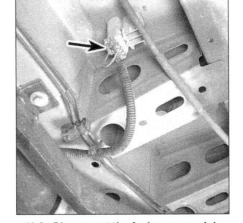

11.6c Disconnect the fuel pump module electrical connector

8 Support the fuel tank with two floor jacks.
9 Remove the bolts from the fuel tank straps; let the straps hang down.
10 Slowly lower the fuel tank and remove it from under the vehicle.
11 Installation is the reverse of removal. Tighten the fuel tank strap bolts securely.
12 Reconnect the cable to the negative battery terminal (see Chapter 5), then start the engine and check for fuel leaks.

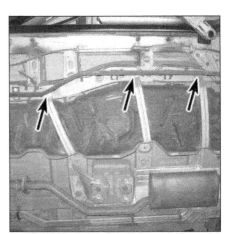

11.9 Remove the fuel tank strap bolts

12 Air filter housing - removal and installation

Air intake duct

1 Disconnect the PCV fresh air hose from the valve cover **(see illustration)**.
2 Loosen the clamps at the air filter housing and the throttle body and lift off the duct.
3 Installation is the reverse of removal.

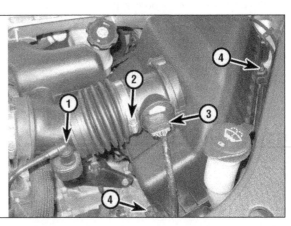

12.1 Air intake duct and air filter housing cover details

1 PCV fresh air hose
2 Hose clamp
3 MAF/IAT sensor electrical connector
4 Air filter housing cover screws (2 of 6 shown)

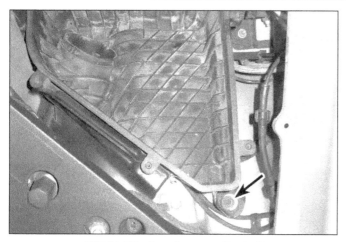

12.7 Air filter housing mounting bolt

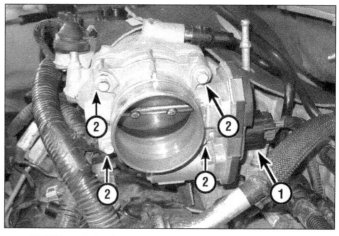

13.3 Throttle body electrical connector (1) and mounting fasteners (2)

Air filter housing

4   Relieve the fuel system pressure (see Section 3).

5   Disconnect the electrical connector from the MAF sensor **(see illustration 12.1)**.

6   Loosen the clamp and detach the duct from the air filter housing.

7   Remove the air filter housing cover and filter element.

8   Remove the hood release cable mounting clip from the air filter housing.

9   Disconnect the fuel supply line from the fuel rail (see Chapter 4).

10   Disconnect the EVAP line from the purge solenoid.

11   Release the Powertrain Control Module (PCM) from the bracket (see Chapter 6) and position it away from the air filter housing.

Note: *Don't disconnect the electrical connectors from the PCM.*

12   Remove the air filter housing mounting fastener and lift off the housing.

13   Installation is the reverse of removal.

13   Throttle body - removal and installation

Warning: *Wait until the engine is completely cool before beginning this procedure.*

Caution: *The manufacturer recommends that an idle-learn procedure be done any time the throttle body is removed. This is done to ensure that the idle is stable and that no Diagnostic Trouble Codes (DTCs) occur. Refer to Chapter 6 for information on this procedure.*

1   Disconnect the cable from the negative battery terminal (see Chapter 5).

2   Remove the air intake duct (see Section 12).

3   Disconnect the electrical connector from the throttle body **(see illustration)**.

4   Remove the throttle body mounting fasteners **(see illustration 13.3)**, and carefully lift off the throttle body. Discard the gasket; it should be replaced with a new one.

5   Cover the intake manifold opening with a clean shop towel.

14.4 Location of the fuel supply line fitting at the fuel rail

6   Installation is the reverse of removal. Use a new gasket and tighten the throttle body fasteners to the torque listed in this Chapter's Specifications.

Idle relearn procedure

7   Start the engine and allow it to idle for three minutes.

Note: *If a scanner is available, connect the scanner and view or record the actual engine idle speeds.*

8   After the three minutes, the PCM start to learn the new idle patterns and begin adjusting the idle speed. The engine should begin to slow to almost normal.

9   When that happens, turn the engine off for one minute, then start the engine and allow it to idle again for three minutes. At the end of the second three minute cycle, the idle should be normal.

Note: *If the Check Engine light comes on during the cycles, clear the codes after the second three minute cycle ends.*

10   If the engine does not idle properly after the relearn procedure, drive the vehicle above 44 mph, then decelerate and accelerate several times, allowing it to idle in between. Stop the vehicle and turn the

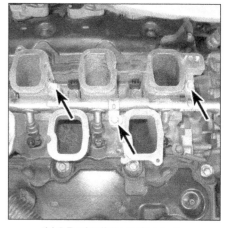

14.7 Fuel rail mounting bolts

engine off for one minute; the engine should now idle normally.

11   Clear any trouble codes that may have been set (see Chapter 6).

14   Fuel rail and injectors - removal and installation

Warning: *Gasoline is extremely flammable. See* **Fuel system warnings** *in Section 1.*

2008 and earlier models

Removal

Note: *Even if you only removed the fuel rail assembly to replace a single injector or a leaking O-ring, it's a good idea to remove all of the injectors from the fuel rail and replace all of the O-rings at the same time.*

1   Relieve the fuel system pressure (see Section 3).

2   Disconnect the cable from the negative battery terminal (see Chapter 5).

3   Remove the upper intake manifold (see Chapter 2A).

4   Disconnect the fuel supply line at the fuel rail **(see illustration)**.

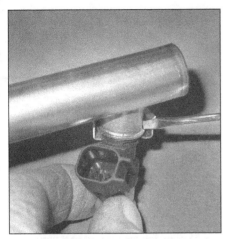

14.8 Carefully pry off each injector retaining clip with a small screwdriver or nose pliers, then pull the injector out of the fuel rail

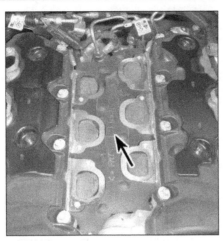

14.21 Remove the foam insulator covering the fuel rails

14.24 With the outer bolts removed and the inner bolts loosened, evenly pry the injectors out of the cylinder head . . .

14.25 . . . and remove the fuel rail and injectors as an assembly

14.28 Use the pliers to open up the retaining clip, then remove it

5   Disconnect the electrical connector for the main fuel injector harness.
6   Disconnect the electrical connector from each injector.
7   Remove the fuel rail mounting fasteners, then remove both fuel rails and their injectors as an assembly **(see illustration)**.
8   Release the fuel injector retaining clips **(see illustration)**. Remove each injector from its bore in the fuel rail. Remove and discard the upper O-ring and the lower O-ring. Repeat this procedure for each injector.

Installation

9   Coat the new upper O-rings with clean engine oil and slide them into place on each of the fuel injectors.
10   Coat the new lower O-rings with clean engine oil and install them on the lower ends of the injectors.
11   Coat the outside surface of each upper O-ring with clean engine oil, insert each injector into its bore in the fuel rail, then install the retaining clips. Install the isolators with new seals on the injectors, if equipped.

12   Install the injectors and fuel rail assembly on the intake manifold. Tighten the fuel rail mounting bolts to the torque listed in this Chapter's Specifications.
13   The remainder of installation is the reverse of removal.
14   Reconnect the cable to the negative battery terminal (see Chapter 5).
15   Turn the ignition switch to ON (but don't operate the starter). This activates the fuel pump for about two seconds, which builds up fuel pressure in the fuel lines and the fuel rail.
16   Repeat Step 15 two or three times, then check the fuel lines, fuel rails and injectors for fuel leaks.

2009 and later models

17   Disconnect the cable from the negative terminal of the battery (see Chapter 5).
18   Relieve the fuel system pressure (see Section 3).
19   Disconnect the fuel supply line at the fuel rail (see Section 5).

20   Remove the intake manifold (see Chapter 2A).
21   Remove the foam insulator **(see illustration)**.
22   Remove the fuel pressure sensor (see Section 8). Also remove the high-pressure fuel line from the pump and fuel rails (see Section 10).
23   Remove the mounting bolts from both fuel rails in reverse of the tightening sequence **(see illustration 14.42)**.
24   Install the center bolts, leaving several threads exposed, then evenly pry the Bank 2 (front side) fuel rail and injectors out of the cylinder head **(see illustration)**.
25   Remove the bolts and work the fuel rail and injectors from the cylinder head **(see illustrations)**.
26   Repeat the procedure on the bank 1 injectors.
27   Disconnect the injector harness from the injectors.
28   Using external snap-ring pliers, carefully spread the retaining clip open then remove the clip **(see illustration)**.

14.29 Remove the injectors from the fuel rail

14.30 To remove the Teflon sealing ring, cut it off with a hobby knife (be careful not to scratch the injector groove)

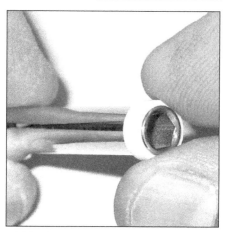

14.35 Slide the new Teflon seal onto the end of a socket that's the same diameter as the end of the fuel injector . . .

29   Remove each injector from its bore in the fuel rail **(see illustration)**.

30   Remove the old combustion chamber Teflon sealing ring and the upper O-ring and support ring from each injector **(see illustration)**. **Caution:** *Be extremely careful not to damage the groove for the seal or the rib in the floor of the groove. If you damage the groove or the rib, you must replace the injector.*

31   Before installing the new Teflon seal on each injector, thoroughly clean the groove for the seal and the injector shaft. Remove all combustion residue and varnish with a clean shop rag.

Teflon seal installation using the special tools

32   The manufacturer recommends that you use the tools included in the special injector tool set to install the Teflon lower seals on the injectors: Install the special seal assembly cone on theinjector, install the special sleeve on the injector and use the sleeve to pushon the assembly cone, which pushes the Teflon

seal into place on its groove. Do NOT use any lubricants to do so.

33   Pushing the Teflon seal into place in its groove expands it slightly. There are three sizing sleeves in the special tool set with progressively smaller inside diameters. Using a clockwise rotating motion of about 180 degrees, install the slightly larger sleeve onto the injector and over the Teflon seal until the sleeve hits its stop, then carefully turn the sleeve counterclockwise as you pull it off the injector. Use the slightly smaller sizing sleeve the same way, followed by the smallest sizing ring. The seal is now sized. Repeat this step for each injector.

Teflon seal installation without special tools

34   If you don't have the special injector tool set, the Teflon seal can be installed using this method: First, find a socket that is equal or very close in diameter to the diameter of the end of the fuel injector.

35   Work the new Teflon seal onto the end of

the socket **(see illustration)**.

36   Place the socket against the end of the injector **(see illustration)** and slide the seal from the socket onto the injector. Do NOT use any lubricants to do so. Continue pushing the seal onto the injector until it seats into its mounting groove.

37   Because the inside diameter of the seal has to be stretched open to fit over the bore of the socket and the injector, its outside diameter is now slightly too large - it is no longer flush with the surface of the injector. It must be shrunk it back to its original size. To do so, push a piece of plastic tubing with an interference fit onto the end of the socket; a plastic straw that fits tightly on the injector will work. After pushing the plastic tubing onto the socket about an inch, snip off the rest of the tubing, then use the socket to push the tubing onto the end of the injector **(see illustration)** and slide it onto the injector until it completely covers the new seal **(see illustration)**. Leave the tubing on for a few hours, then remove it. The seal should now be shrunk back its original outside diameter, or close to it.

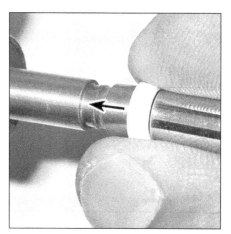

14.36 . . . align the socket with the end of the injector and slide the seal onto the injector and into its mounting groove

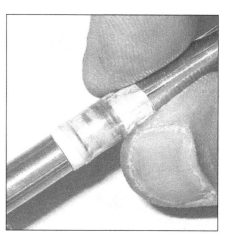

14.37a Use the socket to push a short section of plastic tubing onto the end of the injector and over the new seal . . .

14.37b . . . then leave the plastic tubing in place for several hours to compress the new seal

14.38 Note that the upper O-ring (1) is installed above the support ring (2)

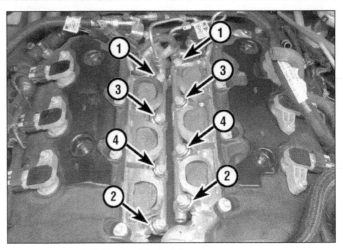

14.42 Fuel rail bolt tightening sequence

Injector and fuel rail installation

38   Install the new support ring at the upper end of the injector. Lubricate the new upper O-ring with clean engine oil and install it on the injector. Do NOT oil the new Teflon seal. Note that the seal is installed above the spacer **(see illustration)**.

39   Thoroughly clean the injector bores with a small nylon brush.

40   Install the new compensation element and retaining ring to the bottom of each injector.

41   Install the fuel injectors in the cylinder head (NOT in the fuel rail). You should be able to push each assembled injector into its bore in the cylinder head. The bore is slightly tapered, so you will encounter some resistance as the Teflon seal nears the bottom of the bore. Press the injector into its bore until it stops, making sure to properly align the injector.

42   Install a new spring steel retainer on each injector. Install the fuel rail, starting with the front injector and working toward the rear, then install the mounting bolts (outer bolts first, then the inner bolts). Tighten the bolts in sequence **(see illustration)** to the torque listed in this Chapter's Specifications.

43   The remainder of installation is the reverse of removal. Clean the high pressure fuel line fittings on the fuel rails, then apply a little clean engine oil to the threads. Install the new high-pressure fuel line, tightening the fittings to the torque listed in this Chapter's Specifications. Install a new fuel pressure sensor (see Section 8).

Notes

Chapter 5
Engine electrical systems

Contents

	Section		Section
Alternator - removal and installation	7	General information and precautions	1
Battery - disconnection	3	Ignition coils - replacement	6
Battery - removal and installation	4	Starter motor - removal and installation	8
Battery cables - replacement	5	Troubleshooting	2

1 General information and precautions

General information

Ignition system

1 The electronic ignition system consists of the Crankshaft Position (CKP) sensor, the Camshaft Position (CMP) sensor, the Knock Sensor (KS), the Powertrain Control Module (PCM), the ignition switch, the battery, the individual ignition coils, and the spark plugs. For more information on the CKP, CMP and KS sensors, as well as the PCM, refer to Chapter.

Charging system

2 The charging system includes the alternator (with an integral voltage regulator), the Powertrain Control Module (PCM), the Body Control Module (BCM), a charge indicator light on the dash, the battery, a fuse or fusible link and the

wiring connecting all of these components. The charging system supplies electrical power for the ignition system, the lights, the radio, etc. The alternator is driven by a drivebelt.

Starting system

3 The starting system consists of the battery, the ignition switch, the starter relay, the Powertrain Control Module (PCM), the Body Control Module (BCM), the Transmission Range (TR) switch, the starter motor and solenoid assembly, and the wiring connecting all of the components.

Precautions

4 Always observe the following precautions when working on the electrical system:

a) *Be extremely careful when servicing engine electrical components. They are easily damaged if checked, connected or handled improperly.*

b) *Never leave the ignition switched on for long periods of time when the engine is not running.*

c) *Never disconnect the battery cables while the engine is running.*

d) *Maintain correct polarity when connecting battery cables from another vehicle during jump starting - see "Booster Battery - jump starting" at the beginning of this manual.*

e) *Always disconnect the cable from the negative battery terminal before working on the electrical system, but read the battery disconnection procedure first (see Section 3).*

5 It's also a good idea to review the safety-related information regarding the engine electrical systems located in "Safety first!" at the beginning of this manual, before beginning any operation included in this Chapter.

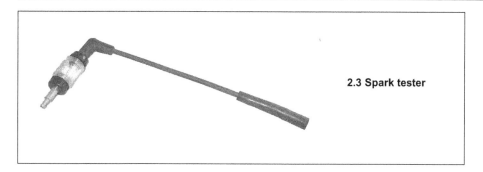

2.3 Spark tester

2   Troubleshooting

Ignition system

1   If a malfunction occurs in the ignition system, do not immediately assume that any particular part is causing the problem. First, check the following items:

a) *Make sure that the cable clamps at the battery terminals are clean and tight.*
b) *Test the condition of the battery (see Steps 21 through 24). If it doesn't pass all the tests, replace it.*
c) *Check the ignition coil or coil pack connections.*
d) *Check any relevant fuses in the engine compartment fuse and relay box (see Chapter 12). If they're burned, determine the cause and repair the circuit.*

Check

Warning: *Because of the high voltage generated by the ignition system, use extreme care when performing a procedure involving ignition components.*
Note: *The ignition system components on these vehicles are difficult to diagnose. In the event of ignition system failure that you can't diagnose, have the vehicle tested at a dealer service department or other qualified auto repair facility.*
Note: *You'll need a spark tester for the following test. Spark testers are available at most auto supply stores.*
2   If the engine turns over but won't start, verify that there is sufficient ignition voltage to fire the spark plugs as follows.
3   On models with a coil-over-plug type ignition system, remove a coil and install the tester between the boot at the lower end of the coil and the spark plug **(see illustration)**. On models with spark plug wires, disconnect a spark plug wire from a spark plug and install the tester between the spark plug wire boot and the spark plug.
Caution: *Do NOT crank the engine or allow it to run for more than five seconds; running the engine for more than five seconds may set a Diagnostic Trouble Code (DTC) for a cylinder misfire.*
4   Crank the engine and note whether or not the tester flashes.

Models with a coil-over-plug type ignition system

5   If the tester flashes during cranking, the coil is delivering sufficient voltage to the spark plug to fire it. Repeat this test for each cylinder to verify that the other coils are OK.
6   If the tester doesn't flash, remove a coil from another cylinder and swap it for the one being tested. If the tester now flashes, you know that the original coil is bad. If the tester still doesn't flash, the PCM or wiring harness is probably defective. Have the PCM checked out by a dealer service department or other qualified repair shop (testing the PCM is beyond the scope of the do-it-yourselfer because it requires expensive special tools).
7   If the tester flashes during cranking but a misfire code (related to the cylinder being tested) has been stored, the spark plug could be fouled or defective.

Models with spark plug wires

8   If the tester flashes during cranking, sufficient voltage is reaching the spark plug to fire it.
9   Repeat this test on the remaining cylinders.
10   Proceed on this basis until you have verified that there's a good spark from each spark plug wire. If there is, then you have verified that the coils in the coil pack are functioning correctly and that the spark plug wires are OK.
11   If there is no spark from a spark plug wire, then either the coil is bad, the plug wire is bad or a connection at one end of the plug wire is loose. Assuming that you're using new plug wires or known good wires, then the coil is probably defective. Also inspect the coil pack electrical connector. Make sure that it's clean, tight and in good condition.
12   If all the coils are firing correctly, but the engine misfires, then one or more of the plugs might be fouled. Remove and check the spark plugs or install new ones (see Chapter 1).
13   No further testing of the ignition system is possible without special tools. If the problem persists, have the ignition system tested by a dealer service department or other qualified repair shop.

Charging system

14   If a malfunction occurs in the charging system, do not automatically assume the alternator is causing the problem. First check the following items:

a) *Check the drivebelt tension and condition (see Chapter 1). Replace it if it's worn or deteriorated.*
b) *Make sure the alternator mounting bolts are tight.*
c) *Inspect the alternator wiring harness and the connectors at the alternator and voltage regulator. They must be in good condition, tight and have no corrosion.*
d) *Check the fusible link (if equipped) or main fuse in the underhood fuse/relay box. If it is burned, determine the cause, repair the circuit and replace the link or fuse (the vehicle will not start and/or the accessories will not work if the fusible link or main fuse is blown).*
e) *Start the engine and check the alternator for abnormal noises (a shrieking or squealing sound indicates a bad bearing).*
f) *Check the battery. Make sure it's fully charged and in good condition (one bad cell in a battery can cause overcharging by the alternator).*
g) *Disconnect the battery cables (negative first, then positive). Inspect the battery posts and the cable clamps for corrosion. Clean them thoroughly if necessary (see Chapter 1). Reconnect the cables (positive first, negative last).*

Alternator - check

15   Use a voltmeter to check the battery voltage with the engine off. It should be at least 12.6 volts **(see illustration 2.21)**.
16   Start the engine and check the battery voltage again. It should now be approximately 13.5 to 15 volts.
17   If the voltage reading is more or less than the specified charging voltage, the voltage regulator is probably defective, which will require replacement of the alternator (the voltage regulator is not replaceable separately). Remove the alternator and have it bench tested (most auto parts stores will do this for you).
18   The charging system (battery) light on the instrument cluster lights up when the ignition key is turned to ON, but it should go out when the engine starts.
19   If the charging system light stays on after the engine has been started, there is a problem with the charging system. Before replacing the alternator, check the battery condition, alternator belt tension and electrical cable connections.
20   If replacing the alternator doesn't restore voltage to the specified range, have the charging system tested by a dealer service department or other qualified repair shop.

Battery - check

21   Check the battery state of charge. Visually inspect the indicator eye on the top of the battery (if equipped with one); if the indicator eye is black in color, charge the battery (see Chapter 1). Then perform an open circuit voltage test using a digital voltmeter. With the engine and all accessories Off, touch the negative probe of the voltmeter to the negative terminal of the battery and the positive probe to the positive terminal of the battery

2.21 To test the open circuit voltage of the battery, touch the black probe of the voltmeter to the negative terminal and the red probe to the positive terminal of the battery; a fully charged battery should be at least 12.6 volts

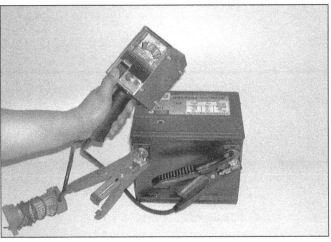

2.23 Connect a battery load tester to the battery and check the battery condition under load following the tool manufacturer's instructions

(see illustration). The battery voltage should be 12.6 volts or slightly above. If the battery is less than the specified voltage, charge the battery before proceeding to the next test. Do not proceed with the battery load test unless the battery charge is correct.

Note: *The battery's surface charge must be removed before accurate voltage measurements can be made. Turn on the high beams for ten seconds, then turn them off and let the vehicle stand for two minutes.*

22 Disconnect the negative cable, then the positive cable from the battery.

23 Perform a battery load test. An accurate check of the battery condition can only be performed with a load tester **(see illustration).** This test evaluates the ability of the battery to operate the starter and other accessories during periods of high current draw. Connect the load tester to the battery terminals. Load test the battery according to the tool manufacturer's instructions. This tool increases the load demand (current draw) on the battery.

24 Maintain the load on the battery for 15 seconds and observe that the battery voltage does not drop below 9.6 volts. If the battery condition is weak or defective, the tool will indicate this condition immediately.

Note: *Cold temperatures will cause the minimum voltage reading to drop slightly. Follow the chart given in the manufacturer's instructions to compensate for cold climates. Minimum load voltage for freezing temperatures (32 degrees F) should be approximately 9.1 volts.*

Starting system

The starter rotates, but the engine doesn't

25 Remove the starter (see Section 8). Check the overrunning clutch and bench test the starter to make sure the drive mechanism extends fully for proper engagement with the flywheel ring gear. If it doesn't, replace the starter.

26 Check the flywheel ring gear for missing teeth and other damage. With the ignition turned off, rotate the flywheel so you can check the entire ring gear.

The starter is noisy

27 If the solenoid is making a chattering noise, first check the battery (see Steps 21 through 24). If the battery is okay, check the cables and connections.

28 If you hear a grinding, crashing metallic sound when you turn the key to Start, check for loose starter mounting bolts. If they're tight, remove the starter and inspect the teeth on the starter pinion gear and flywheel ring gear. Look for missing or damaged teeth.

29 If the starter sounds fine when you first turn the key to Start, but then stops rotating the engine and emits a zinging sound, the problem is probably a defective starter drive that's not staying engaged with the ring gear. Replace the starter.

The starter rotates slowly

30 Check the battery (see Steps 21 through 24).

31 If the battery is okay, verify all connections (at the battery, the starter solenoid and motor) are clean, corrosion-free and tight. Make sure the cables aren't frayed or damaged.

32 Check that the starter mounting bolts are tight so it grounds properly. Also check the pinion gear and flywheel ring gear for evidence of a mechanical bind (galling, deformed gear teeth or other damage).

The starter does not rotate at all

33 Check the battery (see Steps 21 through 24).

34 If the battery is okay, verify all connections (at the battery, the starter solenoid and

motor) are clean, corrosion-free and tight. Make sure the cables aren't frayed or damaged.

35 Check all of the fuses in the underhood fuse/relay box.

36 Check that the starter mounting bolts are tight so it grounds properly.

37 Check for voltage at the starter solenoid "S" terminal when the ignition key is turned to the start position. If voltage is present, replace the starter/solenoid assembly. If no voltage is present, the problem could be the starter relay, the Transmission Range (TR) switch (see Chapter 6), or with an electrical connector somewhere in the circuit (see the wiring diagrams at the end of Chapter 12). Also, on many modern vehicles, the Powertrain Control Module (PCM) and the Body Control Module (BCM) control the voltage signal to the starter solenoid; on such vehicles a special scan tool is required for diagnosis.

---

3   Battery - disconnection

Warning: *If the vehicle is equipped with On-Star, make absolutely sure the ignition key is in the Off position and Retained Accessory Power (RAP) has been depleted before disconnecting the cable from the negative battery terminal. Also, never remove the OnStar fuse with the ignition key in any position other than Off. If these precautions are not taken, the OnStar system's back-up battery will be activated, and remain activated, until it goes dead. If this happens, the OnStar system will not function as it should in the event that the main vehicle battery power is cut off (as might happen during a collision).*

Caution: *Always disconnect the cable from the negative battery terminal FIRST and hook it up LAST or the battery may be shorted by the tool being used to loosen the cable clamps.*

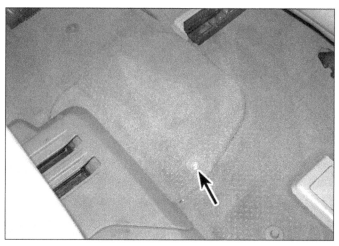

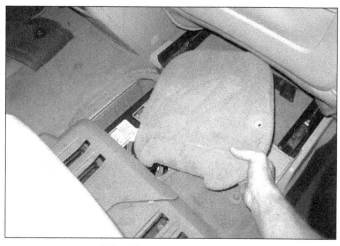

4.3a Remove the fastener securing the battery cover...

4.3b... then remove the cover

1   Some systems on the vehicle require battery power to be available at all times, either to maintain continuous operation (alarm system, power door locks, etc.), or to maintain control unit memory (radio station presets, Powertrain Control Module and other control units). When the battery is disconnected, the power that maintains these systems is cut. So, before you disconnect the battery, please note that on a vehicle with power door locks, it's a wise precaution to remove the key from the ignition and to keep it with you, so that it does not get locked inside if the power door locks should engage accidentally when the battery is reconnected!

2   Devices known as memory-savers can be used to avoid some of these problems. Precise details vary according to the device used. The typical memory saver is plugged into the cigarette lighter and is connected to a spare battery. Then the vehicle battery can be disconnected from the electrical system. The memory saver will provide sufficient current to maintain audio unit security codes, PCM memory, etc. and will provide power to always hot circuits such as the clock and radio memory circuits. **Warning:** *Some memory savers deliver a considerable amount of current in order to keep vehicle systems operational after the main battery is disconnected. If you're using a memory saver, make sure that the circuit concerned is actually open before servicing it.*

Warning: *If you're going to work near any of the airbag system components, the battery MUST be disconnected and a memory saver must NOT be used. If a memory saver is used, power will be supplied to the airbag, which means that it could accidentally deploy and cause serious personal injury.*

3   To disconnect the battery for service procedures requiring power to be cut from the vehicle, first open the driver's door to disable Retained Accessory Power (RAP), then remove the battery cover (see Section 4), loosen the cable end bolt and disconnect the cable from the negative battery terminal. Isolate the cable end to prevent it from coming into accidental contact with the battery terminal.

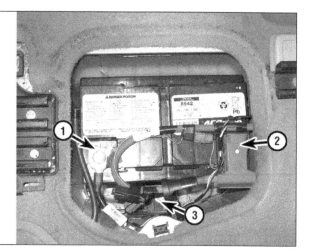

4.4 Battery details:

1   Negative battery cable
2   Positive battery cable
3   Battery hold-down block

---

4   Battery - removal and installation

---

1   Move the front passenger seat as far forward as possible. Move the right-side second-row seat as far rearward as possible.
2   Pull up the floor mat behind the passenger seat.
Note: *There is one tab holding the mat in place.*
3   Remove the battery cover fastener, then remove the cover and set it aside **(see illustrations).**
4   Disconnect the cable from the negative battery terminal first, then disconnect the cable from the positive battery terminal **(see illustration).**
Note: *On later models, it will be necessary to disconnect the electrical connectors to the current sensor and the body harness connectors to the negative and positive cables to remove the battery.*
5   Remove the battery hold-down block.
6   Lift out the battery. Be careful - it's heavy.
Note: *Battery straps and handlers are available at most auto parts stores for reasonable prices. They make it easier to remove and carry the battery.*

7   If you are replacing the battery, make sure you get one that's identical, with the same dimensions, amperage rating, cold cranking rating, etc.
8   Installation is the reverse of removal. Always connect the positive cable first and the negative cable last.

---

5   Battery cables - replacement

---

1   When removing the cables, always disconnect the cable from the negative battery terminal first and hook it up last, or you might accidentally short out the battery with the tool you're using to loosen the cable clamps. Even if you're only replacing the cable for the positive terminal, be sure to disconnect the negative cable from the battery first.
2   Disconnect the old cables from the battery, then trace each of them to their opposite ends and disconnect them. Note the routing of each cable before disconnecting it to ensure correct installation.
3   If you are replacing any of the old cables, take them with you when buying new cables. It is vitally important that you replace the cables

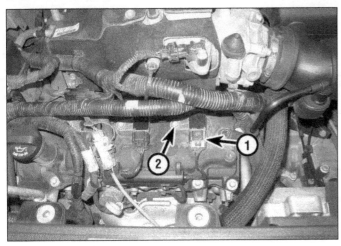

6.4 Ignition coil electrical connector (1) and mounting fastener (2)

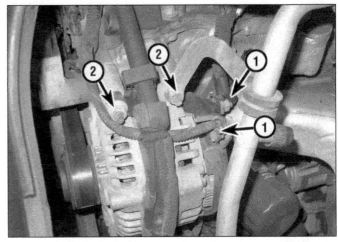

7.7 Alternator electrical connectors (1) and mounting fasteners (2)

with identical parts.

4   Clean the threads of the solenoid or ground connection with a wire brush to remove rust and corrosion. Apply a light coat of battery terminal corrosion inhibitor or petroleum jelly to the threads to prevent future corrosion.

5   Attach the cable to the solenoid or ground connection and tighten the mounting nut/bolt securely.

6   Before connecting a new cable to the battery, make sure that it reaches the battery post without having to be stretched.

7   Connect the cable to the positive battery terminal first, *then* connect the ground cable to the negative battery terminal.

6   Ignition coils - replacement

1   Disconnect the cable from the negative battery terminal (see Section 3). Remove the engine cover.

2   On 2008 and earlier models, remove the underhood fuse/relay box (see Chapter 12) to access the rear bank ignition coils.

3   On 2009 and later models, remove the intake manifold (see Chapter 2A) to access the front bank ignition coils.

4   Disconnect the electrical connector from coil **(see illustration)**.

5   Remove the ignition coil mounting fastener **(see illustration 6.4)**.

6   Pull up and twist to remove the coil.

7   Apply a little silicone dielectric compound to the inside of each boot before installing it.

8   Installation is the reverse of removal.

7   Alternator - removal and installation

1   Disconnect the cable from the negative battery terminal (see Section 3).

2   Remove the drivebelt (see Chapter 1).

3   Remove the engine cover.

4   Remove the coolant reservoir (see Chapter 3).

5   Remove the engine mount strut bracket (see Chapter 2A).

6   Remove the air conditioning line clamp bolt from upper radiator support.

7   Disconnect the battery cable and the other wiring connections from the alternator terminals **(see illustration)**.

8   Unscrew the idler pulley mounting bolt **(see illustration)**. Slide the pulley out as far as possible.

9   Remove the air conditioning compressor hose bracket from alternator.

10   Remove the power steering fluid reservoir hose bracket from the alternator.

11   Remove the alternator mounting fasteners and maneuver it, along with the idler pulley, out of the vehicle **(see illustration 7.7)**. Then while out of the vehicle, separate the idler pulley from the alternator.

12   If you're replacing the alternator, take the old one with you when purchasing the replacement unit. Make sure that the new/rebuilt unit looks identical to the old alternator. Look at

the electrical terminals on the backside of the alternator. They should be the same in number, size and location as the terminals on the old alternator. Finally, look at the identification numbers. They will be stamped into the housing or printed on a tag attached to the housing. Make sure that the I.D. numbers are the same on both alternators.

13   Many new/rebuilt alternators DO NOT have a pulley installed, so you might have to swap the pulley from the old unit to the new/rebuilt one. When buying an alternator, find out the store's policy regarding pulley swaps. Some stores perform this service free of charge. If your local auto parts store doesn't offer this service, you'll have to purchase a puller for removing the pulley and do it yourself.

14   Installation is the reverse of removal. Tighten the alternator mounting bolts securely.

15   Reconnect the cable to the negative terminal of the battery. Check the charging voltage (see Section 2) to verify that the alternator is operating correctly.

7.8 Unscrew the idler pulley bolt as far as possible

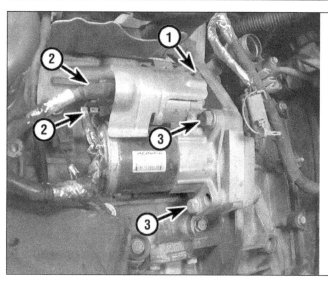

8.5 Starter motor details (shown from the front with the radiator removed)

1   *Starter motor heat shield mounting fastener*
2   *Starter motor wiring*
3   *Starter motor mounting bolts*

8   Starter motor - removal and installation

1   Detach the cable from the negative terminal of the battery (see Section 3).
2   Raise the vehicle and support it securely on jackstands.
3   Remove the front bank catalytic converter (see Chapter 6).
4   Remove the starter motor heat shield mounting fastener, then remove the heat shield.
5   Disconnect the wiring from the starter **(see illustration)**.
6   Unscrew the starter mounting bolts and remove the starter.
7   Installation is the reverse of removal. Tighten the starter mounting bolts securely, then reconnect the cable to the negative terminal of the battery (see Section 4).

Chapter 6
Emissions and engine control systems

Contents

	Section
Accelerator Pedal Position (APP) sensor - replacement	4
Camshaft Position Actuator Solenoid Valve - replacement	6
Camshaft Position (CMP) sensor(s) - replacement	5
Catalytic converter - replacement	16
Crankshaft Position (CKP) sensor - replacement	7
Engine Coolant Temperature (ECT) sensor - replacement	8
Evaporative Emissions Control (EVAP) system - component replacement	17
General information	1
Intake Manifold Tuning Valve (2007 and 2008 models) - removal and installation	18

	Section
Knock sensor - replacement	11
Manifold Absolute Pressure (MAP) sensor (2007 and 2008 models) - replacement	10
Mass Air Flow/Intake Air Temperature (MAF/IAT) sensor - replacement	9
Obtaining and clearing Diagnostic Trouble Codes (DTCs)	3
On Board Diagnosis (OBD) system	2
Oxygen sensors - replacement	12
Powertrain Control Module (PCM) - removal and installation	15
Transmission range switch - removal and installation	13
Transmission speed sensors - replacement	14

Specifications

Torque specifications

Knock sensor	17 ft-lbs	23 Nm

1 General information

1   To prevent pollution of the atmosphere from incompletely burned and evaporating gases, and to maintain good driveability and fuel economy, a number of emission control systems are incorporated. They include the:

Catalytic converter

2   A catalytic converter is an emission control device in the exhaust system that reduces certain pollutants in the exhaust gas stream. There are two types of converters: oxidation converters and reduction converters.

3   Oxidation converters contain a monolithic substrate (a ceramic honeycomb) coated with the semi-precious metals platinum and palladium. An oxidation catalyst reduces unburned hydrocarbons (HC) and carbon monoxide (CO) by adding oxygen to the exhaust stream as it passes through the substrate, which, in the presence of high temperature and the catalyst materials, converts the HC and CO to water vapor (H_2O) and carbon dioxide (CO_2).

4   Reduction converters contain a monolithic substrate coated with platinum and rhodium. A reduction catalyst reduces oxides of nitrogen (NOx) by removing oxygen, which in the presence of high temperature and the catalyst material produces nitrogen (N) and carbon dioxide (CO_2).

5   Catalytic converters that combine both types of catalysts in one assembly are known as "three-way catalysts" or TWCs. A TWC can reduce all three pollutants.

Evaporative Emissions Control (EVAP) system

6   The Evaporative Emissions Control (EVAP) system prevents fuel system vapors (which contain unburned hydrocarbons) from escaping into the atmosphere. On warm days, vapors trapped inside the fuel tank expand until the pressure reaches a certain threshold. Then the fuel vapors are routed from the fuel tank through the fuel vapor vent valve and the fuel vapor control valve to the EVAP canister, where they're stored temporarily until the next time the vehicle is operated. When the conditions are right (engine warmed up, vehicle up to speed, moderate or heavy load on the engine, etc.), the PCM opens the canister purge valve, which allows fuel vapors to be drawn from the canister into the intake manifold. Once in the intake manifold, the fuel vapors mix with incoming air before being drawn through the intake ports into the combustion chambers where they're burned up with the rest of the air/fuel mixture. The EVAP system is complex and virtually impossible to troubleshoot without the right tools and training.

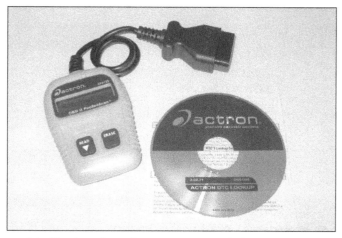

2.4a Simple code readers are an economical way to extract trouble codes when the CHECK ENGINE light comes on

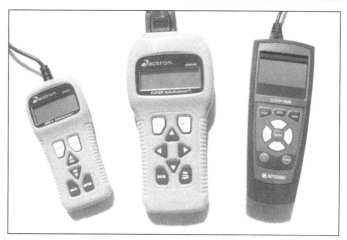

2.4b Hand-held scan tools like these can extract computer codes and also perform diagnostics

Secondary Air Injection (AIR) system

7   Some models are equipped with a Secondary Air Injection (AIR) system. The AIR system is used to reduce tailpipe emissions on initial engine start-up. The system uses an electric motor/pump assembly, relay, vacuum valve/solenoid, air shut-off valve, check valves and tubing to inject fresh air directly into the exhaust manifolds. The fresh air (oxygen) reacts with the exhaust gas in the catalytic converter to reduce HC and CO levels. The air pump and solenoid are controlled by the PCM through the AIR relay. During initial start-up, the PCM energizes the AIR relay, the relay supplies battery voltage to the air pump and the vacuum valve/solenoid, engine vacuum is applied to the air shut-off valve which opens and allows air to flow through the tubing into the exhaust manifolds. The PCM will operate the air pump until closed loop operation is reached (approximately four minutes). During normal operation, the check valves prevent exhaust backflow into the system.

Powertrain Control Module (PCM)

8   The Powertrain Control Module (PCM) is the brain of the engine management system. It also controls a wide variety of other vehicle systems. In order to program the new PCM, the dealer needs the vehicle as well as the new PCM. If you're planning to replace the PCM with a new one, there is no point in trying to do so at home because you won't be able to program it yourself.

Positive Crankcase Ventilation (PCV) system

9   The Positive Crankcase Ventilation (PCV) system reduces hydrocarbon emissions by scavenging crankcase vapors, which are rich in unburned hydrocarbons. A PCV valve or orifice regulates the flow of gases into the intake manifold in proportion to the amount of intake vacuum available.

10   The PCV system generally consists of the fresh air inlet hose, the PCV valve or orifice and the crankcase ventilation hose (or PCV hose). The fresh air inlet hose connects the air intake duct to a pipe on the valve cover. The crankcase ventilation hose (or PCV hose) connects the PCV valve or orifice in the valve cover to the intake manifold.

11   On the vehicles covered by this manual, there are no serviceable components in the PCV system.

---

2   On Board Diagnosis (OBD) system

General description

1   All models are equipped with the second generation OBD-II system. This system consists of an on-board computer known as the Powertrain Control Module (PCM), and information sensors, which monitor various functions of the engine and send data to the PCM. This system incorporates a series of diagnostic monitors that detect and identify fuel injection and emissions control system faults and store the information in the computer memory. This system also tests sensors and output actuators, diagnoses drive cycles, freezes data and clears codes.

2   The PCM is the brain of the electronically controlled fuel and emissions system. It receives data from a number of sensors and other electronic components (switches, relays, etc.). Based on the information it receives, the PCM generates output signals to control various relays, solenoids (fuel injectors) and other actuators. The PCM is specifically calibrated to optimize the emissions, fuel economy and driveability of the vehicle.

3   It isn't a good idea to attempt diagnosis or replacement of the PCM or emission control components at home while the vehicle is under warranty. Because of a federally-mandated warranty which covers the emissions system components and because any owner-induced damage to the PCM, the sensors and/or the control devices may void this warranty, take the vehicle to a dealer service department if the PCM or a system component malfunctions.

Scan tool information

4   Because extracting the Diagnostic Trouble Codes (DTCs) from an engine management system is now the first step in troubleshooting many computer-controlled systems and components, a code reader, at the very least, will be required **(see illustration)**. More powerful scan tools can also perform many of the diagnostics once associated with expensive factory scan tools **(see illustration)**. If you're planning to obtain a generic scan tool for your vehicle, make sure that it's compatible with OBD-II systems. If you don't plan to purchase a code reader or scan tool and don't have access to one, you can have the codes extracted by a dealer service department or an independent repair shop.

Note: *Some auto parts stores even provide this service.*

Information Sensors

Accelerator Pedal Position (APP) sensor - as you press the accelerator pedal, the APP sensor alters its voltage signal to the PCM in proportion to the angle of the pedal, and the PCM commands a motor inside the throttle body to open or close the throttle plate accordingly

Camshaft Position (CMP) sensor - produces a signal that the PCM uses to identify the number 1 cylinder and to time the firing sequence of the fuel injectors

Crankshaft Position (CKP) sensor - produces a signal that the PCM uses to calculate engine speed and crankshaft position, which enables it to synchronize ignition timing with fuel injector timing, and to detect misfires

Engine Coolant Temperature (ECT) sensor - a thermistor (temperature-sensitive variable resistor) that sends a voltage signal to the PCM, which uses this data to determine the temperature of the engine coolant

Fuel tank pressure sensor - measures the fuel tank pressure and controls fuel tank pressure by signaling the EVAP system to purge the fuel tank vapors when the pressure becomes excessive

Intake Air Temperature (IAT) sensor - monitors the temperature of the air entering the engine and sends a signal to the PCM to determine injector pulse-width (the duration of each injector's on-time) and to adjust spark timing (to prevent spark knock)

Knock sensor - a piezoelectric crystal that oscillates in proportion to engine vibration which produces a voltage output that is monitored by the PCM. This retards the ignition timing when the oscillation exceeds a certain threshold

Manifold Absolute Pressure (MAP) sensor - monitors the pressure or vacuum inside the intake manifold. The PCM uses this data to determine engine load so that it can alter the ignition advance and fuel enrichment

Mass Air Flow (MAF) sensor - measures the amount of intake air drawn into the engine. It uses a hot-wire sensing element to measure the amount of air entering the engine

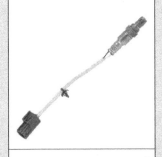

Oxygen sensors - generates a small variable voltage signal in proportion to the difference between the oxygen content in the exhaust stream and the oxygen content in the ambient air. The PCM uses this information to maintain the proper air/fuel ratio. A second oxygen sensor monitors the efficiency of the catalytic converter

Throttle Position (TP) sensor - a potentiometer that generates a voltage signal that varies in relation to the opening angle of the throttle plate inside the throttle body. Works with the PCM and other sensors to calculate injector pulse width (the duration of each injector's on-time)

Photos courtesy of Wells Manufacturing, except APP and MAF sensors.

3   Obtaining and clearing Diagnostic Trouble Codes (DTCs)

1   All models covered by this manual are equipped with on-board diagnostics. When the PCM recognizes a malfunction in a monitored emission or engine control system, component or circuit, it turns on the Malfunction Indicator Light (MIL) on the dash. The PCM will continue to display the MIL until the problem is fixed and the Diagnostic Trouble Code (DTC) is cleared from the PCM's memory. You'll need a scan tool to access any DTCs stored in the PCM.

2   Before outputting any DTCs stored in the PCM, thoroughly inspect ALL electrical connectors and hoses. Make sure that all electrical connections are tight, clean and free of corrosion. And make sure that all hoses are correctly connected, fit tightly and are in good condition (no cracks or tears).

Accessing the DTCs

3   The Diagnostic Trouble Codes (DTCs) can only be accessed with a code reader or scan tool. Professional scan tools are expensive, but relatively inexpensive generic code readers or scan tools **(see illustrations 2.4a and 2.4b)** are available at most auto parts stores. Simply plug the connector of the scan tool into the diagnostic connector **(see illus-**

3.3 The Data Link Connector (DLC) is located under the lower edge of the dash, to the left of the steering column

tration). Then follow the instructions included with the scan tool to extract the DTCs.

4   Once you have outputted all of the stored DTCs, look them up on the accompanying DTC chart.

5   After troubleshooting the source of each DTC, make any necessary repairs or replace the defective component(s).

Clearing the DTCs

6   Clear the DTCs with the code reader or scan tool in accordance with the instructions provided by the tool's manufacturer.

Diagnostic Trouble Codes

7   The accompanying tables are a list of the Diagnostic Trouble Codes (DTCs) that can be accessed by a do-it-yourselfer working at home (there are many, many more DTCs available to professional mechanics with proprietary scan tools and software, but those codes cannot be accessed by a generic scan tool). If, after you have checked and repaired the connectors, wire harness and vacuum hoses (if applicable) for an emission-related system, component or circuit, the problem persists, have the vehicle checked by a dealer service department or other qualified repair shop.

OBD-II trouble codes

Code	Probable cause
P0008	Engine position system performance (bank 1)
P0009	Engine position system performance (bank 2)
P0010	Intake camshaft position actuator circuit open (bank 1)
P0011	"A" Camshaft position - timing over-advanced (bank 1)
P0013	"B" Camshaft position - actuator circuit malfunction (bank 1)
P0014	"B" Camshaft position - timing over-advanced (bank 1)
P0016	Crankshaft position/camshaft position, bank 1, sensor A - correlation
P0017	Crankshaft position/camshaft position, bank 1, sensor B - correlation
P0018	Crankshaft position/camshaft position, bank 2, sensor A - correlation
P0019	Crankshaft position/camshaft position, bank 2, sensor B - correlation
P0020	Intake camshaft position actuator circuit open (bank 2)
P0021	Intake camshaft position-timing over-advanced (bank 2)
P0023	"B" Camshaft position - actuator circuit (bank 2)

Code	Probable cause
P0024	"B" Camshaft position - timing over-advanced or system performance problem (bank 2)
P0030	HO2S heater control circuit (bank 1, sensor 1)
P0031	HO2S heater control circuit low (bank 1, sensor 1)
P0032	HO2S heater control circuit high (bank 1, sensor 1)
P0036	HO2S heater control circuit (bank 1 sensor 2)
P0037	HO2S heater control circuit low (bank 1, sensor 2)
P0038	HO2S heater control circuit high (bank 1, sensor 2)
P0050	HO2S heater control circuit (bank 2, sensor 1)
P0051	HO2S heater control circuit low (bank 2, sensor 1)
P0052	HO2S heater control circuit high (bank 2, sensor 1)
P0053	HO2S heater resistance (bank 1, sensor 1)
P0054	HO2S heater resistance (bank 1, sensor 2)
P0056	HO2S heater control circuit malfunction (bank 2, sensor 2)
P0057	HO2S heater control circuit low (bank 2, sensor 2)
P0058	HO2S heater control circuit high (bank 2, sensor 2)
P0068	Throttle Position (TP) sensor inconsistent with Mass Air Flow (MAF) sensor
P0100	Mass air flow or volume air flow circuit malfunction
P0101	Mass air flow or volume air flow circuit, range or performance problem
P0102	Mass air flow or volume air flow circuit, low input
P0103	Mass air flow or volume air flow circuit, high input
P0106	Manifold absolute pressure or barometric pressure circuit, range or performance problem
P0107	Manifold absolute pressure or barometric pressure circuit, low input
P0108	Manifold absolute pressure or barometric pressure circuit, high input
P0111	Intake air temperature circuit, range or performance problem
P0112	Intake air temperature circuit, low input
P0113	Intake air temperature circuit, high input
P0115	Engine coolant temperature circuit
P0116	Engine coolant temperature circuit range/performance problem
P0117	Engine coolant temperature circuit, low input
P0118	Engine coolant temperature circuit, high input
P0119	Engine coolant temperature circuit, intermittent

OBD-II trouble codes (continued)

Code	Probable cause
P0120	Throttle position or pedal position sensor/switch circuit malfunction
P0121	Throttle position or pedal position sensor/switch circuit, range or performance problem
P0122	Throttle position or pedal position sensor/switch circuit, low input
P0123	Throttle position or pedal position sensor/switch circuit, high input
P0125	Insufficient coolant temperature for closed loop fuel control
P0128	Coolant thermostat (coolant temperature below thermostat regulating temperature)
P0130	O2 sensor circuit malfunction (bank 1, sensor 1)
P0131	O2 sensor circuit, low voltage (bank 1, sensor 1)
P0132	O2 sensor circuit, high voltage (bank 1, sensor 1)
P0133	O2 sensor circuit, slow response (bank 1, sensor 1)
P0134	O2 sensor circuit - no activity detected (bank 1, sensor 1)
P0135	O2 sensor heater circuit malfunction (bank 1, sensor 1)
P0137	O2 sensor circuit, low voltage (bank 1, sensor 2)
P0138	O2 sensor circuit, high voltage (bank 1, sensor 2)
P0139	O2 sensor circuit, slow response (bank 1, sensor 2)
P0140	O2 sensor circuit - no activity detected (bank 1, sensor 2)
P0141	O2 sensor heater circuit malfunction (bank 1, sensor 2)
P0151	O2 sensor circuit, low voltage (bank 2, sensor 1)
P0152	O2 sensor circuit, high voltage (bank 2, sensor 1)
P0153	O2 sensor circuit, slow response (bank 2, sensor 1)
P0154	O2 sensor circuit - no activity detected (bank 2, sensor 1)
P0155	O2 sensor heater circuit malfunction (bank 2, sensor 1)
P0157	O2 sensor circuit, low voltage (bank 2, sensor 2)
P0158	O2 sensor circuit, high voltage (bank 2, sensor 2)
P0160	O2 sensor circuit - no activity detected (bank 2, sensor 2)
P0161	O2 sensor heater circuit malfunction (bank 2, sensor 2)
P0171	System too lean (bank 1)
P0172	System too rich (bank 1)
P0201	Injector circuit malfunction - cylinder no. 1
P0202	Injector circuit malfunction - cylinder no. 2

Code	Probable cause
P0203	Injector circuit malfunction - cylinder no. 3
P0204	Injector circuit malfunction - cylinder no. 4
P0205	Injector circuit malfunction - cylinder no. 5
P0206	Injector circuit malfunction - cylinder no. 6
P0218	Transmission overheating condition
P0220	Throttle position or pedal position sensor/switch B circuit malfunction
P0221	Throttle position or pedal position sensor/switch B, range or performance problem
P0222	Throttle position or pedal position sensor/switch B circuit, low input
P0223	Throttle position or pedal position sensor/switch B circuit, high input
P0230	Fuel pump primary circuit malfunction
P0261	Cylinder no. 1 injector circuit, low
P0262	Cylinder no. 1 injector circuit, high
P0264	Cylinder no. 2 injector circuit, low
P0265	Cylinder no. 2 injector circuit, high
P0267	Cylinder no. 3 injector circuit, low
P0268	Cylinder no. 3 injector circuit, high
P0270	Cylinder no. 4 injector circuit, low
P0271	Cylinder no. 4 injector circuit, high
P0273	Cylinder no. 5 injector circuit, low
P0274	Cylinder no. 5 injector circuit, high
P0276	Cylinder no. 6 injector circuit, low
P0277	Cylinder no. 6 injector circuit, high
P0300	Random/multiple cylinder misfire detected
P0301	Cylinder no. 1 misfire detected
P0302	Cylinder no. 2 misfire detected
P0303	Cylinder no. 3 misfire detected
P0304	Cylinder no. 4 misfire detected
P0305	Cylinder no. 5 misfire detected
P0306	Cylinder no. 6 misfire detected
P0315	Crankshaft position system - variation not learned
P0324	Knock control system error

OBD-II trouble codes (continued)

Code	Probable cause
P0325	Knock sensor no. 1 circuit malfunction (bank 1 or single sensor)
P0326	Knock sensor no. 1 circuit, range or performance problem (bank 1 or single sensor)
P0327	Knock sensor no. 1 circuit, low input (bank 1 or single sensor)
P0328	Knock sensor no. 1 circuit, high input (bank 1 or single sensor)
P0330	Knock sensor no. 2 circuit malfunction (bank 2)
P0331	Knock sensor no. 2 circuit, range or performance problem (bank 2)
P0332	Knock sensor no. 2 circuit, low input (bank 2)
P0333	Knock sensor no. 2 circuit, high input (bank 2)
P0335	Crankshaft position sensor "A" - circuit malfunction
P0336	Crankshaft position sensor "A" - range or performance problem
P0338	Crankshaft position sensor "A" - high input
P0340	Camshaft position sensor "A" - circuit malfunction (bank 1)
P0341	Camshaft position sensor "A" - range or performance problem (bank 1)
P0342	Camshaft position sensor "A" - low input (bank 1)
P0343	Camshaft position sensor "A" - high input (bank 1)
P0346	Camshaft position sensor "A" - range/performance problem (bank 2)
P0347	Camshaft position sensor "A" - low input (bank 2)
P0348	Camshaft position sensor "A" - range/performance problem (bank 2)
P0351	Ignition coil 1 primary or secondary circuit malfunction
P0352	Ignition coil 2 primary or secondary circuit malfunction
P0353	Ignition coil 3 primary or secondary circuit malfunction
P0354	Ignition coil 4 primary or secondary circuit malfunction
P0355	Ignition coil 5 primary or secondary circuit malfunction
P0356	Ignition coil 6 primary or secondary circuit malfunction
P0366	Camshaft position sensor "B" - range/performance problem (bank 1)
P0367	Camshaft position sensor "B" - low input (bank 1)
P0368	Camshaft position sensor "B" circuit high input (bank 1)
P0391	Camshaft position sensor "B" - range/performance problem (bank 2)
P0392	Camshaft position sensor "B" - low input (bank 2)
P0393	Camshaft position sensor "B" - high input (bank 2)

Code	Probable cause
P0401	Exhaust gas recirculation - insufficient flow detected
P0403	Exhaust gas recirculation - circuit malfunction
P0404	Exhaust gas recirculation - range or performance problem
P0405	Exhaust gas recirculation valve position sensor A - circuit low
P0406	Exhaust gas recirculation valve position sensor A - circuit high
P0420	Catalyst system efficiency below threshold (bank 1)
P0430	Catalyst system efficiency below threshold (bank 2)
P0442	Evaporative emission control system, small leak detected
P0443	Evaporative emission control system, purge control valve circuit malfunction
P0446	Evaporative emission control system, vent control circuit malfunction
P0449	Evaporative emission control system, vent valve/solenoid circuit malfunction
P0450	Evaporative emission control system, pressure sensor malfunction
P0451	Evaporative emission control system, pressure sensor range or performance problem
P0452	Evaporative emission control system, pressure sensor low input
P0453	Evaporative emission control system, pressure sensor high input
P0454	Evaporative emission control system, pressure sensor intermittent
P0455	Evaporative emission (EVAP) control system leak detected (no purge flow or large leak)
P0458	Evaporative emission control system, purge control valve - circuit low
P0459	Evaporative emission control system, purge control valve - circuit high
P0461	Fuel level sensor circuit, range or performance problem
P0462	Fuel level sensor circuit, low input
P0463	Fuel level sensor circuit, high input
P0464	Fuel level sensor circuit, intermittent
P0480	Cooling fan no. 1, control circuit malfunction
P0481	Cooling fan no. 2, control circuit malfunction
P0496	Evaporative emission system - high purge flow
P0497	Evaporative emission system - low purge flow
P0498	Evaporative emission system, vent control - circuit low
P0499	Evaporative emission system, vent control - circuit high
P0506	Idle control system, rpm lower than expected
P0507	Idle control system, rpm higher than expected

OBD-II trouble codes (continued)

Code	Probable cause
P0513	Incorrect immobilizer key
P0520	Engine oil pressure sensor/switch circuit malfunction
P0532	A/C refrigerant pressure sensor, low input
P0533	A/C refrigerant pressure sensor, high input
P0562	System voltage low
P0563	System voltage high
P0571	Cruise control/brake switch A, circuit malfunction
P0572	Cruise control/brake switch A, circuit low
P0573	Cruise control/brake switch A, circuit high
P0575	Cruise control system - input circuit malfunction
P0601	Internal control module, memory check sum error
P0602	Control module, programming error
P0603	Internal control module, keep alive memory (KAM) error
P0604	Internal control module, random access memory (RAM) error
P0606	PCM processor fault
P0607	Control module performance
P0615	Starter relay - circuit malfunction
P0616	Starter relay - circuit low
P0617	Starter relay - circuit high
P0621	Alternator L terminal circuit malfunction
P0622	Alternator F terminal circuit malfunction
P0625	Alternator field terminal - circuit low
P0626	Alternator field terminal - circuit high
P0627	Fuel pump control - circuit open
P0628	Fuel pump control - circuit low
P0629	Fuel pump control - circuit high
P0633	Immobilizer key not programmed - ECM
P0634	ECM/TCM - internal temperature too high
P0638	Throttle actuator control range/performance problem (bank 1)
P0641	Sensor reference voltage A - circuit open

Code	Probable cause
P0642	Engine control module (ECM), knock control - defective
P0643	Sensor reference voltage A - circuit high
P0644	Driver display, serial communication - circuit malfunction
P0645	A/C clutch relay control circuit
P0650	Malfunction indicator lamp (MIL), control circuit malfunction
P0651	Sensor reference voltage B - circuit open
P0653	Sensor reference voltage B - circuit high
P0667	ECM/TCM internal temperature sensor - circuit range/performance problem
P0668	ECM/TCM internal temperature sensor - circuit low
P0669	ECM/TCM internal temperature sensor - circuit high
P0685	ECM power relay, control - circuit open
P0686	ECM power relay control - circuit low
P0687	Engine, control relay - short to ground
P0689	ECM power relay sense - circuit low
P0690	ECM power relay sense - circuit high
P0691	Engine coolant blower motor 1 - short to ground
P0700	Transmission control system malfunction
P0703	Torque converter/brake switch B, circuit malfunction
P0705	Transmission range sensor, circuit malfunction (PRNDL input)
P0711	Transmission fluid temperature sensor circuit, range or performance problem
P0712	Transmission fluid temperature sensor circuit, low input
P0713	Transmission fluid temperature sensor circuit, high input
P0717	Input/turbine speed sensor circuit, no signal
P0722	Output speed sensor circuit, no signal
P0727	Engine speed input circuit, no signal
P0730	Incorrect gear ratio
P0731	Incorrect gear ratio, first gear
P0732	Incorrect gear ratio, second gear
P0733	Incorrect gear ratio, third gear
P0734	Incorrect gear ratio, fourth gear
P0735	Incorrect gear ratio, fifth gear

OBD-II trouble codes (continued)

Code	Probable cause
P0736	Incorrect gear ratio, reverse gear
P0741	Torque converter clutch, circuit performance problem or stuck in Off position
P0742	Torque converter clutch circuit, stuck in On position
P0762	Shift solenoid C, stuck in On position
P0962	Pressure control (PC) solenoid A - control circuit low
P0963	Pressure control (PC) solenoid A - control circuit high
P0966	Pressure control (PC) solenoid B - control circuit low
P0967	Pressure control (PC) solenoid B - control circuit high
P0970	Pressure control (PC) solenoid C - control circuit low
P0971	Pressure control (PC) solenoid C - control circuit high
P0973	Shift solenoid (SS) A - control circuit low
P0974	Shift solenoid (SS) A - control circuit high
P0976	Shift solenoid (SS) B - control circuit low
P0977	Shift solenoid (SS) B - control circuit high
P0979	Shift solenoid (SS) C - control circuit low
P0980	Shift solenoid (SS) C - control circuit high
P0982	Shift solenoid (SS) D - control circuit low
P0983	Shift solenoid (SS) D - control circuit high
P0985	Shift solenoid (SS) E - control circuit low
P0986	Shift solenoid (SS) E - control circuit high

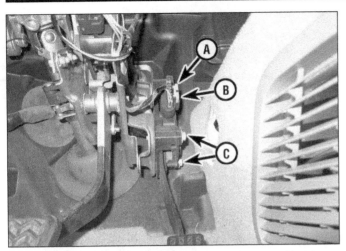

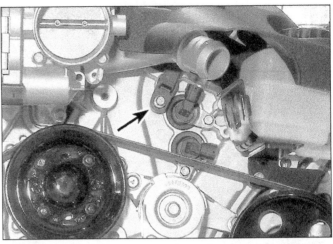

4.2 The APP sensor is located at the top of the accelerator pedal. To unplug the connector, slide the white lock (A) out, then depress the tab on the side (B) and pull off the connector. (C) are two of the three mounting bolts

5.1 The camshaft position sensors are located on the front of the cylinder heads, just below the valve cover (left side [front bank cylinder head] intake sensor shown, right side [rear bank cylinder head] sensors similar)

4 Accelerator Pedal Position (APP) sensor - replacement

1   Remove the left-side under-dash panel and the driver's knee bolster (see Chapter 11).
2   Disconnect the electrical connector from the APP sensor **(see illustration)**.
3   Remove the APP sensor module mounting bolts and remove the sensor module.
Note: *The top bolt must be left in the sensor until it's removed from the vehicle. Make sure to place it in its hole before installing the sensor.*
4   Installation is the reverse of removal.

5 Camshaft Position (CMP) sensor(s) - replacement

Warning: *The engine must be completely cool before beginning this procedure.*
1   These engines have four CMP sensors **(see illustration)**.
2   If you're replacing a left-side (front cylinder bank) sensor, drain the engine coolant (see Chapter 1), support the engine from below with

a floor jack and block of wood, remove the engine mount strut and bracket, the radiator inlet hose, and the power steering reservoir bolts. Position the reservoir out of the way.
3   Disconnect the CMP sensor electrical connector.
4   Remove the CMP sensor mounting bolt, and remove the sensor.
5   Installation is the reverse of removal.
6   Refill the cooling system if it was drained (see Chapter 1).

6 Camshaft Position Actuator Solenoid Valve - replacement

Warning: *The engine must be completely cool before beginning this procedure.*
1   The 3.6L engine has four camshaft position actuator solenoid valves **(see illustration)**.
2   If you're replacing a left-side (front cylinder bank) solenoid valve, drain the engine coolant (see Chapter 1), remove the engine mount strut and bracket, the radiator inlet hose, and the power steering reservoir

bolts. Position the reservoir out of the way.
3   Disconnect the solenoid valve electrical connector.
4   Remove the solenoid valve mounting bolt and the solenoid valve.
5   Installation is the reverse of removal.
6   Refill the cooling system if it was drained (see Chapter 1).

7 Crankshaft Position (CKP) sensor - replacement

1   Raise the front of the vehicle and support it securely on jackstands.
2   Remove the heat shield covering the wiring harness and sensor **(see illustration 11.2)**. Disconnect the electrical connector from the CKP sensor **(see illustration)**.
3   Unscrew the mounting fastener and remove the CKP sensor.
4   Before installing the sensor, lubricate the O-ring with clean engine oil. Installation is otherwise the reverse of removal.

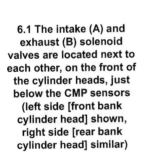

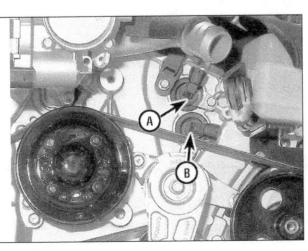

6.1 The intake (A) and exhaust (B) solenoid valves are located next to each other, on the front of the cylinder heads, just below the CMP sensors (left side [front bank cylinder head] shown, right side [rear bank cylinder head] similar)

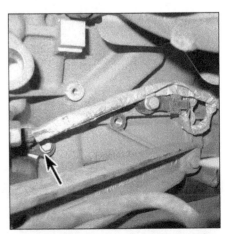

7.2 The CKP sensor is located at the rear of the engine block on the firewall side

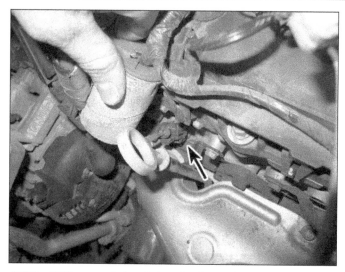

8.3 The ECT sensor is located on the front cylinder head near the engine oil dipstick, covered by a protective shield

9.2 The MAF/IAT sensor is located on the air filter housing cover (2011 and earlier models shown; on 2012 and later models, the sensor slides into the top of the cover and is retained by two screws)

8   Engine Coolant Temperature (ECT) sensor - replacement

Warning: *Wait until the engine has cooled completely before beginning this procedure.*
Note: *The ECT sensor is located on the front-bank cylinder head, near the engine oil dipstick, just below the valve cover.*
1   Remove the engine cover.
2   Drain the engine coolant to a point lower than that of the sensor (see Chapter 1).
3   Disconnect the electrical connector from the ECT sensor **(see illustration).**
4   Unscrew and remove the ECT sensor.
5   Before installing the ECT sensor, apply some sealant to the threads.
6   Installation is the reverse of removal. Refill the cooling system (see Chapter 1).

9   Mass Air Flow/Intake Air Temperature (MAF/IAT) sensor - replacement

Note: *The IAT sensor is an integral component of the Mass Air Flow (MAF) sensor.*
1   On 2011 and earlier models, remove the air intake duct (see Chapter 4).
2   Disconnect the electrical connector from the MAF/IAT sensor **(see illustration)**.
3   Remove the fasteners and detach the sensor from the air filter housing.
4   On 2011 and earlier models, remove the MAF/IAT sensor seal and inspect it for damage. If it is OK, it can be re-used.
5   Installation is the reverse of removal.

10   Manifold Absolute Pressure (MAP) sensor (2007 and 2008 models) - replacement

1   Remove the engine cover.
2   Disconnect the electrical connector from the MAP sensor **(see illustration)**.
3   Remove the MAP sensor bolt and pull the sensor from the upper intake manifold.
4   Installation is the reverse of removal. Use a new MAP sensor O-ring seal, and lubricate it with a little clean engine oil.

11   Knock sensor - replacement

1   Raise the vehicle and support it securely on jackstands.

10.2 MAP sensor electrical connector and retaining bolt

11.2 Knock sensor heat shield bolts (rear cylinder bank [Bank 1])

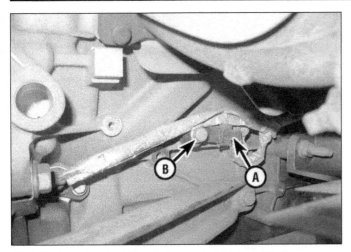

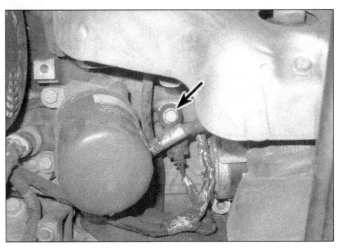

11.4a Depress the tab (A) and disconnect the electrical connector, then remove the knock sensor retaining bolt (B)

11.4b Location of the knock sensor on the front cylinder bank (Bank 2), shown from the front with the radiator removed

2 If you're replacing the knock sensor on the rear cylinder bank (Bank 1), remove the heat shield from the engine block **(see illustration)**.

3 If you're replacing the knock sensor on the front cylinder bank (Bank 2), remove the engine oil filter (see Chapter 1).

Note: *On 2009 and later models, it will be necessary to remove the catalytic converter to access the Bank 2 knock sensor (see Section 16).*

4 Disconnect the knock sensor electrical connector **(see illustrations)**.

5 Unscrew the knock sensor retaining bolt and detach the sensor from the engine block.

6 Installation is the reverse of removal. Tighten the knock sensor bolt to the torque listed in this Chapter's Specifications.

12 Oxygen sensors - replacement

Note: *Because it is installed in the exhaust manifold or pipe, both of which contract when cool, an oxygen sensor might be very difficult*

to loosen when the engine is cold. Rather than risk damage to the sensor or its mounting threads, start and run the engine for a minute or two, then shut it off. Be careful not to burn yourself during the following procedure.

1 Be particularly careful when servicing an oxygen sensor:

a) Oxygen sensors have a permanently attached pigtail and an electrical connector that cannot be removed. Damaging or removing the pigtail or electrical connector will render the sensor useless.

b) Keep grease, dirt and other contaminants away from the electrical connector and the louvered end of the sensor.

c) Do not use cleaning solvents of any kind on an oxygen sensor.

d) Oxygen sensors are extremely delicate. Do not drop a sensor or handle it roughly.

e) Make sure that the silicone boot on the sensor is installed in the correct position. Otherwise, the boot might melt and it might prevent the sensor from operating correctly.

Replacement

Note: *These sensors are either in the exhaust pipe or screwed directly into the exhaust manifold(s). There are four oxygen sensors; one in each exhaust manifold, and one just after each catalytic converter.*

Upstream oxygen sensors

2 Remove the engine cover.

3 Disconnect the upstream oxygen sensor electrical connector **(see illustrations)**. Unclip it from any retainers.

4 Remove the upstream oxygen sensor.

5 If you're going to install the old sensor, apply anti-seize compound to the threads of the sensor to facilitate future removal. If you're going to install a new oxygen sensor, it's not necessary to apply anti-seize compound to the threads; the threads on new sensors already have anti-seize compound on them.

6 Installation is the reverse of removal. Tighten the oxygen sensor securely.

Downstream oxygen sensors

7 Raise the vehicle and support it securely on jackstands.

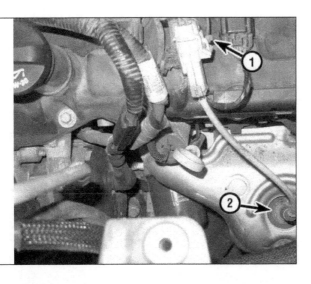

12.3a There are two upstream oxygen sensors, one in each exhaust manifold. This is the sensor for Bank 2 (sensor 2/1) . . .

1 Electrical connector (slide out the lock then lift the tab to unplug)
2 Oxygen sensor

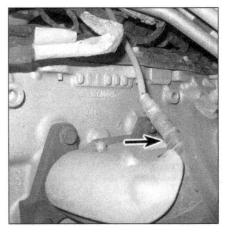

12.3b . . . and this is the sensor for Bank 1 (sensor 1/1)

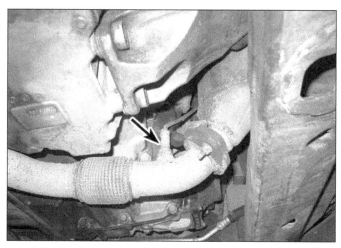

12.8a The downstream oxygen sensor(s) are located in the exhaust pipe, just after the catalytic converter(s). This is the sensor for Bank 2 (sensor 2/2)...

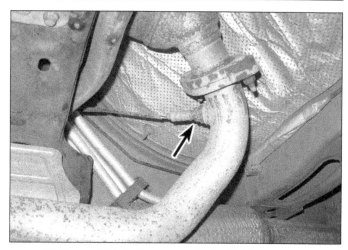

12.8b... and this is the sensor for Bank 1 (sensor 1/2)

8 Locate the downstream oxygen sensor **(see illustrations)**, then trace the lead up to the electrical connector and disconnect the connector.

9 Unscrew the downstream oxygen sensor.

10 If you're going to install the old sensor, apply anti-seize compound to the threads of the sensor to facilitate future removal. If you're going to install a new oxygen sensor, it's not necessary to apply anti-seize compound to the threads. The threads on new sensors already have anti-seize compound on them.

11 Installation is the reverse of removal. Tighten the oxygen sensor securely.

---

13 Transmission range switch - removal and installation

The transmission range switch is internally mounted, requiring valve body removal; we don't recommend replacing it at home.

---

14 Transmission speed sensors - replacement

The speed sensors on these transaxles are internally mounted and we don't recommend replacing them at home.

---

15 Powertrain Control Module (PCM) - removal and installation

Caution: *To avoid electrostatic discharge damage to the PCM, handle the PCM only by its case. Do not touch the electrical terminals during removal and installation. If available, ground yourself to the vehicle with an anti-static ground strap, available at computer supply stores.*

Note: *This procedure applies only to disconnecting, removing and installing the PCM that is already installed in your vehicle. If the PCM is defective and has to be replaced, it must be programmed with new software and calibrations. This procedure requires the use of GM's TECH-2 scan tool and GM's latest PCM-programming software, so you WILL NOT BE ABLE TO REPLACE THE PCM AT HOME.*

1 Disconnect the cable from the negative battery terminal (see Chapter 5).

2 Disconnect the electrical connectors from the PCM **(see illustration).**

3 Release the retaining tab and detach the PCM from the fan shroud.

4 Installation is the reverse of removal.

---

16 Catalytic converter - replacement

Warning: *Replace catalytic converters only after enough time has elapsed after driving the vehicle to allow the system components to cool completely. Also, when working under*

the vehicle, make sure it is securely supported on jackstands.

Note: *The following is a generalized procedure for most catalytic converters. The specifics vary among the engines used in these vehicles.*

Note: *Many exhaust specialist shops are able to replace catalytic converters at a lower cost than what you might pay for a new one from a dealer.*

1 Raise the vehicle and support it securely on jackstands.

2 Disconnect the electrical connector from the related oxygen sensor and remove the sensor (see Section 12).

3 Support the other sections of the exhaust system as necessary.

4 Remove the retaining nuts from the front and rear catalyst mounting flanges.

5 Remove the catalyst and pull off the gaskets.

6 Installation is the reverse of removal. Replace any rusted or damaged fasteners along with the gaskets.

15.2 The PCM is located on the fan shroud. To disconnect the electrical connectors, slide the lock (A) to the right, depress the tab (B), then swing the locking lever (C) open. (D) is the PCM retaining tab

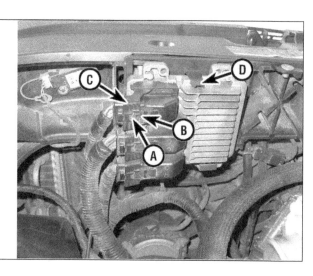

17.1 The EVAP purge control solenoid is located on the top of the intake manifold, next to the throttle body

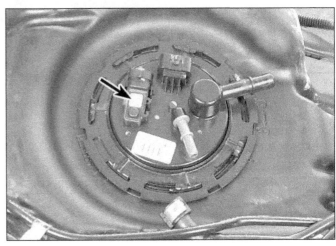

17.6 Location of the fuel tank pressure sensor

17.11 The EVAP canister is located on top of the fuel tank, at the rear

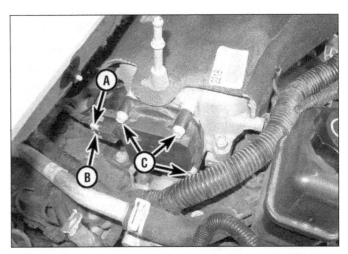

18.4 Slide out the connector lock (A), depress the tab (B) and disconnect the electrical connector, then remove the mounting bolts (C) and pull the Intake Manifold Tuning Valve from the upper intake manifold

17 Evaporative Emissions Control (EVAP) system - component replacement

EVAP purge control solenoid valve

1   Disconnect the electrical connector from the valve **(see illustration)**.
2   Disconnect the hose. See Chapter 4 for information on quick-connect fittings.
3   Unscrew the bolt and remove the purge valve from the upper intake manifold.
4   Installation is the reverse of removal.

Fuel tank pressure sensor

Warning: *Gasoline is extremely flammable.* *See* **Fuel system warnings** *in Chapter 4.*

5   Remove the fuel tank (see Chapter 4).
6   Disconnect the electrical connector and pull the fuel tank pressure sensor from the fuel pump module **(see illustration)**.
7   Installation is the reversal of removal.

EVAP canister

8   Raise the vehicle and support it securely on jackstands.
9   Remove the fuel tank (see Chapter 4).
10   Disconnect the electrical connector from the canister vent solenoid valve.
11   Disconnect the hoses from the canister **(see illustration)**. See Chapter 4 for information on quick-connect fittings.
12   Remove the retainer securing the canister on top of the fuel tank and remove the canister.
13   Installation is the reversal of removal.

18 Intake Manifold Tuning Valve (2007 and 2008 models) - removal and installation

1   Disconnect the cable from the negative terminal of the battery (see Chapter 5).
2   Remove the engine cover.
3   Unbolt the underhood fuse/relay box and position it aside for clearance.
4   Disconnect the electrical connector from the valve **(see illustration)**.
5   Unscrew the mounting bolts and remove the valve from the upper intake manifold.
6   Inspect the O-ring; if it's in good condition it can be re-used.
7   Installation is the reverse of removal. Tighten the fasteners securely.

Notes

Chapter 7 Part A
Automatic transaxle

Contents

	Section
Automatic transaxle - removal and installation	8
Automatic transaxle overhaul - general information	9
Brake Transmission Shift Interlock (BTSI) system - description, component replacement and adjustment	6
Diagnosis - general	2
Driveaxle oil seals - replacement	7

	Section
General information	1
Shift cable - replacement and adjustment	5
Shift lever - replacement	3
Shift lever knob - replacement	4
Transaxle mount - replacement	10

Specifications

General
Fluid type and capacity .... See Chapter 1

Torque specifications

	Ft-lbs	Nm
Driveplate inspection cover bolts (FWD)	55	75
Engine/transaxle subframe bolts	See Chapter 10	
Starter bolts	See Chapter 5	
Transaxle brace bolts (FWD)	37	50
Torque converter-to-driveplate bolts	46	62
Transaxle-to-engine mounting bolts		
Upper	55	75
Lower	43	58
Transaxle mount fasteners		
To transaxle*	44	60
To frame*	55	75
Mount bracket-to-transaxle		
Long bolt	44	60
Short bolts	66	89

* These fasteners must be replaced with new ones whenever they have been loosened or removed.

1 General information

1 Information on the automatic transaxle is included in this Part of Chapter 7. Information on the transfer case used on all-wheel drive models can be found in Chapter 7B.

2 Because of the complexity of the automatic transaxles and the specialized equipment necessary to perform most service operations, this Chapter contains only those procedures related to general diagnosis, routine maintenance, adjustment and removal and installation.

3 If the transaxle requires major repair work, it should be left to a dealer service department or an automotive or transmission repair shop. Once properly diagnosed you can, however, remove and install the transaxle yourself and save the expense, even if the repair work is done by a transmission shop. Keep in mind, however, that transaxle removal is difficult on these models. Transmission shops are generally equipped with vehicle hoists and other specialized equipment that is necessary for transaxle removal.

2 Diagnosis - general

1 Automatic transaxle malfunctions may be caused by five general conditions:

a) *Poor engine performance*
b) *Improper adjustments*
c) *Hydraulic malfunctions*
d) *Mechanical malfunctions*
e) *Malfunctions in the computer or its signal network*

2 Diagnosis of these problems should always begin with a check of the easily repaired items: fluid level and condition (see Chapter 1), shift cable adjustment and shift lever installation. Next, perform a road test to determine if the problem has been corrected or if more diagnosis is necessary. If the problem persists after the preliminary tests and corrections are completed, additional diagnosis should be performed by a dealer service department or other qualified transmission repair shop. Refer to "Troubleshooting" at the beginning of this manual for information on symptoms of transaxle problems.

Preliminary checks

3 Drive the vehicle to warm the transaxle to normal operating temperature.

4 Check the fluid level (see Chapter 1):

a) *If the fluid level is unusually low, add enough fluid to bring the level within the designated area of the dipstick, then check for external leaks (see following).*

b) *If the fluid level is abnormally high, drain off the excess, then check the drained fluid for contamination by coolant. The presence of engine coolant in the automatic transmission fluid indicates that a failure has occurred in the internal radiator oil cooler walls that separate the coolant from the transmission fluid (see Chapter 3).*

c) *If the fluid is foaming, drain it and refill the transaxle, then check for coolant in the fluid, or a high fluid level.*

5 Check the engine idle speed.
Note: *If the engine is malfunctioning, do not proceed with the preliminary checks until it has been repaired and runs normally.*

6 Check and adjust the shift cable, if necessary (see Section 5).

7 If hard shifting is experienced, inspect the shift cable under the steering column and at the manual lever on the transaxle (see Section 5).

Fluid leak diagnosis

8 Most fluid leaks are easy to locate visually. Repair usually consists of replacing a seal or gasket. If a leak is difficult to find, the following procedure may help.

9 Identify the fluid. Make sure it's transmission fluid and not engine oil or brake fluid (automatic transmission fluid is a deep red color).

10 Try to pinpoint the source of the leak. Drive the vehicle several miles, then park it over a large sheet of cardboard. After a minute or two, you should be able to locate the leak by determining the source of the fluid dripping onto the cardboard.

11 Make a careful visual inspection of the suspected component and the area immediately around it. Pay particular attention to gasket mating surfaces. A mirror is often helpful for finding leaks in areas that are hard to see.

12 If the leak still cannot be found, clean the suspected area thoroughly with a degreaser or solvent, then dry it thoroughly.

13 Drive the vehicle for several miles at normal operating temperature and varying speeds. After driving the vehicle, visually inspect the suspected component again.

14 Once the leak has been located, the cause must be determined before it can be properly repaired. If a gasket is replaced but the sealing flange is bent, the new gasket will not stop the leak. The bent flange must be straightened.

15 Before attempting to repair a leak, check to make sure that the following conditions are corrected or they may cause another leak.
Note: *Some of the following conditions cannot be fixed without highly specialized tools and expertise. Such problems must be referred to a qualified transmission shop or a dealer service department.*

Gasket leaks

16 Check the pan periodically. Make sure the bolts are tight, no bolts are missing, the gasket is in good condition and the pan is flat (dents in the pan may indicate damage to the valve body inside).

17 If the pan gasket is leaking, the fluid level or the fluid pressure may be too high, the vent may be plugged, the pan bolts may be too tight, the pan sealing flange may be warped, the sealing surface of the transaxle housing may be damaged, the gasket may be damaged or the transaxle casting may be cracked

or porous. If sealant instead of gasket material has been used to form a seal between the pan and the transaxle housing, it may be the wrong type of sealant.

Seal leaks

18 If a transaxle seal is leaking, the fluid level or pressure may be too high, the vent may be plugged, the seal bore may be damaged, the seal itself may be damaged or improperly installed, the surface of the shaft protruding through the seal may be damaged or a loose bearing may be causing excessive shaft movement.

19 Make sure the dipstick tube seal is in good condition and the tube is properly seated. Periodically check the area around the sensors for leakage. If transmission fluid is evident, check the seals for damage.

Case leaks

20 If the case itself appears to be leaking, the casting is porous and will have to be repaired or replaced.

21 Make sure the oil cooler hose fittings are tight and in good condition.

Fluid comes out vent pipe or fill tube

22 If this condition occurs, the possible causes are: the transaxle is overfilled, there is coolant in the fluid, the case is porous, the dipstick is incorrect, the vent is plugged or the drain-back holes are plugged.

3 Shift lever - replacement

Warning: *These models are equipped with Supplemental Restraint Systems (SRS), more commonly known as airbags. Always disable the airbag system before working in the vicinity of any airbag system component to avoid the possibility of accidental deployment of the airbag(s), which could cause personal injury (see Chapter 12).*

1 Set the parking brake and chock the wheels.

2 Remove the components of the upper bezel and floor console to expose the shift lever assembly (see Chapter 11).

3 Disconnect the shift cable end from the ball of the shift lever arm by prying away from the arm **(see illustration 5.3)**.

4 Unclip the shift cable housing from the shift lever bracket **(see illustration 5.4)**.

5 Remove the nuts and bolts attaching the shift lever assembly to the center console structure **(see illustration)**.

6 Disconnect any electrical connectors for the shift lever assembly.

7 Remove the shift lever assembly.

8 Installation is the reverse of removal.

4 Shift lever knob - replacement

Warning: *These models are equipped with Supplemental Restraint Systems (SRS), more*

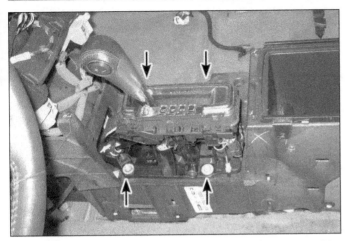

3.5 Remove the nuts and bolts attaching the shift lever assembly

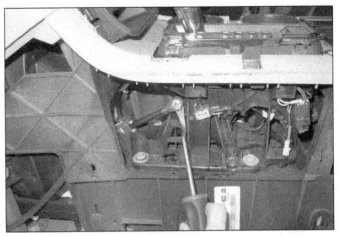

5.3 Pry the shift cable end from the ball of the lever arm

commonly known as airbags. Always disable the airbag system before working in the vicinity of any airbag system component to avoid the possibility of accidental deployment of the airbag(s), which could cause personal injury (see Chapter 12).

1   Set the parking brake and chock the wheels.

2   Remove the driver's side floor console extension panel (see Chapter 11).

3   Disconnect the electrical connector for the harness attached to the shift knob for the tap up/down shifting option (if equipped).

Note: *Make note of the harness routing for installation.*

4   Move the shift lever towards the rear of the vehicle.

5   Using a small, flat-tipped screwdriver, separate the shift knob trim ring from the knob by carefully prying the trim ring down to expose the retaining screw.

6   Remove the shift knob retaining screw and remove the shift knob.

7   Installation is reverse of removal.

Caution: *Ensure the shift knob harness is routed in its original position to prevent interference during shift lever operation.*

5   Shift cable - replacement and adjustment

Warning: *These models are equipped with Supplemental Restraint Systems (SRS), more commonly known as airbags. Always disable the airbag system before working in the vicinity of any airbag system component to avoid the possibility of accidental deployment of the airbag(s), which could cause personal injury (see Chapter 12).*

Replacement

1   Set the parking brake and black chocks on the wheels.

2   Remove the driver's side floor console extension panel (see Chapter 11).

3   Disconnect the shift cable end from the ball of the shift lever arm by prying away from the arm **(see illustration)**.

4   Unclip the shift cable housing from the shift lever bracket **(see illustration)**.

5   Remove the air filter housing air inlet duct and the air filter housing (see Chapter 4).

6   Disconnect the shift cable end from the ball of the transaxle shift lever arm by prying

5.4 Unclip the cable housing from the bracket

away from the arm **(see illustration)**.

7   Remove the shift cable housing retainer, pinch the locking tabs in, and remove the cable from the shift cable bracket **(see illustration)**.

8   Remove the shift cable grommet from the engine compartment firewall.

5.6 Pry the cable end off the ballstud

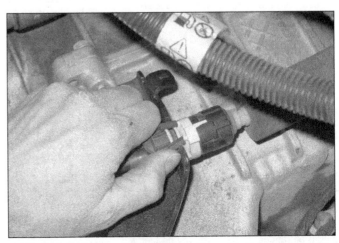

5.7 Pinch the locking tabs to remove the cable from the bracket

5.13 Pull the adjustment clip (A) up and slide the cable and cable end together until all freeplay is removed

6.5 Remove the BTSI terminals from the connector

6.6 Remove the screw and detach the solenoid from the shifter base

9 Remove the shift cable from the vehicle; pull out from the engine compartment end to remove the cable from the vehicle.
Caution: *DO NOT attempt to separate the two cable halves at the metal retainer. The shift cable is not designed to be disassembled. If the cable is separated, it must be replaced.*
10 Installation is reverse of removal.

Adjustment
11 Set the parking brake and chock the wheels.
12 Place the shift lever in park and ensure the transaxle shift lever is in the park position as well.
13 At the transaxle end of the cable, release the cable adjustment clip **(see illustration)**. Pull the shift cable towards the transaxle lever until all freeplay is removed from the cable.
14 Once all freeplay is removed, push in the adjustment clip.
15 Pull on the shift cable to ensure the adjustment clip is fully locked.
16 Check the shift lever for correct operation in all ranges. If there are any problems, perform the adjustment again.

6   Brake Transmission Shift Interlock (BTSI) system - description, component replacement and adjustment

Warning: *These models are equipped with Supplemental Restraint Systems (SRS), more commonly known as airbags. Always disable the airbag system before working in the vicinity of any airbag system component to avoid the possibility of accidental deployment of the airbag(s), which could cause personal injury (see Chapter 12).*

Description
1 This system prevents the transmission from being shifted out of Park unless the key is turned On and the brake pedal is pressed. When the vehicle is started, the BTSI is energized, locking the shift lever in Park; when the brake pedal is pressed, the solenoid is de-energized, so the shift lever can be moved to another gear.
2 Your vehicle is equipped with either a push or pull type of BTSI solenoid. Once the console extension panels are removed, you will be able to identify the type to proceed to the proper instructions.

Replacement
3 Remove the console extension panels (see Chapter 11). Check to see what kind of solenoid your vehicle has: Push-to-release type is attached to the passenger side of the shift lever assembly by a screw and has a harness attached. Pull-to-release type clips into the passenger side of the shift lever assembly and has a connector.
4 If your vehicle has a pull-to-release style solenoid, remove the shift lever assembly (see Section 3).

Push-to-release type
5 Remove the BTSI solenoid electrical connector terminal pins from the shared connector **(see illustration)**.

Note: *This will require a special tool (or equivalent) that is inserted into the back side of the connector to dislodge the terminal retaining tangs.*
Note: *Note the harness routing and location of the terminal pins in the connector for installation.*
6 Remove the retaining screw and the solenoid from the shift lever assembly **(see illustration)**.
7 Installation is the reverse of removal.
Note: *Ensure the harness is routed properly and terminal pins are installed correctly.*

Pull-to-release type
8 Depress the connector lock tab and disconnect the electrical connector from the solenoid.
9 Press in on the tabs to release the solenoid and remove it from the shift lever base.
10 Installation is reverse of removal, noting the following:
   a) *When pressing the new solenoid into the shift lever assembly, press on the solenoid housing and not the white plunger.*
   b) *Ensure the solenoid connector is fully seated and locked.*

7   Driveaxle oil seals - replacement

Note: *This procedure does not apply to the transfer case seals on the right side of AWD models.*
1 Oil leaks frequently occur due to wear of the driveaxle oil seals. Replacement of these seals is relatively easy since the repairs can be performed without removing the transaxle from the vehicle.
2 The seals are located on the sides of the transaxle where the driveaxles are attached. If leakage of a seal is suspected, raise the vehicle and support it securely on jackstands. If the seal is leaking, fluid will be found on the sides of the transaxle.
3 Remove the driveaxle (see Chapter 8).
4 Note how deep the seal is installed, then

use a screwdriver to carefully pry the seal from its bore. If it can't be removed with a screwdriver, a special seal removal tool (available at most auto supply stores) will be needed.

5   Compare the old seal to the new one to be sure it's correct.

6   Coat the inside and outside diameters of the new seal with transaxle fluid.

7   Using a seal installation tool or a large socket, install the new seal. Drive it into the bore squarely and make sure it's seated to its original depth.

8   Install the driveaxle (see Chapter 8).

9   Installation is the reverse of removal. Check the transaxle fluid level, adding if necessary (see Chapter 1).

8   Automatic transaxle - removal and installation

Removal

1   Remove the air filter duct and filter housing (see Chapter 4).

2   Disconnect all wiring from the transaxle and transfer case (if so equipped). This includes the ground wires, the transaxle range switch, the transaxle speed sensors and the transaxle control module.

3   Disconnect the shift cable from the shift lever on the transaxle and remove the cable from the bracket (see Section 5).

4   Disconnect the transmission cooler lines from the transaxle. Plug and cap the lines and openings to prevent contamination.

5   Disconnect any transaxle or transfer case vent tubes.

6   Install an engine support fixture and connect the chains solidly to the lifting brackets on top of the engine. If lifting brackets are not present, connect the support chains to substantial parts of the engine (such as the threaded holes at the left ends of the cylinder heads, the exhaust manifolds and the engine mount strut bracket.

Note: *Engine support fixtures are available at most equipment rental yards.*

7   Remove the four uppermost transaxle-to-engine mounting bolts.

8   Loosen the front wheel lug nuts and the driveaxle/hub nuts, then raise the front of the vehicle and support it securely on jackstands.

9   Drain the transaxle fluid (see Chapter 1).

10   Remove the front wheels and the inner fender splash shields (see Chapter 11).

11   Remove the frame brace from between the lower control arms, disconnect the oxygen sensors and remove the flexible exhaust pipe.

12   On AWD models, remove the driveshaft (see Chapter 8).

13   Remove the subframe (see Chapter 10).

14   Remove the driveaxles (see Chapter 8).

15   On FWD models, remove the intermediate shaft (see Chapter 8) and transaxle-to-engine brace.

16   On AWD models, remove the transfer case (see Chapter 7B).

17   Remove the transaxle mount and bracket (see Section 10).

18   Remove the starter (see Chapter 5). Make matchmarks on the torque converter and driveplate so they can be assembled in the same position.

19   Remove the four torque converter bolts through the starter opening.

20   Support the transaxle with a jack - preferably one designed for this purpose. Transmission/transaxle adapters are available for most heavy duty floor jacks. If the jack is equipped with safety chains or straps, use them to secure the transaxle to the jack.

21   On FWD models, remove the driveplate inspection cover.

22   Remove the lowermost transaxle-to-engine bolts, then pull the transaxle away from the engine until it can be lowered using the jack.

Caution: *Make sure the torque converter doesn't become detached from the transaxle. Clamping a pair of locking pliers to the transaxle case will prevent this.*

23   Lower the transaxle until it can be safely set on the ground.

Installation

24   Installation is the reverse of removal, noting the following points:

a) *As the torque converter is reinstalled, ensure that the drive tangs at the center of the torque converter hub engage with the recesses in the automatic transaxle fluid pump inner gear. This can be confirmed by turning the torque converter while pushing it towards the transaxle. If it isn't fully engaged, it will clunk into place. Align the matchmarks you made on the driveplate and torque converter.*

b) *Replace the steering shaft pinch bolt with a new one and tighten it to the torque listed in the Chapter 10 Specifications.*

c) *Install all of the driveplate-to-torque converter nuts before tightening any of them.*

d) *Tighten the driveplate-to-torque converter bolts to the torque listed in this Chapter's Specifications.*

e) *Tighten the transaxle mounting bolts to the torque listed in this Chapter's Specifications.*

f) *Replace all O-rings discarded previously.*

g) *Tighten the new subframe bolts to the torque values listed in the Chapter 10 Specifications.*

h) *Tighten the wheel lug nuts to the torque listed in the Chapter 1 Specifications.*

i) *Fill the transaxle with the correct type and amount of automatic transmission fluid (see Chapter 1).*

j) *Adjust the shift cable (see Section 5).*

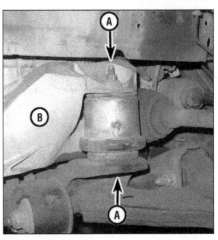

10.3 Remove the upper and lower mount nuts (A) and transaxle mount bracket (B)

9   Automatic transaxle overhaul - general information

1   In the event of a problem occurring, it will be necessary to establish whether the fault is electrical, mechanical or hydraulic in nature, before repair work can be contemplated. Diagnosis requires detailed knowledge of the transaxle's operation and construction, as well as access to specialized test equipment, and so is deemed to be beyond the scope of this manual. It is therefore essential that problems with the automatic transaxle are referred to a dealer service department or other qualified repair facility for assessment.

2   Note that a faulty transaxle should not be removed before the vehicle has been diagnosed by a knowledgeable technician equipped with the proper tools, as troubleshooting must be performed with the transaxle installed in the vehicle.

10   Transaxle mount - replacement

1   Raise the vehicle and support it securely on jackstands.

2   Place a floor jack under the transaxle with a block of wood between the jack and transaxle to prevent damage.

3   Remove the upper and lower mount nuts attaching the mount to the frame and transaxle mount bracket **(see illustration)**.

4   Remove the transaxle mount-to-transaxle mounting bolts. To access and remove the lower rear bolt, you may need use the jack to lift the transaxle high enough to tilt the mount.

5   Remove the transaxle mount and bracket.

6   Installation is the reverse of removal. Tighten all bolts to the torque listed in this Chapter's Specifications.

Notes

Chapter 7 Part B
Transfer case

Contents

	Section		Section
General information	1	Transfer case - removal and installation	2

Specifications

Transfer case fluid type	See Chapter 1	

Torque specifications

	Ft-lbs	Nm
Transfer case-to-transaxle bolts	37	50
Rear engine mount nuts	55	75
Rear engine mount bracket bolts	43	58

1   General information

1   Due to the complexity of the transfer case covered in this manual and the need for specialized equipment to perform most service operations, this Chapter contains only removal and installation procedures.

2   If the transfer case requires major repair work, it should be taken to a dealer service department or an automotive or transmission repair shop. You can, however, remove and install the transfer case yourself and save the expense of that labor, even if the repair work is done by a transmission shop.

2   Transfer case - removal and installation

1   Loosen the right front wheel lug nuts and driveaxle/hub nut. Raise the vehicle and support it securely on jackstands. Remove the wheel.

2   Drain the transfer case lubricant (see Chapter 1).

3   Drain the transaxle fluid (see Chapter 1).

4   Remove the driveshaft (see Chapter 8).

5   Remove the right driveaxle (see Chapter 8).

6   Disconnect the front flexible exhaust pipe at both front and rear catalytic converters and the muffler pipe and remove the front portion of the exhaust system. Support the rear exhaust pipe with wire.

7   Support the transaxle with a jack, jackstand or equivelent.

8   Remove the right-side (rear) engine mount (see Chapter 2A)

Note: *If equipped, disconnect the active engine mount vacuum line before removing the rear engine mount.*

9   Remove the rear engine mount bracket bolts and remove the bracket from the engine.

10   Remove the transfer case mounting bolts.

11   Disconnect the intermediate shaft by pulling the transfer case away from the transaxle, and remove the transfer case.

12   Installation is the reverse of removal, noting the following points:

a) *Install a new O-ring seal between the transfer case and transaxle.*

b) *Use two guide pins during installation of the transfer case to the transmission. Failure to do so may result in damage to the O-ring seal, causing a fluid leak. Guide pins can be made by cutting the heads off of 2 M12x1.75x75 bolts. Hand thread the guide pins into the upper and lower transfer case mounting bolt holes. Once the transfer case is installed, remove the guide pins and install the mounting bolts.*

c) *Tighten the exhaust system fasteners to the torque listed in the Chapter 4 Specifications.*

d) *Tighten the driveshaft fasteners to the torque listed in the Chapter 8 Specifications.*

e) *Tighten the transfer case mounting bolts to the torque listed in this Chapter's Specifications.*

f) *Refill the transfer case and transaxle with the proper type and amount of lubricant (see Chapter 1).*

g) *Tighten the driveaxle/hub nut to the torque listed in the Chapter 8 Specifications.*

h) *Tighten the wheel lug nuts to the torque listed in the Chapter 1 Specifications.*

Chapter 8 Driveline

Contents

	Section
Clutch control module - replacement	12
Differential clutch drum (AWD models) - removal and installation	11
Driveaxle boot - replacement	3
Driveaxles - removal and installation	2
Driveshaft (AWD models) - removal and installation	4
Driveshaft center support bearing (AWD models) - replacement	6
Driveshaft universal and constant velocity joints (AWD models) - general information and check	5

	Section
General information	1
Rear differential assembly (AWD models) - removal and installation	9
Rear differential pinion seal (AWD models) - replacement	8
Rear driveaxle oil seals (AWD models) - replacement	7
Torque tube (AWD models) - removal and installation	10

Specifications

Torque specifications	Ft-lbs	Nm
Driveaxles		
Driveaxle/hub nut*		
Front	173	234
Rear (AWD models)	151	205
Intermediate shaft bearing bracket fasteners	43	58
* Nut must be replaced		
Driveshaft (AWD models)		
Center bearing mounting nuts	43	58
Driveshaft-to-transfer case bolts	25	34
Driveshaft-to-rear differential bolts	43	58
Rear differential assembly (AWD models)		
Side cover bolts	21	28
Differential drain/fill plugs	26	35
Torque tube-to-differential bolts	40	54
Torque tube front mounting bolt	137	186
Differential rear support bushing bolts	118	160
Differential front mounting nuts	21	28
Pinion housing mounting bolts	21	28

2.1 Loosen the driveaxle/hub nut with a long breaker bar

2.11 Carefully pry the inner end of the driveaxle from the outer end of the intermediate shaft, or from the transaxle

1   General information

1   The information in this Chapter deals with the components from the rear of the engine to the drive wheels, except for the transaxle (and transfer case on AWD models), which is dealt with in Chapters 7A and 7B.

2   Since nearly all the procedures covered in this Chapter involve working under the vehicle, make sure it's securely supported on sturdy jackstands or on a hoist where the vehicle can be easily raised and lowered.

2   Driveaxles - removal and installation

Front

Removal

1   Remove the wheel center cover. Break the driveaxle/hub nut loose with a socket and large breaker bar **(see illustration)**.

Note: *If the opening in the wheel is too small to accommodate your socket, wait until after the wheel is removed to loosen the nut. You can prevent the brake disc from turning by inserting a long punch into a disc cooling vane and letting it come to rest against the caliper mounting bracket.*

2   Loosen the wheel lug nuts, raise the vehicle and support it securely on jackstands. Remove the wheel.

3   Remove the front inner fender splash shields if necessary (see Chapter 11).

4   Disconnect the tie-rod end from the steering knuckle (see Chapter 10).

5   Disconnect the stabilizer bar link from the stabilizer bar (see Chapter 10).

6   Remove the driveaxle/hub nut and discard it.

7   Separate the control arm balljoint from the steering knuckle (see Chapter 10).

8   Swing the knuckle/hub assembly out (away from the vehicle) until the end of the driveaxle is free of the hub.

Note: *If the driveaxle splines stick in the hub, tap on the end of the driveaxle with a plastic hammer. Support the outer end of the drive-axle with a piece of wire to avoid unnecessary strain on the inner CV joint.*

9   Secure the strut/knuckle/hub assembly out of the way to make room to work.

10   If you're removing the right driveaxle, carefully pry the inner CV joint off the intermediate shaft using a large screwdriver or prybar positioned between the CV joint housing and the intermediate shaft bearing support.

11   If you're removing the left driveaxle, pry the inner CV joint out of the transaxle using a large screwdriver or prybar positioned between the transaxle and the CV joint housing **(see illustration)**. Be careful not to damage the differential seal.

12   Support the CV joints and carefully remove the driveaxle from the vehicle. If equipped, remove the washer from the outer CV end of the driveaxle and save it for installation.

Installation

13   Pry the old spring clip from the inner end of the driveaxle (left side) or outer end of the intermediate shaft (right side) and install a new one. Lubricate the differential or intermediate shaft seal with multi-purpose grease and raise the driveaxle into position while supporting the CV joints.

Note: *Position the spring clip with the opening facing down; this will ease insertion of the driveaxle and prevent damage to the clip.*

14   Apply a light coat of multi-purpose grease to the intermediate shaft splines.

15   Push the splined end of the inner CV joint into the differential side gear (left side) or onto the intermediate shaft (right side) and make sure the spring clip locks in its groove.

16   Grasp the inner CV joint housing (not the driveaxle) and try to pull out to ensure the driveaxle retaining ring has seated securely in the transaxle or on the intermediate shaft.

17   Ensure the washer removed from the old driveaxle is installed on the outer CV end of the driveaxle.

18   Apply a light coat of multi-purpose

grease to the outer CV joint splines, pull out on the steering knuckle assembly and install the stub axle into the hub.

19   Insert the balljoint stud into the steering knuckle and tighten the bolt to the torque listed in the Chapter 10 Specifications. Reconnect the tie-rod end and the stabilizer bar link (see Chapter 10). Install a new cotter pins if required.

20   Install the front inner fender splash shields, if removed (see Chapter 11).

21   Install a *new* driveaxle/hub nut. Tighten the hub nut securely, but don't try to tighten it to the actual torque specification until you've lowered the vehicle to the ground.

Note: *If your socket won't fit through the opening in the wheel, tighten the driveaxle/hub nut now, using the technique described in Step 1 to prevent the brake disc from turning.*

22   Install the wheel and lug nuts, then lower the vehicle. Tighten the lug nuts to the torque listed in the Chapter 1 Specifications.

23   Tighten the driveaxle/hub nut to the torque listed in this Chapter's Specifications.

Intermediate shaft

Removal

24   Remove the right front driveaxle (see Steps 1 through 12).

25   Remove the bolts mounting the intermediate shaft bearing retainer to the engine bracket.

26   Pull the intermediate shaft from the transaxle.

Installation

27   Lubricate the lips of the transaxle seal with multi-purpose grease. Carefully guide the intermediate shaft into the transaxle side gear, then install the mounting bolts hand tight for the bearing support. Make sure that the shaft is positively engaged into the transaxle.

28   Tighten the nuts to the torque listed in this Chapter's Specifications.

29   Install a new spring clip on the right front driveaxle (see Step 13).

30   The remainder of installation is the reverse of removal.

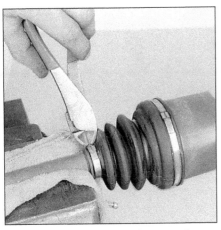

3.3 Cut off the boot clamps and discard them

3.4 Mark the relationship of the tri-pod assembly to the outer race

3.6 Mark the tri-pod and axleshaft to ensure proper installation

Rear (AWD models)

Removal

31 Remove the wheel center cover. Break the driveaxle/hub nut loose with a socket and large breaker bar **(see illustration 2.1)**.

Note: *If the opening in the wheel is too small to accommodate your socket, wait until after the wheel is removed to loosen the nut. You can prevent the brake disc from turning by inserting a long punch into a disc cooling vane and letting it come to rest against the caliper mounting bracket.*

32 Block the front wheels to prevent the vehicle from rolling. Loosen the wheel lug nuts, raise the rear of the vehicle and support it securely on jackstands. Remove the wheel.

33 Remove the driveaxle/hub nut and discard it. Use a brass punch or a block of wood with a hammer to break the end of the driveaxle loose from the wheel hub.

Caution: *The manufacturer recommends that all suspension nuts and cotter pins be replaced with new ones any time they are removed.*

34 Remove the hub and bearing assembly (see Chapter 10).

35 If removing the right rear driveaxle, remove the muffler assembly (see Chapter 4).

36 Pry the driveaxle out of the differential and remove it through the opening in the knuckle. Remove the retaining ring - it will have to be replaced with a new one.

Installation

37 Install a new retaining ring on the inner end of the driveaxle. Apply a light film of grease to the area on the inner CV joint stub shaft where the seal rides, guide the driveaxle through the opening in the knuckle, then insert the splined end of the inner CV joint into the differential. Make sure the spring clip locks in its groove.

Note: *Position the spring clip with the opening facing down; this will ease insertion of the driveaxle and prevent damage to the clip.*

38 Install the hub and bearing assembly (see Chapter 10).

3.7a Spread the ends of the stop-ring apart and slide it towards the center of the shaft . . .

39 Install a *new* driveaxle/hub nut. Tighten the hub nut securely, but don't try to tighten it to the actual torque specification until you've lowered the vehicle to the ground.

40 Install the wheel and lug nuts, then lower the vehicle. Tighten the lug nuts to the torque listed in the Chapter 1 Specifications.

41 Tighten the driveaxle/hub nut to the torque listed in this Chapter's Specifications. Install the wheel center cover.

---

3   Driveaxle boot - replacement

Note: *If the CV joints are worn, indicating the need for an overhaul (usually due to torn boots), explore all options before beginning the job. Complete rebuilt driveaxles are available on an exchange basis, which eliminates much time and work.*

1 Remove the driveaxle from the vehicle (see Section 2).

2 Mount the driveaxle in a vise. The jaws

3.7b . . . then slide the tri-pod assembly back and remove the retainer clip

of the vise should be lined with wood or rags to prevent damage to the driveaxle.

Inner CV joint boot (front, all models; rear, 2007 and 2008 models)/Outer CV joint boot (rear, 2007 and 2008 models)

Removal

3 Remove the boot clamps **(see illustration)**.

4 Pull the boot back from the inner CV joint and slide off the joint housing. Mark the relationship of the tri-pod to the outer race **(see illustration)**.

5 On rear CV joints, not the orientation of the spring and guide, then remove them from the housing

6 Use a center punch to mark the relationship between the tri-pod and the axleshaft to ensure proper installation **(see illustration)**.

7 Remove the stop-ring and retainer clip from the end of the axleshaft **(see illustrations)**.

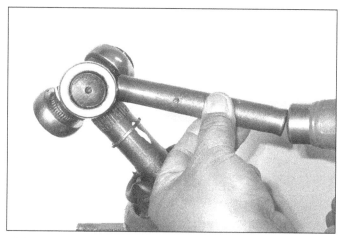

3.8 Drive the tri-pod joint from the axleshaft with a brass punch and hammer, making sure not to damage the bearing surfaces or the splines on the shaft

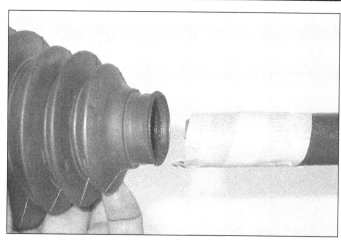

3.11 Wrap the splined area of the axleshaft with tape to prevent damage to the boot(s) when installing it

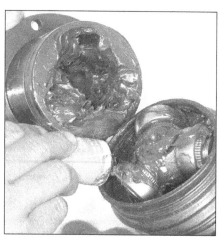

3.13 Pack the outer race with CV joint grease and slide it over the tri-pod assembly - make sure the match marks on the CV joint housing and tri-pod line up

Note: *2007 and 2008 models don't have a stop-ring.*

8   Use a hammer and a brass punch to drive the tri-pod joint from the driveaxle if necessary **(see illustration)**. On 2009 and later models, remove the stop-ring from the axleshaft.

9   Slide the boot off the axleshaft.

Inspection

10   Clean the old grease from the outer race and the tri-pod bearing assembly. Carefully disassemble each section of the tri-pod assembly, one at a time so as not to mix up the parts, and clean the needle bearings with solvent. Inspect the rollers, tri-pod, bearings and outer race for scoring, pitting or other signs of abnormal wear, which will warrant the replacement of the inner CV joint.

Reassembly

11   Slide the clamps and boot onto the axleshaft. It's a good idea to wrap the axleshaft

splines with tape to prevent damaging the boot **(see illustration)**.

12   On 2009 and later models, spread the stop-ring and install it on the shaft, past its groove. On all models, place the tri-pod on the shaft (aligning the previously-made matchmarks) and install a new retainer clip. On 2009 and later models, seat the stop-ring in its groove.

13   Apply CV joint grease to the tri-pod assembly, the inside of the joint housing and the inside of the boot **(see illustration)**. On rear driveaxles, install the spring and guide into the housing. On all models, slide the housing over the tri-pod.

14   Slide the boot into place.

15   Position the CV joint mid-way through its travel, then equalize the pressure in the boot **(see illustration)**.

16   Tighten the boot clamps **(see illustrations)**.

17   Install the driveaxle assembly (see Section 2).

3.15 Equalize the pressure inside the boot by inserting a small, dull screwdriver between the boot and the outer race

3.16a To install new fold-over type clamps, bend the tang down...

3.16b... and flatten the tabs to hold it in place

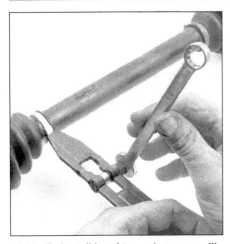

3.16c To install band-type clamps, you'll need a special tool; install the band with its end pointing in the direction of axle rotation and tighten it securely, then pivot the tool up 90-degrees and tap the center of the clip with a center punch...

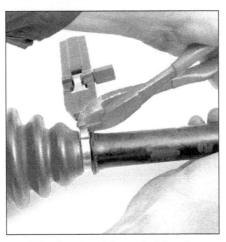

3.16d... then bend the end of the clamp back over the clip and cut off the excess

3.16e If you're installing crimp-type boot clamps, you'll need a pair of special crimping pliers (available at most auto parts stores)

Outer CV joint boot (front, all models; rear, 2009 and later models)

Removal

18 Remove the boot clamps (see illustration 3.3).

19 Strike the edge of the CV joint housing sharply with a soft-face hammer to dislodge the outer CV joint from the axleshaft (see illustration). Remove and discard the bearing retainer clip from the axleshaft.

20 Slide the outer CV joint boot off the axleshaft.

21 Place marks on the inner race and cage so they both can be installed facing out when reassembling the joint.

22 Press down on the inner race far enough

3.19 Strike the edge of the CV joint housing sharply with a soft-faced hammer to dislodge the CV joint from the shaft

to allow a ball bearing to be removed. If it's difficult to tilt, tap the inner race with a brass punch and hammer (see illustration).

23 Pry the balls out of the cage, one at a time, with a blunt screwdriver or wooden tool (see illustration).

3.22 Gently tap the inner race with a brass punch to tilt it enough to allow ball bearing removal

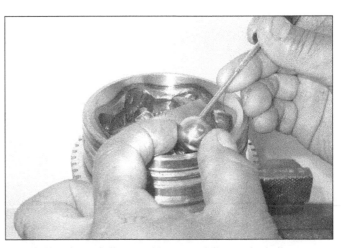

3.23 Using a dull screwdriver, carefully pry the balls out of the cage

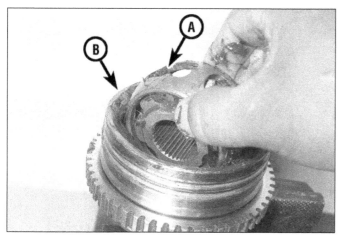

3.24 Tilt the inner race and cage 90-degrees, then align the windows in the cage (A) with the lands (B) and rotate the inner race up and out of the outer race

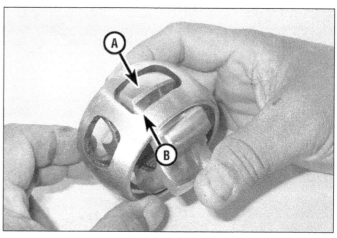

3.25 Align the inner race lands (A) with the cage windows (B) and rotate the inner race out of the cage

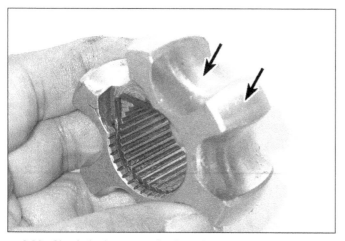

3.26a Check the inner race lands and grooves for pitting and score marks

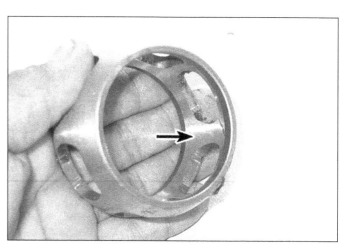

3.26b Check the cage for cracks, pitting and score marks - shiny spots are normal and don't affect operation

24 With all of the balls removed from the cage and the cage/inner race assembly tilted 90-degrees, align the cage windows with the outer race lands and remove the assembly from the outer race **(see illustration)**.

25 Remove the inner race from the cage by turning the inner race 90-degrees in the cage, aligning the inner lands with the cage windows and rotating the inner race out of the cage **(see illustration)**.

Inspection

26 Clean the components with solvent to remove all traces of grease. Inspect the cage and races for pitting, score marks, cracks and other signs of wear and damage. Shiny, polished spots are normal and won't adversely affect CV joint operation **(see illustrations)**.

Reassembly

27 Install the inner race in the cage by reversing the removal technique.

28 Install the inner race and cage assembly in the outer race by reversing the removal technique. The marks that were previously applied to the inner race and cage must both be visible after the assembly is installed in the outer race.

29 Press the balls into the cage windows **(see illustration)**.

30 Slide a new sealing boot clamp and sealing boot onto the axleshaft. It's a good idea to wrap the axleshaft splines with tape to prevent damaging the boot **(see illustration 3.10)**.

31 Place a new bearing retainer clip onto the axleshaft.

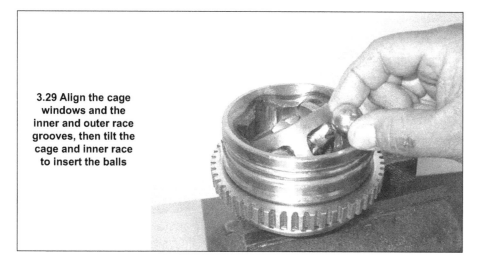

3.29 Align the cage windows and the inner and outer race grooves, then tilt the cage and inner race to insert the balls

3.32a Pack the outer CV joint assembly with CV grease...

3.32b... then apply grease to the inside of the boot...

3.32c... until the level is up to the end of the axle

4.2 Scribe a line between the driveshaft CV joint flange and the transfer case yoke

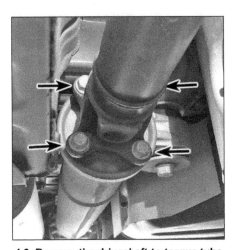

4.3 Remove the driveshaft-to-torque tube flange mounting bolts

4.4 Remove the driveshaft constant velocity (CV) mounting bolts with spaced washers – five of six bolts shown

32 Place half the grease provided in the sealing boot kit into the outer CV joint assembly housing **(see illustration)**. Put the remaining grease into the sealing boot **(see illustrations)**.

33 Align the splines on the axleshaft with the splines on the outer CV joint assembly and gently drive the CV joint onto the axleshaft using a soft-faced hammer until the CV joint is seated to the axleshaft.

34 Position the CV joint mid-way through its travel, then equalize the pressure in the boot **(see illustration 3.14)**.

35 Tighten the boot clamps **(see illustrations 3.15a through 3.15e)**.

36 Install the driveaxle (see Section 2).

---

4   Driveshaft (AWD models) - removal and installation

Note: *The manufacturer recommends replacing driveshaft fasteners with new ones when installing the driveshaft.*

1   Raise the rear of the vehicle and support it securely on jackstands. Block the front

wheels to prevent the vehicle from rolling. Place the shift selector lever in Neutral.

2   Use chalk or a scribe to index the relationship of the driveshaft to the differential pinion yoke and the flange at the transfer case **(see illustration)**.

3   Disconnect the rear of the driveshaft from the torque tube flange. Support the rear of the driveshaft **(see illustration)**.

4   Remove the front Constant Velocity (CV) joint bolts and washers from the transfer case. Support the front of the driveshaft **(see illustration)**.

Note: *If the front CV joint will not break free from the transfer case flange, use a brass drift and a hammer to carefully drive the CV joint out from the flange.*

5   Remove the mounting bolts from the center support bearing **(see illustration)**.

6   Remove the driveshaft from the vehicle.

7   Installation is the reverse of removal, noting the following points:

a) Make sure the marks you made previously are aligned.

b) Tighten the fasteners to the torques listed in this Chapter's Specifications.

4.5 Remove the driveshaft center support bearing mounting bolts and lower the driveshaft from the vehicle

5   Driveshaft universal and constant velocity joints (AWD models) - general information and check

Universal joints

1   Universal joints are mechanical couplings which connect two rotating components that meet each other at different angles.

2   These joints are composed of a yoke on each side connected by a crosspiece called a trunnion. Cups at each end of the trunnion contain needle bearings which provide smooth transfer of the torque load. Snap-rings, either inside or outside of the bearing cups, hold the assembly together.

3   Wear in the needle roller bearings is characterized by vibration in the driveline, noise during acceleration, and in extreme cases of lack of lubrication, metallic squeaking and ultimately grating and shrieking sounds as the bearings disintegrate.

4   It is easy to check if the needle bearings are worn with the driveshaft in position, by trying to turn the shaft with one hand, the other hand holding the rear differential pinion flange when the rear universal joint is being checked, and the front half coupling when the front universal joint is being checked. Any movement between the driveshaft and the front half couplings, and around the rear half couplings, is indicative of considerable wear. Another method of checking for universal joint wear is to use a prybar inserted into the gap between the universal joint and the driveshaft or flange. Leave the vehicle in gear and try to pry the joint both radially and axially. Any looseness should be apparent with this method. A final test for wear is to attempt to lift the shaft and note any movement between the yokes of the joints.

5   If any of the above conditions exist, replace the driveshaft.

Constant velocity joint

6   The constant velocity joint on the forward end of the driveshaft is not serviceable either. If the joint wears out or the boot becomes damaged, the front portion of the driveshaft must be replaced.

6   Driveshaft center support bearing (AWD models) - replacement

The center support bearing is not replaceable separately; if it is in need of replacement, the entire driveshaft must be replaced.

7   Rear driveaxle oil seals (AWD models) - replacement

1   Raise the rear of the vehicle and support it securely on jackstands. Place the transaxle in Neutral with the parking brake off. Block the front wheels to prevent the vehicle from rolling.

2   If working on the right rear driveaxle, remove the muffler assembly.

3   Remove the driveaxle(s) (see Section 2).

4   Carefully pry out the driveaxle oil seal with a seal removal tool or a large screwdriver. Be careful not to damage or scratch the seal bore.

5   Using a seal installer or a large deep socket as a drift, install the new oil seal. Drive it into the bore squarely and make sure it's completely seated.

6   Lubricate the lip of the new seal with multi-purpose grease, then install the driveaxle. Be careful not to damage the lip of the new seal.

7   Check the differential lubricant level and add some, if necessary, to bring it to the appropriate level (see Chapter 1).

8   Rear differential pinion seal (AWD models) - replacement

1   Raise the vehicle and support it securely on jackstands.

2   Remove the torque tube (see Section 10).

3   Thread self-tapping screws into the seal and use pliers or a slide hammer to remove the pinion seal from the differential.

4   Using a seal installer or a large deep socket as a drift, install the new oil seal. Drive it into the bore squarely and make sure it's completely seated.

5   Lubricate the lip of the new seal with multi-purpose grease.

6   Installation is the reverse of removal.
Caution: *DO NOT reuse the gasket between the torque tube and differential.*

7   Check the differential lubricant level and add some, if necessary, to bring it to the appropriate level (see Chapter 1).

9   Rear differential assembly (AWD models) - removal and installation

Removal

1   Loosen the rear wheel lug nuts, raise the rear of the vehicle and support it securely on jackstands. Remove the rear wheels. Block the front wheels to prevent the vehicle from rolling.

2   Remove the driveshaft (see Section 4).

3   Remove the driveaxles (see Section 2).

4   Remove the torque tube front mounting bolt. Lower the front of the torque tube down to gain access to the clutch control module and disconnect the electrical connector.

5   Support the torque tube with a jackstand and place a floor jack under the differential assembly.

6   Remove the differential rear mounting bolts.

7   Carefully lower and remove the differential from the vehicle.

8   Separate the torque tube from the differential (see Section 10).
Caution: *DO NOT reuse the gasket between the torque tube and differential.*

Installation

9   Installation is the reverse of removal, noting the following points:

   a) *Don't tighten any of the mounting fasteners until all of them have been installed.*

   b) *Use a new gasket between the torque tube and differential.*

   c) *Tighten all fasteners to the torque listed in this Chapter's Specifications.*

10   Torque tube (AWD models) - removal and installation

Removal

1   Raise the vehicle and support it securely on jackstands.

2   Remove the driveshaft (see Section 4).

3   Remove the torque tube front mounting bolt.

4   Remove the clutch control module (see Section 12).

5   Remove the bolts attaching the torque tube to the rear differential and remove the torque tube from the vehicle.
Caution: *DO NOT reuse the gasket between the torque tube and differential.*

Installation

6   Inspect the torque tube and differential for signs of pinion seal leakage and replace as necessary.

7   Using a new gasket, install the torque tube to the differential alignment pins and hand tighten the mounting bolts.

8   Support the front of the torque tube using a jackstand and tighten the torque tube-to-differential bolts to the torque listed in this Chapter's Specifications.

9   Install the clutch control module (see Section 12).

10   Install the torque tube mounting bolt, tightening it to the torque listed in this Chapter's Specifications.

11   Install the driveshaft (see Section 4).

11   Differential clutch drum (AWD models) - removal and installation

Caution: *When removing and installing the clutch drum from the torque tube, use care not to damage the electrical harness, as it is not available for purchase separately if damaged.*

1   Raise the vehicle and support it securely on jackstands.

2   Remove the torque tube (see Section 10).

3   Disconnect the clutch control module connector and detach the harness from the torque tube.

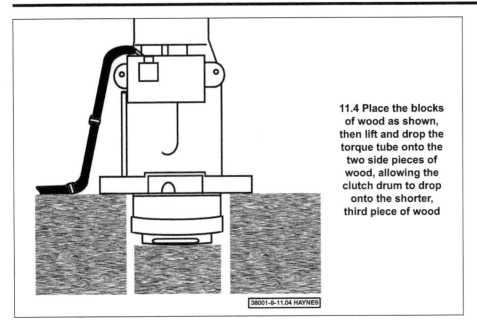

11.4 Place the blocks of wood as shown, then lift and drop the torque tube onto the two side pieces of wood, allowing the clutch drum to drop onto the shorter, third piece of wood

4   Using two pieces of wood and one thinner piece of wood, place the wood under the torque tube as shown **(see illustration)**. Lift the torque tube and drop it onto the side pieces of wood to remove the clutch drum. The clutch drum will drop onto the third piece of wood when it releases from the torque tube.

Caution: *The third piece of wood must be used to prevent damage to the clutch drum during removal.*

5   Remove the clutch drum from the torque tube.

6   Installation is the reverse of removal, noting the following points:

a) *DO NOT reuse the clutch drum-to-torque tube retaining clip. Use a NEW retaining clip.*

b) *DO NOT reuse the clutch drum spline O-ring. Use a NEW O-ring.*

12 Clutch control module - replacement

1   Raise the vehicle and support it securely on jackstands.

2   Remove the driveshaft (see Section 4).

3   Remove the torque tube front mounting bolt.

4   Lower the front of the torque tube down to gain access to the clutch control module.

5   Disconnect the electrical connector.

6   Remove the two mounting screws for the clutch control module and remove the module.

7   Installation is the reverse of removal. Tighten all fasteners to the torque listed in this Chapter's Specifications.

Notes

Chapter 9   Brakes

Contents

	Section
Anti-lock Brake System (ABS) - general information	3
Brake disc - inspection, removal and installation	6
Brake hoses and lines - inspection and replacement	8
Brake hydraulic system - bleeding	9
Disc brake caliper and caliper mounting bracket - removal and installation	5
Disc brake pads - replacement	4

	Section
General information	1
Master cylinder - removal and installation	7
Parking brake - adjustment	12
Parking brake shoes - replacement	13
Power brake booster - check, removal and installation	10
Troubleshooting	2
Vacuum auxiliary pump	11

Specifications

General
Brake fluid type .... See Chapter 1

Disc brakes
Minimum pad lining thickness	See Chapter 1
Disc lateral runout limit	0.002 inch (0.06 mm)
Disc minimum thickness	Cast into disc

Torque specifications

Note: *One foot-pound (ft-lb) of torque is equivalent to 12 inch-pounds (in-lbs) of torque. Torque values below approximately 15 foot-pounds are expressed in inch-pounds, because most foot-pound torque wrenches are not accurate at these smaller values.*

	Ft-lbs (unless otherwise indicated)	Nm
Brake line banjo bolt		
Front	30	40
Rear	37	50
Caliper mounting bracket bolts		
Front	129	175
Rear		
2007 through 2009 models	151	205
2010 and later models	129	175
Caliper guide pin bolts (tighten the bolt closest to the bleed valve first)		
Front	47	64
Rear	20	27
Brake disc retaining screw	106 in-lbs	12
Master cylinder mounting nuts	15	20
Brake line-to-master cylinder nuts	15	20
Brake Pressure Modulator Valve (BPMV) bracket nuts	15	20
Power brake booster nuts	18	25
Brake booster vacuum pump mounting bolts	89 in-lbs	10
Wheel lug nuts	See Chapter 1	

1   General information

1   The vehicles covered by this manual are equipped with hydraulically operated front and rear brake systems. The front and rear brakes are disc type. Both the front and rear brakes are self adjusting. The disc brakes automatically compensate for pad wear.

Hydraulic system

2   The hydraulic system consists of two separate circuits. The master cylinder has separate reservoirs for the two circuits, and, in the event of a leak or failure in one hydraulic circuit, the other circuit will remain operative. Brake balance between the front and rear brakes is monitored and maintained by the Dynamic Rear Proportioning (DRP) function of the Anti-lock Brake System (ABS).

Power brake booster

3   The power brake booster, utilizing engine manifold vacuum and atmospheric pressure to provide assistance to the hydraulically operated brakes, is mounted on the firewall in the engine compartment. On some models, additional vacuum is provided by an electrically powered vacuum pump.

Parking brake

4   The parking brake operates the rear brakes only, through cable actuation. It's activated by a pedal in the driver's foot well. Parking brake shoes are pressed against the inside of the drum portion of the rear brake discs, thereby preventing the rear wheels from rotating.

Service

5   After completing any operation involving disassembly of any part of the brake system, always test drive the vehicle to check for proper braking performance before resuming normal driving. When testing the brakes, perform the tests on a clean, dry, flat surface. Conditions other than these can lead to inaccurate test results.
6   Test the brakes at various speeds with both light and heavy pedal pressure. The vehicle should stop evenly without pulling to one side or the other.
7   Tires, vehicle load and wheel alignment are factors which also affect braking performance.

Precautions

8   There are some general cautions and warnings involving the brake system on this vehicle:

a)   *Use only brake fluid conforming to DOT 3 specifications. Adding DOT 5 fluid or any other type of hydraulic oil, will contaminate the entire system and require a complete flushing and replacement of many components.*

b)   *The brake pads contain fibers which are hazardous to your health if inhaled. Whenever you work on brake system components, clean all parts with brake system cleaner. Do not allow the fine dust to become airborne. Also, wear an approved filtering mask.*

c)   *Safety should be paramount whenever any servicing of the brake components is performed. Do not use parts or fasteners which are not in perfect condition, and be sure that all clearances and torque specifications are adhered to. If you are at all unsure about a certain procedure, seek professional advice. Upon completion of any brake system work, test the brakes carefully in a controlled area before putting the vehicle into normal service. If a problem is suspected in the brake system, don't drive the vehicle until it's fixed.*

2   Troubleshooting

PROBABLE CAUSE	CORRECTIVE ACTION

No brakes - pedal travels to floor

1 Low fluid level 2 Air in system	1 and 2 Low fluid level and air in the system are symptoms of another problem - a leak somewhere in the hydraulic system. Locate and repair the leak
3 Defective seals in master cylinder	3 Replace master cylinder
4 Fluid overheated and vaporized due to heavy braking	4 Bleed hydraulic system (temporary fix). Replace brake fluid (proper fix)

Brake pedal slowly travels to floor under braking or at a stop

1 Defective seals in master cylinder	1 Replace master cylinder
2 Leak in a hose, line, caliper or wheel cylinder	2 Locate and repair leak
3 Air in hydraulic system	3 Bleed the system, inspect system for a leak

Brake pedal feels spongy when depressed

1 Air in hydraulic system	1 Bleed the system, inspect system for a leak
2 Master cylinder or power booster loose	2 Tighten fasteners
3 Brake fluid overheated (beginning to boil)	3 Bleed the system (temporary fix). Replace the brake fluid (proper fix)
4 Deteriorated brake hoses (ballooning under pressure)	4 Inspect hoses, replace as necessary (it's a good idea to replace all of them if one hose shows signs of deterioration)

Brake pedal feels hard when depressed and/or excessive effort required to stop vehicle

1 Power booster faulty	1 Replace booster
2 Engine not producing sufficient vacuum, or hose to booster clogged, collapsed or cracked	2 Check vacuum to booster with a vacuum gauge. Replace hose if cracked or clogged, repair engine if vacuum is extremely low
3 Defective auxiliary vacuum pump (if equipped)	3 Replace auxiliary vacuum pump
4 Brake linings contaminated by grease or brake fluid	4 Locate and repair source of contamination, replace brake pads or shoes
5 Brake linings glazed	5 Replace brake pads or shoes, check discs and drums for glazing, service as necessary
6 Caliper piston(s) or wheel cylinder(s) binding or frozen	6 Replace calipers or wheel cylinders
7 Brakes wet	7 Apply pedal to boil-off water (this should only be a momentary problem)
8 Kinked, clogged or internally split brake hose or line	8 Inspect lines and hoses, replace as necessary

Excessive brake pedal travel (but will pump up)

1 Drum brakes out of adjustment	1 Adjust brakes
2 Air in hydraulic system	2 Bleed system, inspect system for a leak

Excessive brake pedal travel (but will not pump up)

1 Master cylinder pushrod misadjusted	1 Adjust pushrod
2 Master cylinder seals defective	2 Replace master cylinder
3 Brake linings worn out	3 Inspect brakes, replace pads and/or shoes
4 Hydraulic system leak	4 Locate and repair leak

Brake pedal doesn't return

1 Brake pedal binding	1 Inspect pivot bushing and pushrod, repair or lubricate
2 Defective master cylinder	2 Replace master cylinder

Troubleshooting (continued)

PROBABLE CAUSE	CORRECTIVE ACTION

Brake pedal pulsates during brake application

PROBABLE CAUSE	CORRECTIVE ACTION
1 Brake drums out-of-round	1 Have drums machined by an automotive machine shop
2 Excessive brake disc runout or disc surfaces out-of-parallel	2 Have discs machined by an automotive machine shop
3 Loose or worn wheel bearings	3 Adjust or replace wheel bearings
4 Loose lug nuts	4 Tighten lug nuts

Brakes slow to release

PROBABLE CAUSE	CORRECTIVE ACTION
1 Malfunctioning power booster	1 Replace booster
2 Pedal linkage binding	2 Inspect pedal pivot bushing and pushrod, repair/lubricate
3 Malfunctioning proportioning valve	3 Replace proportioning valve
4 Sticking caliper or wheel cylinder	4 Repair or replace calipers or wheel cylinders
5 Kinked or internally split brake hose	5 Locate and replace faulty brake hose

Brakes grab (one or more wheels)

PROBABLE CAUSE	CORRECTIVE ACTION
1 Grease or brake fluid on brake lining	1 Locate and repair cause of contamination, replace lining
2 Brake lining glazed	2 Replace lining, deglaze disc or drum

Vehicle pulls to one side during braking

PROBABLE CAUSE	CORRECTIVE ACTION
1 Grease or brake fluid on brake lining	1 Locate and repair cause of contamination, replace lining
2 Brake lining glazed	2 Deglaze or replace lining, deglaze disc or drum
3 Restricted brake line or hose	3 Repair line or replace hose
4 Tire pressures incorrect	4 Adjust tire pressures
5 Caliper or wheel cylinder sticking	5 Repair or replace calipers or wheel cylinders
6 Wheels out of alignment	6 Have wheels aligned
7 Weak suspension spring	7 Replace springs
8 Weak or broken shock absorber	8 Replace shock absorbers

Brakes drag (indicated by sluggish engine performance or wheels being very hot after driving)

PROBABLE CAUSE	CORRECTIVE ACTION
1 Brake pedal pushrod incorrectly adjusted	1 Adjust pushrod
2 Master cylinder pushrod (between booster and master cylinder) incorrectly adjusted	2 Adjust pushrod
3 Obstructed compensating port in master cylinder	3 Replace master cylinder
4 Master cylinder piston seized in bore	4 Replace master cylinder
5 Contaminated fluid causing swollen seals throughout system	5 Flush system, replace all hydraulic components
6 Clogged brake lines or internally split brake hose(s)	6 Flush hydraulic system, replace defective hose(s)
7 Sticking caliper(s) or wheel cylinder(s)	7 Replace calipers or wheel cylinders
8 Parking brake not releasing	8 Inspect parking brake linkage and parking brake mechanism, repair as required
9 Improper shoe-to-drum clearance	9 Adjust brake shoes
10 Faulty proportioning valve	10 Replace proportioning valve

PROBABLE CAUSE

CORRECTIVE ACTION

Brakes fade (due to excessive heat)

1 Brake linings excessively worn or glazed	1 Deglaze or replace brake pads and/or shoes
2 Excessive use of brakes	2 Downshift into a lower gear, maintain a constant slower speed (going down hills)
3 Vehicle overloaded	3 Reduce load
4 Brake drums or discs worn too thin	4 Measure drum diameter and disc thickness, replace drums or discs as required
5 Contaminated brake fluid	5 Flush system, replace fluid
6 Brakes drag	6 Repair cause of dragging brakes
7 Driver resting left foot on brake pedal	7 Don't ride the brakes

Brakes noisy (high-pitched squeal)

1 Glazed lining	1 Deglaze or replace lining
2 Contaminated lining (brake fluid, grease, etc.)	2 Repair source of contamination, replace linings
3 Weak or broken brake shoe hold-down or return spring	3 Replace springs
4 Rivets securing lining to shoe or backing plate loose	4 Replace shoes or pads
5 Excessive dust buildup on brake linings	5 Wash brakes off with brake system cleaner
6 Brake drums worn too thin	6 Measure diameter of drums, replace if necessary
7 Wear indicator on disc brake pads contacting disc	7 Replace brake pads
8 Anti-squeal shims missing or installed improperly	8 Install shims correctly

Brakes noisy (scraping sound)

1 Brake pads or shoes worn out; rivets, backing plate or brake shoe metal contacting disc or drum	1 Replace linings, have discs and/or drums machined (or replace)

Brakes chatter

1 Worn brake lining	1 Inspect brakes, replace shoes or pads as necessary
2 Glazed or scored discs or drums	2 Deglaze discs or drums with sandpaper (if glazing is severe, machining will be required)
3 Drums or discs heat checked	3 Check discs and/or drums for hard spots, heat checking, etc. Have discs drums machined or replace them
4 Disc runout or drum out-of-round excessive	4 Measure disc runout and/or drum out-of-round, have discs or drums machined or replace them
5 Loose or worn wheel bearings	5 Adjust or replace wheel bearings
6 Loose or bent brake backing plate (drum brakes)	6 Tighten or replace backing plate
7 Grooves worn in discs or drums	7 Have discs or drums machined, if within limits (if not, replace them)
8 Brake linings contaminated (brake fluid, grease, etc.)	8 Locate and repair source of contamination, replace pads or shoes
9 Excessive dust buildup on linings	9 Wash brakes with brake system cleaner
10 Surface finish on discs or drums too rough after machining (especially on vehicles with sliding calipers)	10 Have discs or drums properly machined
11 Brake pads or shoes glazed	11 Deglaze or replace brake pads or shoes

Troubleshooting (continued)

| PROBABLE CAUSE | CORRECTIVE ACTION |

Brake pads or shoes click

1 Shoe support pads on brake backing plate grooved or excessively worn	1 Replace brake backing plate
2 Brake pads loose in caliper	2 Loose pad retainers or anti-rattle clips
3 Also see items listed under Brakes chatter	

Brakes make groaning noise at end of stop

1 Brake pads and/or shoes worn out	1 Replace pads and/or shoes
2 Brake linings contaminated (brake fluid, grease, etc.)	2 Locate and repair cause of contamination, replace brake pads or shoes
3 Brake linings glazed	3 Deglaze or replace brake pads or shoes
4 Excessive dust buildup on linings	4 Wash brakes with brake system cleaner
5 Scored or heat-checked discs or drums	5 Inspect discs/drums, have machined if within limits (if not, replace discs or drums)
6 Broken or missing brake shoe attaching hardware	6 Inspect drum brakes, replace missing hardware

Rear brakes lock up under light brake application

1 Tire pressures too high	1 Adjust tire pressures
2 Tires excessively worn	2 Replace tires
3 Defective proportioning valve	3 Replace proportioning valve

Brake warning light on instrument panel comes on (or stays on)

1 Low fluid level in master cylinder reservoir (reservoirs with fluid level sensor)	1 Add fluid, inspect system for leak, check the thickness of the brake pads and shoes
2 Failure in one half of the hydraulic system	2 Inspect hydraulic system for a leak
3 Piston in pressure differential warning valve not centered	3 Center piston by bleeding one circuit or the other (close bleeder valve as soon as the light goes out)
4 Defective pressure differential valve or warning switch	4 Replace valve or switch
5 Air in the hydraulic system	5 Bleed the system, check for leaks
6 Brake pads worn out (vehicles with electric wear sensors - small probes that fit into the brake pads and ground out on the disc when the pads get thin)	6 Replace brake pads (and sensors)

Brakes do not self adjust

Disc brakes

1 Defective caliper piston seals	1 Replace calipers. Also, possible contaminated fluid causing soft or swollen seals (flush system and fill with new fluid if in doubt)
2 Corroded caliper piston(s)	2 Same as above

Drum brakes

1 Adjuster screw frozen	1 Remove adjuster, disassemble, clean and lubricate with high-temperature grease
2 Adjuster lever does not contact star wheel or is binding	2 Inspect drum brakes, assemble correctly or clean or replace parts as required
3 Adjusters mixed up (installed on wrong wheels after brake job)	3 Reassemble correctly
4 Adjuster cable broken or installed incorrectly (cable-type adjusters)	4 Install new cable or assemble correctly

PROBABLE CAUSE

CORRECTIVE ACTION

Rapid brake lining wear

1 Driver resting left foot on brake pedal	1 Don't ride the brakes
2 Surface finish on discs or drums too rough	2 Have discs or drums properly machined
3 Also see Brakes drag	

3   Anti-lock Brake System (ABS) - general information

General information

1   The anti-lock brake system is designed to maintain vehicle steerability, directional stability and optimum deceleration under severe braking conditions on most road surfaces. It does so by monitoring the rotational speed of each wheel and controlling the brake line pressure to each wheel during braking. This prevents the wheels from locking up.

2   The ABS system has three main components - the wheel speed sensors, the electronic control unit (ECU) and the hydraulic unit **(see illustration)**. Four wheel-speed sensors - one at each wheel - send a variable voltage signal to the control unit, which monitors these signals, compares them to its program and determines whether a wheel is about to lock up. When a wheel is about to lock up, the control unit signals the hydraulic unit to reduce hydraulic pressure (or not increase it further) at that wheel's brake caliper. Pressure modulation is handled by electrically-operated solenoid valves.

3   If a problem develops within the system, an "ABS" warning light will glow on the dashboard. Sometimes, a visual inspection of the ABS system can help you locate the problem. Carefully inspect the ABS wiring harness. Pay particularly close attention to the harness and connections near each wheel. Look for signs of chafing and other damage caused by incorrectly routed wires. If a wheel sensor harness is damaged, the sensor must be replaced. **Warning:** *Do NOT try to repair an ABS wiring harness. The ABS system is sensitive to even the smallest changes in resistance. Repairing the harness could alter resistance values and cause the system to malfunction. If the ABS wiring harness is damaged in any way, it must be replaced.* **Caution:** *Make sure the ignition is turned off*

3.2 The Brake Pressure Modulation Valve (BPMV) is the hydraulic unit that controls the brake line pressure to each wheel individually and is located below and to the side of the master cylinder

before unplugging or reattaching any electrical connections.

4   Some vehicles come equipped with traction control, which helps prevent wheel spin under acceleration. The Traction Control System (TCS) utilizes some of the same components as the ABS system. The wheel sensors monitor wheel rotation and transmit data through the same wiring to the ABS control module. The control module compares rotation speeds, then controls the amount of power supplied to the wheels by adjusting fuel settings to the engine.

Diagnosis and repair

5   If a dashboard warning light comes on and stays on while the vehicle is in operation, the ABS system requires attention. Although special electronic ABS diagnostic testing tools are necessary to properly diagnose the system, you can perform a few preliminary checks before taking the vehicle to a dealer service department.

   a) Check the brake fluid level in the reservoir.

   b) Verify that the computer electrical connectors are securely connected.

   c) Check the electrical connectors at the hydraulic control unit.

   d) Check the fuses.

   e) Follow the wiring harness to each wheel and verify that all connections are secure and that the wiring is undamaged.

6   If the above preliminary checks do not rectify the problem, the vehicle should be diagnosed by a dealer service department or other qualified repair shop. Due to the complex nature of this system, all actual repair work must be done by a qualified automotive technician.

Wheel speed sensor - removal and installation

7   Loosen the wheel lug nuts, raise the vehicle and support it securely on jackstands. Remove the wheel.

8   Make sure the ignition key is turned to the Off position.

9   Trace the wiring back from the sensor, detaching all brackets and clips while noting

3.10a Front wheel speed sensor mounting bolt

3.10b Rear wheel speed sensor mounting bolt

4.5 Always wash the brakes with brake cleaner before disassembling anything

4.7a Using two large C-clamps, depress the pistons into the bottom of their bores in the caliper to make room for the new pads (rear calipers have only one piston). Tighten the clamps evenly and make sure the fluid in the master cylinder reservoir doesn't overflow

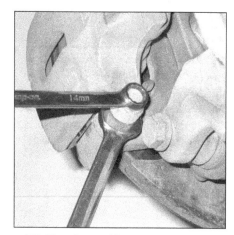

4.7b If you're working on a rear cailper, hold the caliper guide pin with an open-end wrench and unscrew the caliper lower mounting bolt with another wrench

its correct routing, then disconnect the electrical connector.

10 Remove the mounting bolt and carefully pull the sensor out from the knuckle **(see illustrations)**.

11 Installation is the reverse of removal. Tighten the mounting bolt securely.

12 Install the wheel and lug nuts, tightening them securely. Lower the vehicle and tighten the lug nuts to the torque listed in the Chapter 1 Specifications.

---

4   Disc brake pads - replacement

Warning: *Disc brake pads must be replaced on both front or both rear wheels at the same time - never replace the pads on only one wheel. Also, the dust created by the brake system is harmful to your health. Never blow it out with compressed air and don't inhale any of it. An approved filtering mask should be worn when working on the brakes. Do not, under any circumstances, use petroleum-based solvents to clean brake parts. Use*

brake system cleaner only!

1 Remove the cap from the brake fluid reservoir. To prevent brake fluid from possibly overflowing the reservoir when the calipers are fully retracted, use a syringe or suction gun to remove brake fluid until the reservoir is approximately half full.

2 Loosen the wheel lug nuts, raise the vehicle and support it securely on jackstands. Block the wheels at the opposite end.

3 Remove the wheels. Work on one brake assembly at a time, using the assembled brake for reference if necessary.

4 Inspect the brake disc carefully (see Section 6). If machining is necessary, follow the information in that Section to remove the disc, at which time the pads can be removed as well.

5 Before disassembling the brake caliper, wash it thoroughly with brake system cleaner and allow it to dry **(see illustration)**. Position a drain pan under the brake to catch the resi-

due - DO NOT use compressed air to blow off the brake dust.

6 When working on the brakes, keep these points in mind:

   a) *Use only hand tools to remove or install the caliper mounting bolts. Do not use air tools or the caliper could be damaged.*

   b) *When removing or installing the rear caliper mounting bolts, use an open end wrench to hold the guide pin from rotating. Do not allow the open-end wrench to contact the caliper when tightening or it could cause the brakes to pulsate.*

   c) *Do not disconnect the brake line from the caliper unless you are replacing the caliper.*

   d) *If you are removing the caliper but not replacing the pads, note the position of the inner and outer pads so you can reinstall them in the same locations.*

7 For the brake pad replacement sequence, follow the accompanying photos **(illustrations 4.7a through 4.7f)**. Stay in order and read the caption under each illustration, then proceed to Step 8.

4.7c Remove the caliper lower mounting bolt, then rotate the caliper up for access to the pads. Hold the caliper up with a length of wire or a coathanger

4.7d Remove the inner brake pad

8   To re-assemble the brake, replace the pads and reverse the steps above. Coat the guide pin with high-temperature brake grease before re-installing it into the mounting bracket **(see illustrations)**.

9   When reinstalling the caliper, tighten the mounting bolts to the torque listed in this Chapter's Specifications. After the job has been completed, firmly depress the brake pedal a few times to bring the pads into contact with the disc. Check the level of the brake fluid, adding some if necessary. Check the operation of the brakes carefully before placing the vehicle into normal service.

4.7e Remove the outer brake pad

4.7f Remove the shims

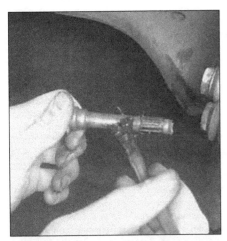

4.8a Clean the caliper guide pins, apply a coat of high-temperature grease to the pins, then reinstall them

4.8b Be sure the guide pin boot is properly seated in this groove on the guide pin

4.8c With the new pads and shims installed in the caliper mounting bracket, swing the caliper down and tighten the caliper mounting bolt to the torque listed in this Chapter's Specifications

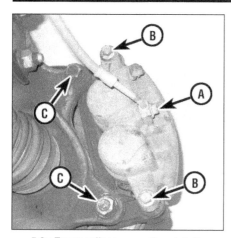

5.2a Front caliper mounting details

A   *Banjo bolt*
B   *Caliper mounting bolts*
C   *Caliper mounting bracket bolts*

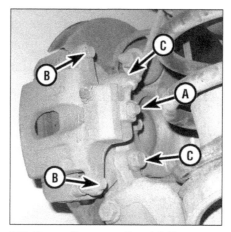

5.2b Rear caliper mounting details

A   *Banjo bolt*
B   *Caliper mounting bolts*
C   *Caliper mounting bracket bolts*

6.3 The brake pads on this vehicle were obviously neglected, as they wore down completely and cut deep grooves into the disc - wear this severe means the disc must be replaced

6.4a To check disc runout, mount a dial indicator as shown and rotate the disc

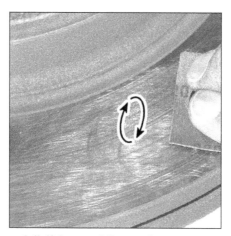

6.4b Using a swirling motion, remove the glaze from the disc surface with sandpaper or emery cloth

Installation

5   Installation is the reverse of removal. Use a high-temperature threadlock on the caliper mounting bracket bolts.

6   When installing the caliper to the bracket, start with the bolt closest to the bleed valve and tighten the guide pin bolts to the torque listed in this Chapter's Specifications.

7   Bleed the brake system (see Section 9). Make sure there are no leaks from the hose connections. Test the brakes carefully before returning the vehicle to normal service.

---

6   **Brake disc - inspection, removal and installation**

Inspection

1   Loosen the wheel lug nuts, raise the vehicle and support it securely on jackstands.

2   Remove the brake caliper (see Section 5). It isn't necessary to disconnect the brake hose. After removing the caliper bolts, suspend the caliper out of the way with a piece of wire **(see illustration 4.7c)**.

3   Visually inspect the disc surface for score marks and other damage. Light scratches and shallow grooves are normal after use and may not always be detrimental to brake operation, but deep scoring requires disc removal and refinishing by an automotive machine shop. Be sure to check both sides of the disc **(see illustration)**. If pulsating has been noticed during application of the brakes, suspect disc runout.

4   To check disc runout, reinstall the lug nuts (inverted) and place a dial indicator at a point about 1/2-inch from the outer edge of the disc **(see illustration)**. Set the indicator to zero and turn the disc. The indicator reading should not exceed the specified allowable runout limit. If it does, the disc should be refinished by an automotive machine shop.

Note: *The discs should be resurfaced regard-*

---

5   Disc brake caliper and caliper mounting bracket - removal and installation

Warning: *Dust created by the brake system is harmful to your health. Never blow it out with compressed air and don't inhale any of it. An approved filtering mask should be worn when working on the brakes. Do not, under any circumstances, use petroleum-based solvents to clean brake parts. Use brake system cleaner only.*

Note: *If replacement is indicated (usually because of fluid leakage or damage to a piston boot), it is recommended that the calipers be replaced, not overhauled. New and factory rebuilt units are available on an exchange basis, which makes this job quite easy. Always replace the calipers in pairs - never replace just one of them.*

Removal

1   Loosen the wheel lug nuts, raise the vehicle (front or rear) and place it securely on

jackstands. Remove the wheels.

2   Remove the banjo fitting bolt and disconnect the brake hose from the caliper **(see illustrations)**. Discard the sealing washers from each side of the hose fitting. Plug the brake hose to keep contaminants out of the brake system and to prevent losing any more brake fluid than is necessary.

Note: *If the caliper is being removed for access to another component, don't disconnect the hose.*

3   Follow **illustrations 4.7a through 4.7c** for the caliper removal procedure. Use only hand tools. Remove the upper guide pin bolt in the same manner as the lower. If the caliper is being removed just to access another component, use a piece of wire to securely hang it out of the way.

Caution: *Do not let the caliper hang by the brake hose.*

4   To remove the caliper mounting bracket, remove the two bracket bolts. Clean any remaining thread locking material from the bracket and the bolts before assembly.

6.5a The minimum thickness dimension is cast into the front or back side of the disc

6.5b Use a micrometer to measure disc thickness

less of the dial indicator reading, as this will impart a smooth finish and ensure a perfectly flat surface, eliminating any brake pedal pulsation or other undesirable symptoms related to questionable discs. At the very least, if you elect not to have the discs resurfaced, remove the glaze from the surface with emery cloth or sandpaper, using a swirling motion (see illustration).

5   It's absolutely critical that the disc not be machined to a thickness under the specified minimum thickness. The minimum (or discard) thickness is cast or stamped into the inside of the disc (see illustration). The disc thickness can be checked with a micrometer (see illustration).

Removal

6   Remove the caliper mounting bracket (the caliper and caliper mounting bracket can be removed as one). Also remove the lug nuts if they were reinstalled for the runout inspection (see illustration).

7   Mark the disc at one of the wheel studs so it can be reinstalled in the same position, then remove the disc retaining screw.

Note: If the rear disc is difficult to remove, it may be necessary to adjust the parking brake shoes away from the inner drum (see Section 12).

Installation

8   While the disc is off, wire-brush the backside of the center portion that contacts the wheel hub. Also clean off any rust or dirt on the hub face.

9   Apply small dots of high-temperature anti-seize around the circumference of the hub, and around the raised center portion.

10   Align your index marks and place the disc in position over the threaded studs. Tighten the disc retaining screw to the torque listed in this Chapter's Specifications.

Note: If the parking brake shoes were adjusted for removal of the rear disc, adjust the shoes back into position and verify the proper operation of the parking brake (see Section 12).

11   Install the caliper mounting bracket and caliper, tightening the bolts to the torque values listed in this Chapter's Specifications.

12   Install the wheel, then lower the vehicle to the ground. Tighten the lug nuts to the torque listed in the Chapter 1 Specifications. Depress the brake pedal a few times to bring the brake pads into contact with the disc. Bleeding won't be necessary unless the brake hose was disconnected from the caliper. Check the operation of the brakes carefully before driving the vehicle.

13   Check the operation of the brakes carefully before driving the vehicle.

---

7   Master cylinder - removal and installation

Removal

1   The master cylinder is located in the engine compartment, mounted to the power brake booster.

2   Be sure the ignition switch is turned to OFF.

3   Remove the air filter element (see Chapter 4). On 2011 and later models, the air filter cover and outlet duct will also need to be removed.

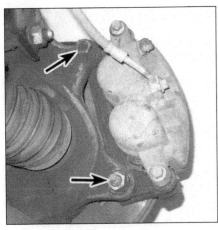

6.6 Caliper mounting bracket-to-knuckle bolts

6.7 Remove the screw that secures the disc to the hub

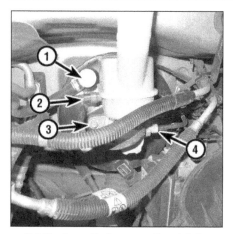

7.5 Master cylinder mounting details

1 Brake booster check valve
2 Fluid level warning switch connector
3 Mounting nut (other nut not visible in photo)
4 Brake line fitting (one of two shown)

4 Remove as much fluid as you can from the reservoir with a syringe, such as an old turkey baster.

Warning: *If a baster is used, never again use it for the preparation of food.*

5 Disconnect the brake booster check valve from the brake booster to remove any vacuum from the booster **(see illustration)**.

6 Disconnect the electrical connector at the brake fluid level switch on the master cylinder reservoir.

7 Remove the windshield washer fluid heater (if equipped).

8 Place rags under the fluid fittings and prepare caps or plastic bags to cover the ends of the lines once they are disconnected.

9 Clean off any dirt or debris from around the brake line fittings to prevent it from entering the system.

Caution: *Brake fluid will damage paint. Cover all body parts and be careful not to spill fluid*

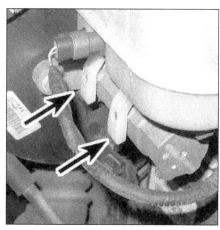

7.12 These tabs must fasten tightly around the posts on the master cylinder (two others are not visible in this photo)

during this procedure.

10 Loosen the fittings at the ends of the brake lines where they enter the master cylinder **(see illustration 7.5)**. To prevent rounding off the corners on these nuts, the use of a flare-nut wrench, which wraps around the nut, is preferred. Pull the brake lines slightly away from the master cylinder and plug the ends to prevent contamination.

11 Remove the nuts attaching the master cylinder to the power booster. Pull the master cylinder off the studs and out of the engine compartment. Again, be careful not to spill the fluid as this is done.

12 If a new master cylinder is being installed, carefully spread the four tabs on the bottom of the reservoir and separate the reservoir from the old master cylinder. Lubricate new reservoir seals with a little brake fluid and install them on the new master cylinder. Secure the reservoir to the master cylinder by locking the four tabs in against the master cylinder housing **(see illustration)**.

Note: *Install new seals when transferring the reservoir.*

Installation

13 Bench bleed the new master cylinder before installing it. Mount the master cylinder in a vise, with the jaws of the vise clamping on the mounting flange.

14 Attach a pair of master cylinder bleeder tubes to the outlet ports of the master cylinder **(see illustration)**.

15 Fill the reservoir with brake fluid of the recommended type (see Chapter 1).

16 Slowly push the pistons into the master cylinder (a large Phillips screwdriver can be used for this) - air will be expelled from the pressure chambers and into the reservoir. Because the tubes are submerged in fluid, air can't be drawn back into the master cylinder when you release the pistons.

17 Repeat the procedure until no more air bubbles are present.

18 Remove the bleed tubes, one at a time, and install plugs in the open ports to prevent fluid leakage and air from entering. Install the reservoir cap.

19 Install the master cylinder over the studs on the power brake booster and tighten the attaching nuts only finger tight at this time.

Note: *Install a new O-ring onto the sleeve of the master cylinder* **(see illustration)**.

20 Thread the brake line fittings into the master cylinder. Since the master cylinder is still a bit loose, it can be moved slightly in order for the fittings to thread in easily. Do not strip the threads as the fittings are tightened.

21 Tighten the mounting nuts to the torque listed in this Chapter's Specifications, then tighten the brake line fittings securely.

22 Fill the master cylinder reservoir with fluid, then bleed the master cylinder and the brake system (see Section 9). To bleed the cylinder on the vehicle, have an assistant depress the brake pedal and hold the pedal to the floor. Loosen the fitting to allow air and fluid to escape. Repeat this procedure on both fittings until the fluid is clear of air bubbles.

Caution: *Have plenty of rags on hand to catch the fluid - brake fluid will ruin painted surfaces. After the bleeding procedure is completed,*

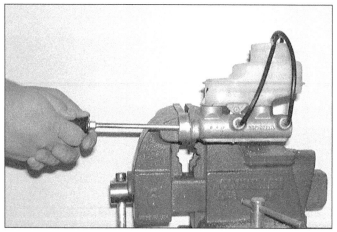

7.14 The best way to bleed air from the master cylinder before installing it on the vehicle is with a pair of bleeder tubes that direct brake fluid into the reservoir during bleeding

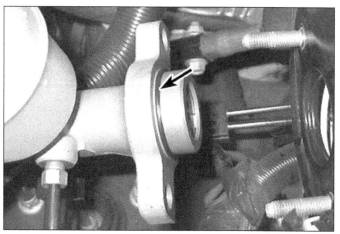

7.19 Install a new O-ring onto the master cylinder sleeve

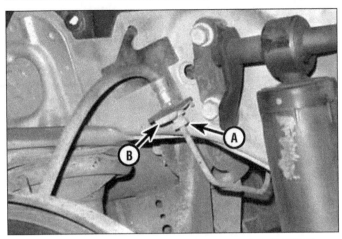

8.3 Unscrew the brake line fitting (A), remove the clip (B), then remove the banjo bolt and detach the hose from the caliper

9.8 When bleeding the brakes, a hose is connected to the bleed screw at the caliper and submerged in brake fluid - air will be seen as bubbles in the tube and container (all air must be expelled before moving to the next wheel)

rinse the area under the master cylinder with clean water.

23   Test the operation of the brake system carefully before placing the vehicle into normal service.

Warning: *Do not operate the vehicle if you are in doubt about the effectiveness of the brake system. It is possible for air to become trapped in the anti-lock brake system hydraulic control unit; if the pedal continues to feel spongy after repeated bleedings or the BRAKE or ANTI-LOCK light stays on, have the vehicle towed to a dealer service department or other qualified shop to be bled with the aid of a scan tool.*

---

8   Brake hoses and lines - inspection and replacement

---

1   About every six months, with the vehicle raised and placed securely on jackstands, the flexible hoses which connect the steel brake lines with the front and rear brake assemblies should be inspected for cracks, chafing of the outer cover, leaks, blisters and other damage. These are important and vulnerable parts of the brake system and inspection should be complete. A light and mirror will be needed for a thorough check. If a hose exhibits any of the above defects, replace it with a new one.

Flexible hoses

2   Clean all dirt away from the ends of the hose.

3   To remove a brake hose, unscrew the tube nut with a flare-nut wrench, if available, to prevent rounding-off the corners of the nut, then remove the bolt(s) or clip(s) securing the hose to the body (and any suspension components) **(see illustration).**

4   Disconnect the hose from the caliper, discarding the sealing washers on either side of the fitting.

5   Using new sealing washers, attach the new brake hose to the caliper. Tighten the

banjo fitting bolt to the torque listed in this Chapter's Specifications.

6   Reverse the removal procedure to install the hose, making sure it isn't twisted.

7   Carefully check to make sure the suspension or steering components don't make contact with the hose. Have an assistant push down on the vehicle and also turn the steering wheel lock-to-lock during inspection.

8   Bleed the brake system (see Section 9).

Metal brake lines

9   When replacing brake lines, be sure to use the correct parts. Don't use copper tubing for any brake system components. Purchase steel brake lines from a dealer parts department or auto parts store.

10   Prefabricated brake line, with the tube ends already flared and fittings installed, is available at auto parts stores and dealer parts departments. These lines can be bent to the proper shapes using a tubing bender.

11   When installing the new line make sure it's well supported in the brackets and has plenty of clearance between moving or hot components.

12   After installation, check the master cylinder fluid level and add fluid as necessary. Bleed the brake system (see Section 9) and test the brakes carefully before placing the vehicle into normal operation.

---

9   Brake hydraulic system - bleeding

Warning: *If air has found its way into the hydraulic control unit, the system must be bled with the use of a scan tool. If the brake pedal feels spongy even after bleeding the brakes, or the ABS light on the instrument panel does not go off, or if you have any doubts whatsoever about the effectiveness of the brake system, have the vehicle towed to a dealer service department or other repair shop equipped with*

the necessary tools for bleeding the system.
Warning: *Wear eye protection when bleeding the brake system. If the fluid comes in contact with your eyes, immediately rinse them with water and seek medical attention.*
Note: *Bleeding the brake system is necessary to remove any air that's trapped in the system when it's opened during removal and installation of a hose, line, caliper, wheel cylinder or master cylinder.*

1   It will probably be necessary to bleed the system at all four brakes if air has entered the system due to low fluid level, or if the brake lines have been disconnected at the master cylinder.

2   If a brake line was disconnected only at a wheel, then only that caliper or wheel cylinder must be bled.

3   If a brake line is disconnected at a fitting located between the master cylinder and any of the brakes, that part of the system served by the disconnected line must be bled.

4   Remove any residual vacuum (or hydraulic pressure) from the brake power booster by applying the brake several times with the engine off.

5   Remove the master cylinder reservoir cap and fill the reservoir with brake fluid. Reinstall the cap.

Note: *Check the fluid level often during the bleeding operation and add fluid as necessary to prevent the fluid level from falling low enough to allow air bubbles into the master cylinder.*

6   Have an assistant on hand, as well as a supply of new brake fluid, an empty clear plastic container, a length of plastic, rubber or vinyl tubing to fit over the bleeder valve and a wrench to open and close the bleeder valve.

7   Beginning at the right rear wheel, loosen the bleeder screw slightly, then tighten it to a point where it's snug but can still be loosened quickly and easily.

8   Place one end of the tubing over the bleeder screw fitting and submerge the other end in brake fluid in the container **(see illustration).**

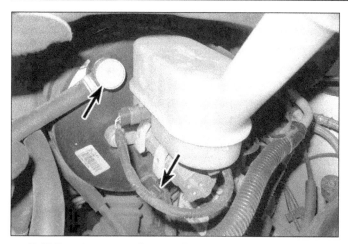

10.10 Booster vacuum hose and sensor electrical connector

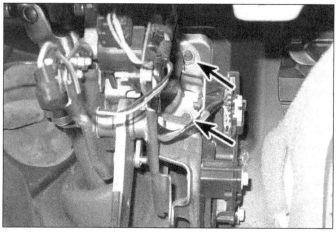

10.17 To detach the power brake booster from the firewall, remove the four nuts (two are not visible here)

9 Have the assistant slowly depress the brake pedal and hold it in the depressed position.

10 While the pedal is held depressed, open the bleeder screw just enough to allow a flow of fluid to leave the valve. Watch for air bubbles to exit the submerged end of the tube. When the fluid flow slows after a couple of seconds, tighten the screw and have your assistant release the pedal.

11 Repeat Steps 9 and 10 until no more air is seen leaving the tube, then tighten the bleeder screw and proceed to the left front wheel, the left rear wheel and the right front wheel, in that order, and perform the same procedure. Check the fluid in the master cylinder reservoir frequently.

12 Never use old brake fluid. It contains moisture which can boil, rendering the brake system inoperative.

13 Refill the master cylinder with fluid at the end of the operation.

14 Check the operation of the brakes. The pedal should feel solid when depressed, with no sponginess. If necessary, repeat the entire process.

Warning: *Do not operate the vehicle if you are in doubt about the effectiveness of the brake system. It is possible for air to become trapped in the anti-lock brake system hydraulic control unit, so, if the pedal continues to feel spongy after repeated bleedings or the BRAKE or ANTI-LOCK light stays on, have the vehicle towed to a dealer service department or other qualified shop to be bled with the aid of a scan tool.*

10   Power brake booster - check, removal and installation

Operating check

1 Depress the brake pedal several times with the engine off and make sure that there is no change in the pedal reserve distance.

2 Depress the pedal and start the engine. If the pedal goes down slightly, operation is normal.

Air tightness check

3 Start the engine and turn it off after one or two minutes. Depress the brake pedal several times slowly. If the pedal goes down farther the first time but gradually rises after the second or third depression, the booster is airtight.

4 Depress the brake pedal while the engine is running, then stop the engine with the pedal depressed. If there is no change in the pedal reserve travel after holding the pedal for 30 seconds, the booster is airtight.

Removal and installation

5 Disassembly of the power unit requires special tools and is not ordinarily performed by the home mechanic. If a problem develops, it's recommended that a new or factory rebuilt unit be installed.

6 Set the parking brake with the vehicle parked on a level surface.

7 With the ignition switch OFF, pump the brake pedal a few times to deplete any remaining vacuum in the booster.

8 Remove the air filter housing assembly.

9 Disconnect the electrical connector at the brake fluid level sensor on the side of the brake fluid reservoir.

10 Disconnect the vacuum hose and the vacuum sensor electrical connector from the booster **(see illustration).**

11 Release the four brake lines from the retaining clip on the left side frame rail under the booster.

Note: *It is not necessary to disconnect any brake hydraulic lines in order to remove the brake booster unit. The brake system will not require bleeding unless brake line fittings are loosened or disconnected.*

12 Locate the Brake Pressure Modulator Valve (BPMV) assembly below and to the side of the master cylinder. Without disconnecting any brake lines from the BPMV, remove the two mounting nuts and lift the assembly off the studs to position it slightly forward.

13 Without disconnecting the brake lines, remove the two master cylinder mounting nuts and position the master cylinder to the side.

Keep the master cylinder upright and support it with mechanic's wire.

14 Under the dash, remove the driver's side instrument panel insulator cover and the driver's side knee bolster (see Chapter 11).

15 Remove the retaining clip and clevis pin, then disconnect the pushrod from the brake pedal.

Note: *2011 and later models utilize a one-piece clevis locking pin to connect the pushrod to the brake pedal. Use a 12-point, 11 mm wrench or socket to release the tabs on the pin. The pin is a one-time-use part and must be discarded and replaced if removed.*

16 Remove the intermediate steering shaft (see Chapter 10).

17 Remove the booster mounting nuts from inside **(see illustration)**, then carefully tilt the booster slightly upward to clear the mounting studs and lift the unit out of the engine compartment.

18 Installation is the reverse of removal. Inspect the gasket seal between the booster and dash panel and replace if necessary. Ensure the O-ring is properly positioned on the master cylinder sleeve. Test the operation of the brakes before placing the vehicle in normal service.

11   Vacuum auxiliary pump

1 With the ignition switch turned to OFF, pump the brake pedal several times to deplete any remaining vacuum.

2 The vacuum pump is located at the left front corner of the engine compartment. Remove the engine cover to gain access to the vacuum pump.

3 Disconnect the vacuum hose from the vacuum pump by releasing the quick connect fitting.

4 Position the power brake booster vacuum hose to the side.

5 Disconnect the electrical connector to the vacuum pump and, if equipped, release the wiring from the bracket.

6 Remove the nuts and pull the pump from

12.6a Remove the cover from the adjusting hole on the
rear brake disc

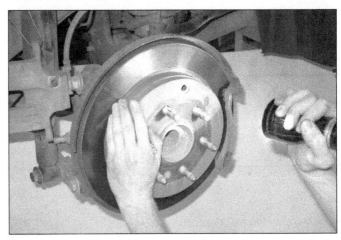

12.6b Shine a flashlight through the hole and align it with the
brake adjuster wheel inside the disc

12.6c Insert a brake adjusting tool or a flat blade screwdriver into
the hole in the brake disc and turn the adjuster wheel

13.5 Unhook the upper spring from the shoes. Remove
the adjuster assembly; turn the adjuster to make it shorter
if necessary

the engine compartment. On some models, it may be necessary to remove the bolts from the pump bracket, then remove the pump and bracket together. Retain the rubber mounting bushings and their sleeves.

7   Installation is the reverse of removal. If the replacement pump did not come with the rubber mounting bushings, transfer the ones from the old pump. Tighten the fasteners securely and check for proper operation of the brake booster system.

12 Parking brake - adjustment

1   The parking brake should be adjusted first at the rear wheels, then at the parking brake equalizer under the vehicle.

2   Block the front wheels to prevent the vehicle from rolling. Verify that the parking brake is fully released and the pedal is against the return stop.

3   Raise the rear of the vehicle and support it securely on jackstands.

4   The rear wheels can remain on the

vehicle. If the brake disc has been removed, install the disc onto the hub and temporarily thread three of the wheel lug nuts onto the studs to hold the disc in place.

5   If you have just replaced the parking brake shoes, you may need to loosen the adjusting nut on the parking brake cable equalizer to allow the new shoes to retract fully.

6   Remove the hole plugs from the brake discs. Use a brake adjusting tool or a screwdriver to turn the adjuster star wheel so that the shoes contact the discs until the discs can't be turned **(see illustrations)**.

7   Turn the adjuster in the opposite direction until you can rotate the disc without resistance (usually about 4 clicks). Install the backing plate hole plugs.

13 Parking brake shoes - replacement

1   Loosen the wheel lug nuts, raise the rear of the vehicle and support it securely on jackstands. Remove both rear wheels but finish one side completely before disassembling

the other in case you need to use the completed side for reference.

2   Remove the caliper bracket (see Section 5).

3   Remove the brake disc (see Section 6). Adjust the parking brake shoes inward (away from the drum) to allow the disc to be removed.

4   Place a drain pan under the wheel you are working on and wash the parking brake assembly with brake system cleaner.

Caution: *Do not use compressed air to blow dust from the brake components. Brake dust is harmful to your health and should not be inhaled.*

5   Remove the upper return spring and adjuster wheel assembly from between the parking brake shoes. **(see illustration)**.

6   On 2012 and later models, compress the hold-down spring and rotate it 1/4 turn. Hold the pin from behind the backing plate while compressing and rotating the spring and retainer cup. Remove the front shoe hold-down spring the same way, then spread the shoes apart at the top and remove both shoes and the lower return spring.

13.7 Press and rotate the hold-down spring retainers or the spring clips. You may need to push on the pin from the back of the backing plate to keep it from rotating

13.8 Lightly grease the parking brake shoe contact areas on the backing plate with high-temperature brake grease

7   On 2011 and earlier models, remove the rear parking brake shoe hold down spring by compressing and rotating the spring 1/4 turn **(see illustration)**. Remove the rear shoe and lower return spring, then release the front shoe hold down spring in the same manner and remove the front shoe.

8   Clean the backing plate and apply a light coating of high-temperature brake grease to the areas on the backing plate where the brake shoes contact. **(see illustration).**

9   Installation is the reverse of removal. Rotate the star wheel of the adjuster assembly to make it as short as possible. When attaching the brake shoes to the backing plate, be sure they are seated properly in the brake actuator.

10   Adjust the parking brakes (see Section 12).

11   Tighten the wheel lug nuts to the torque listed in the Chapter 1 Specifications.

Chapter 10
Suspension and steering systems

Contents

	Section
Balljoints - check and replacement	6
Coil spring (rear) - removal and installation	11
Control arm (front) - removal, inspection and installation	5
General information	1
Hub and bearing assembly (front) - removal and installation	8
Hub and bearing assembly (rear) - removal and installation	12
Power steering pump - removal and installation	20
Power steering system - bleeding	21
Rear knuckle - removal and installation	13
Shock absorber (rear) - removal and installation	9
Stabilizer bar and link (front) - removal and installation	4
Stabilizer bar and links (rear) - removal and installation	14
Steering column - removal and installation	16

	Section
Steering gear - removal and installation	19
Steering gear boots - replacement	18
Steering knuckle - removal and installation	7
Steering wheel - removal and installation	15
Strut/coil spring assembly (front) - removal, inspection and installation	2
Strut/coil spring assembly (front) - replacement	3
Subframe (front) - removal and installation	22
Suspension arms (rear) - removal and installation	10
Tie-rod ends - removal and installation	17
Wheel alignment - general information	24
Wheels and tires - general information	23

Specifications

Torque specifications

Note: *One foot-pound (ft-lb) of torque is equivalent to 12 inch-pounds (in-lbs) of torque. Torque values below approximately 15 foot-pounds are expressed in inch-pounds, because most foot-pound torque wrenches are not accurate at these smaller values.*

	Ft-lbs (unless otherwise indicated)	Nm
Front suspension		
Balljoint-to-control arm mounting fasteners (replacement)	37	50
Subframe mounting bolts		
2009 and earlier models		
Step 1	74	100
Step 2	Tighten an additional 90 degrees	
2010 and later models		
Step 1	81	110
Step 2	Tighten an additional 90 degrees	
Strut		
Upper mounting nuts	33	45
Strut-to-steering knuckle bolt/nuts	144	195
Strut piston rod nut	63	85
Hub and bearing assembly mounting bolts		
2009 and earlier models	103	140
2010 and later models	89	121

Torque specifications

Ft-lbs (unless otherwise indicated)   **Nm**

Note: *One foot-pound (ft-lb) of torque is equivalent to 12 inch-pounds (in-lbs) of torque. Torque values below approximately 15 foot-pounds are expressed in inch-pounds, because most foot-pound torque wrenches are not accurate at these smaller values.*

Front suspension (continued)

Balljoint-to-steering knuckle nut
   2007 models
      Step 1 30   40
      Step 2 .... Tighten an additional 120 degrees
   2008 models
      Step 1 30   40
      Step 2
         Black nut... Tighten an additional 90 degrees
         Silver nut .... Tighten an additional 120 degrees
   2009 through 2013 models
      Step 1 30   40
      Step 2 .... Tighten an additional 90 degrees
   2014 and later models
      Step 1 37   52
      Step 2 .... Tighten an additional 70 degrees
Control arm pivot bolt nuts
   Front 144   195
   Rear 111   150
Control arm bushing-to-subframe bolts/nuts 55   75
Stabilizer bar link nuts
   To stabilizer bar.. 55   75
   To strut
      2009 and earlier models 52   71
      2010 and later models 59   80
Stabilizer bar bushing retainer-to-subframe bolts.................................... 37   50
Driveaxle/hub nut.. See Chapter 8

Rear suspension

Shock absorber
   Lower mounting bolt
      Step 1 74   100
      Step 2
         2009 and earlier models... Tighten an additional 60 degrees
         2010 and later models.. Tighten an additional 90 degrees
   Upper mounting nuts
      2009 and earlier models 52   71
      2010 and later models 54   73
Lower control arm
   To crossmember
      Front bolts/nuts
         Step 1 81   110
         Step 2 .... Tighten an additional 45 degrees
      Rear bolts/nuts
         Step 1 89   121
         Step 2 .... Tighten an additional 60 degrees
   To rear knuckle
      Step 1 74   100
      Step 2
         2009 and earlier models... Tighten an additional 60 degrees
         2010 and later models.. Tighten an additional 90 degrees
Upper control arm
   To crossmember nuts... 103   140
   To knuckle
      Step 1 74   100
      Step 2 .... Tighten an additional 90 degrees
Stabilizer bar
   Link nuts... 16   22
   Bushing bracket bolts.. 37   50
Rear wheel hub/bearing assembly-to-knuckle bolts 103   140
Hub and bearing assembly bolts
   2009 and earlier models... 103   140
   2010 and later models... 96   130

Torque specifications	Ft-lbs (unless otherwise indicated)	Nm

Rear suspension (continued)
Adjuster link
    To knuckle

Step 1	55	75
Step 2	Tighten an additional 60 degrees	
To crossmember	103	140
Driveaxle/hub nut	See Chapter 8	

Steering
Intermediate shaft

Lower pinch bolt	18	25
Upper pinch bolt	16	22
Tie-rod end-to-knuckle nut		
Step 1	22	30
Step 2	Tighten an additional 120 degrees	
Steering column mounting bolts/nuts	20	27
Power steering pump mounting bolts	37	50
Power steering gear inlet hose fitting	25	34
Power steering gear fluid line retaining plate bolt	106 in-lbs	12
Steering gear outlet hose bracket bolt	80 in-lbs	9
Steering wheel bolt	37	50
Steering gear fasteners		
Left side (bolts/nuts)	133	180
Right side (bolts)	74	100
Steering column lower support bracket (if equipped)	89 in-lbs	10

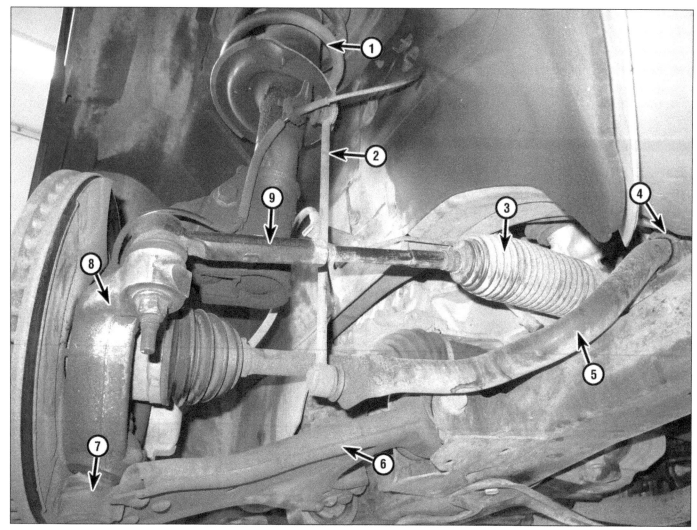

1.1 Front suspension and steering components

1	Strut/coil spring assembly	4	Stabilizer bar bushing	7	Balljoint
2	Stabilizer bar link	5	Stabilizer bar	8	Steering knuckle
3	Steering gear boot	6	Control arm	9	Tie-rod end

1   General information

1   The front suspension is made up of a strut assembly, steering knuckle/hub assembly, a control arm and a stabilizer bar **(see illustration)**. The strut assembly is made up of a shock absorber and a coil spring. The strut combines several functions into one by supporting the weight of the vehicle, function-ing as the spring damper and providing the pivot point for the steering knuckle.

2   The rear suspension employs upper and lower control arms, an adjuster link, an upper-to-lower control arm link, a stabilizer bar and links, and a coil spring and shock absorber on each side **(see illustrations)**.

3   The rack-and-pinion steering gear is located on the front suspension subframe and actuates the tie-rods, which are attached to the steering knuckles. The inner ends of the tie-rods are protected by rubber boots which should be inspected periodically for secure attachment, tears and leaking lubricant (which would indicate a failed rack seal).

4   The power assist system consists of a belt-driven pump and associated lines and hoses. The fluid level in the power steering pump reservoir should be checked periodi-cally (see Chapter 1).

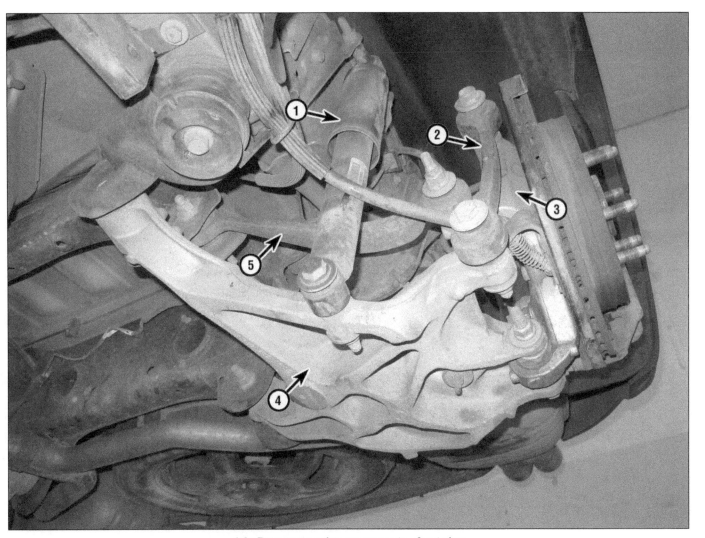

1.2a Rear suspension components - front view

1	Shock absorber	3	Rear knuckle
2	Upper control arm-to-lower	4	Lower control arm
	control arm link	5	Adjuster link

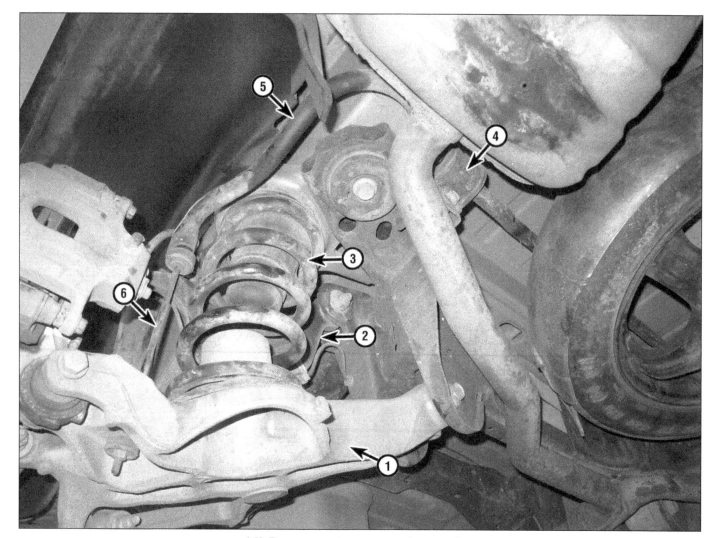

1.2b Rear suspension components - rear view

1	Lower control arm	3	Coil spring	5	Stabilizer bar
2	Upper control arm	4	Stabilizer bar bushing	6	Stabilizer bar link

5 The steering wheel operates the steering shaft, which actuates the steering gear through universal joints. Looseness in the steering can be caused by wear in the steering shaft universal joints, the steering gear, the tie-rod ends and loose retaining bolts.

Precautions

6 Frequently, when working on the suspension or steering system components, you may come across fasteners which seem impossible to loosen. These fasteners on the underside of the vehicle are continually subjected to water, road grime, mud, etc., and can become rusted or frozen, making them extremely difficult to remove. In order to unscrew these stubborn fasteners without damaging them (or other components), use lots of penetrating oil and allow it to soak in for a while. Using a wire brush to clean exposed threads will also ease removal of the nut or bolt and prevent damage to the threads. Sometimes a sharp blow with a hammer and punch will break the bond between a nut and bolt threads, but care must be taken to prevent the punch from slipping off the fastener and ruining the threads. Heating the stuck fastener and surrounding area with a torch sometimes helps too, but isn't recommended because of the obvious dangers associated with fire. Long breaker bars and extension, or cheater, pipes will increase leverage, but never use an extension pipe on a ratchet - the ratcheting mechanism could be damaged. Sometimes tightening the nut or bolt first will help to break it loose. Fasteners that require drastic measures to remove should always be replaced with new ones.

7 Since most of the procedures dealt with in this Chapter involve jacking up the vehicle and working underneath it, a good pair of jackstands will be needed. A hydraulic floor jack is the preferred type of jack to lift the vehicle, and it can also be used to support certain components during various operations.
Warning: *Never, under any circumstances, rely on a jack to support the vehicle while working on it. Whenever any of the suspension or steering fasteners are loosened or removed they*

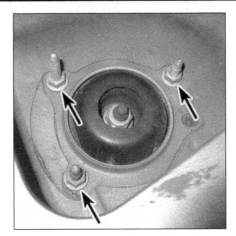

2.1 Upper strut mounting nut locations

must be inspected and, if necessary, replaced with new ones of the same part number or of original equipment quality and design. Torque specifications must be followed for proper reassembly and component retention. Never attempt to heat or straighten any suspension or steering components. Instead, replace any bent or damaged part with a new one.

---

2 Strut/coil spring assembly (front) - removal, inspection and installation

---

Removal

1 Remove the cowl cover (see Chapter 11), then loosen (but don't remove) the strut upper mounting nuts **(see illustration)**.
2 Loosen the wheel lug nuts, raise the vehicle and support it securely on jackstands. Remove the wheel.
3 Detach the wheel speed sensor harness from the strut body.
4 Mark the position of the knuckle to the strut by scribing or drawing a line on the steering knuckle following the line of the strut **(see illustration)**.

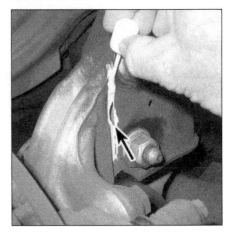

2.4 Mark the relationship of the steering knuckle to the strut

5 Detach the stabilizer bar link from the strut (see Section 4).
6 Support the steering knuckle with a floor jack.
7 Unscrew the nuts securing the strut to the steering knuckle, then tap the bolts out with a hammer **(see illustrations)**.
Caution: *Only turn the nuts; the bolts have serrated shoulders, and turning them could damage the steering knuckle.*
8 Use the floor jack to lower the knuckle to allow room to remove the strut assembly, but be careful not to overextend the inner CV joint. To avoid brake line stress and possible damage, only lower the floor jack enough to clear the strut for removal. Failure to do so could result in fluid leakage and brake system failure.
9 Remove the top nuts (loosened in Step 1), and pull out the strut.

Inspection

10 Check the shock body for leaking fluid, dents, cracks and other obvious damage which would warrant replacement.
11 Check the coil spring for chips or cracks in the spring coating (this will cause prema-

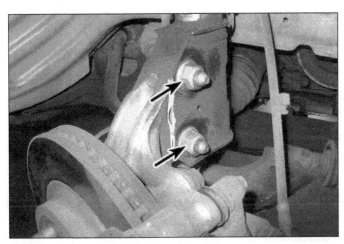

2.7a Unscrew the nuts . . .

2.7b . . . then drive the bolts out with a hammer (don't turn the bolts - they have serrated shoulders)

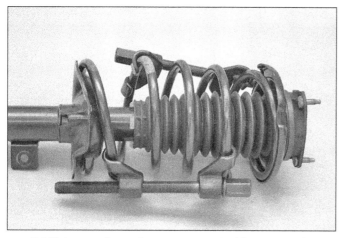

3.3 Make sure the spring compressor tool is on securely

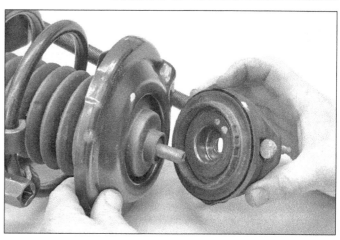

3.5a Remove the upper mount and spring seat . . .

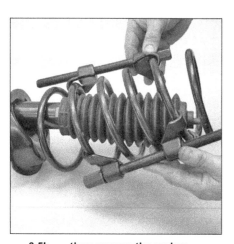

3.5b . . . then remove the spring . . .

3.5c . . . followed by the boot . . .

3.5d . . . and the bump stop

ture spring failure due to corrosion). Inspect the spring seat for cuts, hardness and general deterioration.

12   If any undesirable conditions exist, proceed to the shock absorber/coil spring disassembly procedure (see Section 3).

Installation

13   Guide the assembly up into the fenderwell, and insert the upper mounting studs through the holes in the body. Once the studs protrude, install the nuts so the strut won't fall back through but don't tighten them yet. This is most easily accomplished with the help of an assistant, as the strut is quite heavy and awkward.

14   Use the jack to position the steering knuckle, then install the two lower mounting bolts and nuts. Align the strut with the mark previously made on the knuckle, then tighten the nuts to the torque listed in this Chapter's Specifications. Move to the top nuts and tighten them to the torque listed in this Chapter's Specifications.

15   Reconnect the stabilizer bar link to the strut and tighten the nut to the torque listed in this Chapter's Specifications.

16   Reconnect the wheel speed sensor harness to the strut.

17   Install the wheel and lug nuts, then lower the vehicle and tighten the lug nuts to the torque listed in the Chapter 1 Specifications.

18   Have the front end alignment checked, and if necessary, adjusted.

---

3   Strut/coil spring assembly (front) - replacement

Warning: *Before attempting to disassemble the shock absorber/coil spring assembly, a tool to hold the coil spring in compression must be obtained. Do not attempt to use makeshift methods. Uncontrolled release of the spring could cause damage and personal injury or even death. Use a high-quality spring compressor, and carefully follow the tool manufacturer's instructions provided with it. After removing the coil spring with the compressor still installed, place it in a safe, isolated area. The strut is pressurized; do not apply heat or flame to the assembly or it could explode.*

Warning: *Always replace the struts or coil springs as a set (both sides).*

1   If the front suspension shock absorber/coil springs exhibit signs of wear (leaking fluid, loss of damping capability, sagging or cracked coil springs) then they should be disassembled and overhauled as necessary. The shock absorbers themselves cannot be serviced, and should be replaced if faulty; the springs and related components can be replaced individually. To maintain balanced characteristics on both sides of the vehicle, the components on both sides should be replaced at the same time.

2   With the assembly removed from the vehicle (see Section 2), clean away all external dirt; be careful not to damage or remove the protective coating from the spring, as this helps prevent corrosion damage.

3   Install the coil spring compressor tools (ensuring that they are fully engaged), and compress the spring until all tension is relieved from the upper mount **(see illustration)**.

4   Hold the strut piston rod with a wrench and unscrew the nut. Do not use an impact wrench as this could damage the strut.

5   Remove the upper mount and spring, followed by the boot, bump stop and lower spring seat **(see illustrations)**.

3.9 Ensure the spring is in the channel and resting against the strut body.

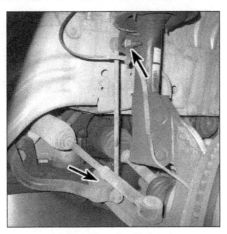

4.4 Stabilizer bar link nuts

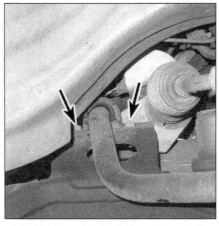

4.11 Stabilizer bar bushing retainer bolts

6   If a new spring is to be installed, the original spring must now be carefully released from the compressor. If it is to be re-used, the spring can be left in compression.

7   With the strut assembly now completely disassembled, examine all the components for wear and damage. Replace components as necessary.

8   Examine the shock for signs of fluid leakage. Check the piston rod for signs of pitting along its entire length, and check the shock body for signs of damage. Test the operation of the shock, while holding it in an upright position, by moving the piston through a full stroke, then through short strokes of 2 to 4 inches. In both cases, the resistance felt should be smooth and continuous. If the resistance is jerky or uneven, or if there is any visible sign of wear or damage, replacement is necessary.

9   To reassemble, lightly lubricate the strut rod and bumpers with a silicone oil or white lithium grease **(see illustration)**. Install the lower spring seat first, then the bump stop and dust boot onto the shock absorber, then install the spring and upper mount/upper seat.

10   With the spring compressed, install the piston rod nut and tighten it to the torque listed in this Chapter's Specifications.

11   Release the tension on the spring and re-install the strut assembly in the vehicle.

4   Stabilizer bar and link (front) - removal and installation

Removal

1   Loosen the front wheel lug nuts. Raise the front of the vehicle and support it securely on jackstands. Apply the parking brake and block the rear wheels to keep the vehicle from rolling off the stands. Remove the front wheels.

2   Unbolt the intermediate driveaxle shaft bracket from the subframe.

3   On all-wheel drive models, remove the

rear driveshaft (see Chapter 8).

4   To detach the stabilizer bar link from the strut or the bar, hold the ballstud with a wrench to prevent rotation, then unscrew the nut **(see illustration)**

5   Detach the tie-rod ends from the steering knuckle (see Section 17).

6   Support the rear of the subframe with a pair of floor jacks (one on each side).

7   Remove the subframe rear reinforcement bolts (see Section 22).

8   Loosen the subframe front reinforcement bolts, then loosen (but don't remove) the subframe front mounting bolts.

9   Loosen the subframe rear mounting bolts a few turns.

10   With the subframe mounting bolts loose, slowly lower the floor jacks. Continue to back the rear bolts out, as necessary, until there is enough room to remove the stabilizer bar.

11   Remove the bolts and detach the stabilizer bar bushing retainers **(see illustration)**

12   While the stabilizer bar is off the vehicle, slide off the bushings and inspect them, replacing them if necessary.

13   Clean the bushing area of the stabilizer bar with a stiff wire brush to remove any rust or dirt. Lubricate the inside and the outside of the new bushings with vegetable oil to facilitate reassembly.

Installation

14   Installation is the reverse of removal. Tighten the fasteners to the torque values listed in this Chapter's Specifications.

15   Raise the subframe using the floor jacks. Tighten the mounting bolts to the specified torque in the following order: Subframe front mounting bolts, subframe front reinforcement bolts, subframe rear mounting bolts, and subframe rear reinforcement mounting bolts.

16   Tighten the intermediate shaft bracket bolts to the torque listed in the Chapter 8 Specifications.

17   Install the wheel and lug nuts. Tighten the lug nuts to the torque listed in the Chapter 1 Specifications.

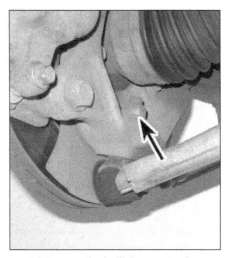

5.3 Loosen the balljoint-to-steering knuckle nut

5   Control arm (front) - removal, inspection and installation

Removal

1   Loosen the wheel lug nuts, raise the front of the vehicle, support it securely on jackstands and remove the wheel.

2   Disconnect the tie-rod end from the steering knuckle and swing it out of the way (see Section 17).

Note: *If you are replacing the balljoint and are going to use a picklefork type balljoint separator, this step is not necessary (it's to enable you to turn the knuckle far enough to install a screw-type balljoint separator).*

3   Loosen the balljoint nut a few turns **(see illustration)**.

Note: *Leaving the nut on the ballstud will prevent the components from separating violently.*

4   Separate the balljoint from the steering knuckle using a puller. If possible, use a screw-type balljoint tool rather than a wedge

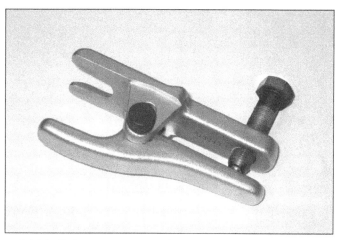

5.4 A balljoint separator tool like this one is available at most auto parts stores and will not damage the balljoint boot when used correctly

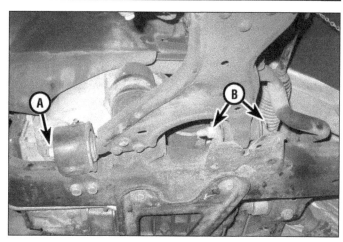

5.5 Control arm front bushing nut (A) and rear pivot bolt/nut (B)

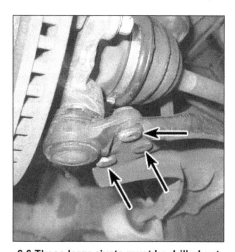

6.6 These large rivets must be drilled out to remove the old balljoint

type tool (picklefork) to avoid damaging the balljoint boot **(see illustration)**.
5   Remove the front bushing nut and the control arm pivot bolt/nut **(see illustration)**.
6   Detach the control arm from the sub-frame.

Inspection

7   Check the control arm for distortion and the bushings for wear, replacing parts as necessary. Do not attempt to straighten a bent control arm.

Installation

8   Installation is the reverse of removal. When installing the control arm, leave the pivot bolt nut and the front bushing nut finger tight until completing the installation of all other components. Before tightening the pivot nuts, raise the outer end of the control arm with a floor jack to simulate normal ride height, then tighten the nuts to the torque listed in this Chapter's Specifications.
9   Have the front wheel alignment checked and, if necessary, adjusted.

6   Balljoints - check and replacement

Check

1   Raise the front of the vehicle and support it securely on jackstands. Apply the parking brake and block the rear wheels to keep the vehicle from rolling off the jackstands.
2   Place a large prybar under the balljoint and resting on the wheel, then try to pry up the balljoint while feeling for movement between the balljoint and steering knuckle. Now, pry between the control arm and the steering knuckle and try to lever the control arm down while feeling for movement between the balljoint and steering knuckle. If excessive movement is evident (over 0.125 inch) in either check, the balljoint is worn. Install a dial indicator and measure the amount of movement to be certain.
3   Have an assistant grasp the tire at the top and bottom and move the top of the tire in-and-out. Touch the balljoint stud nut. If excessive movement is evident (over 0.125 inch), the balljoint or knuckle is worn. Install a dial indicator and measure the amount of movement to be certain.

Replacement

4   Loosen the wheel lug nuts on the side to be disassembled. Apply the parking brake, raise the front of the vehicle, support it securely on jackstands and remove the wheel.
5   Remove the control arm (see Section 5).
6   Drill out the three rivets securing the balljoint **(see illustration)**, remove the balljoint and clean the control arm.
7   Install the new balljoint against the mating surface of the control arm and secure it with the supplied fasteners (nuts and bolts are normally supplied with the new balljoint). Tighten the balljoint fasteners to the torque listed in this Chapter's Specifications, or in the balljoint replacement kit instructions.
8   Reinstall the control arm (see Section 5).

7   Steering knuckle - removal and installation

Warning: *Dust created by the brake system is harmful to your health. Never blow it out with compressed air and don't inhale any of it. Do not, under any circumstances, use petroleum-based solvents to clean brake parts. Use brake system cleaner only.*

Removal

1   Loosen the driveaxle/hub nut (see Chapter 8).
2   Loosen the wheel lug nuts. Raise the vehicle and support it securely on jackstands, then remove the wheel.
3   Remove the brake disc and the ABS wheel speed sensor (see Chapter 9).
4   Loosen the strut-to-knuckle nuts (see Section 2).
5   Separate the control arm from the steering knuckle by separating the control arm balljoint (see Section 5).
6   Separate the tie-rod end from the steering knuckle arm (see Section 17).
7   Detach the steering knuckle from the strut (see Section 2), then remove the knuckle.

Installation

8   Guide the knuckle into position, inserting the driveaxle into the hub.
9   Install the strut-to-knuckle bolts and nuts.
10   Connect the balljoint to the knuckle and tighten the nut to the torque listed in this Chapter's Specifications.
11   Attach the tie-rod end to the steering knuckle arm (see Section 17). Tighten the nut to the torque listed in this Chapter's Specifications.
12   Tighten the strut-to-knuckle bolts to the torque listed in this Chapter's Specifications.
13   Install the wheel speed sensor and harness bracket.
14   Install the brake disc and caliper/mount-

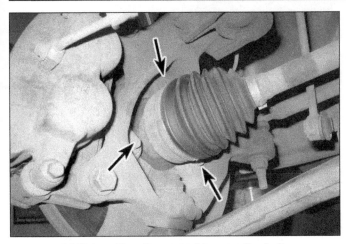

8.6 Hub and bearing assembly mounting bolts

9.2 Shock absorber mounting fasteners

ing bracket assembly (see Chapter 9).

15   Install the wheel and lug nuts. Lower the vehicle and tighten the lug nuts to the torque listed in the Chapter 1 Specifications. Tighten the driveaxle/hub nut to the torque listed in the Chapter 8 Specifications.

16   Have the front-end alignment checked and, if necessary, adjusted.

8   Hub and bearing assembly (front) - removal and installation

Warning: *Dust created by the brake system is harmful to your health. Never blow it out with compressed air and don't inhale any of it. Do not, under any circumstances, use petroleum-based solvents to clean brake parts. Use brake system cleaner only.*

Caution: *Don't allow any magnets near the hub and bearing assembly, and don't drop the assembly - the encoder for the ABS wheel speed sensor will be damaged.*

1   Loosen the driveaxle/hub nut (see Chapter 8).

2   Loosen the wheel lug nuts, raise the vehicle and support it securely on jackstands. Remove the wheel.

3   Remove the driveaxle/hub nut.

4   Remove the brake disc (see Chapter 9).

5   Use wire to support the driveaxle (don't let it hang by the inner CV joint after the hub and bearing has been removed).

6   Remove the hub and bearing assembly mounting bolts **(see illustration)**.

7   Remove the hub and bearing unit. If it's stuck to the driveaxle, tap the driveaxle end with a hammer and a block of wood to avoid damaging the threads, or use a puller to push the driveaxle out of the hub as it is separated from the knuckle.

8   Installation is the reverse of removal. Tighten the bolts to the torque listed in this Chapter's Specifications.

10.3 Upper control arm-to-knuckle nut and bolt

9   Shock absorber (rear) - removal and installation

Warning: *Always replace the shock absorbers as a set - never replace just one of them.*

1   Block the front wheels. Raise the rear of the vehicle and support it securely on jackstands. Remove the wheel.

2   Support the rear lower control arm with a floor jack. Remove the shock absorber upper mounting nut and bolt **(see illustration)**.

3   Remove the shock absorber lower nut and bolt.

4   Remove the shock absorber.

5   Installation is the reverse of removal. Raise the lower control arm with a floor jack to simulate normal ride height, then tighten the fasteners to the torque listed in this Chapter's Specifications.

Caution: *New shock absorbers are gas-filled and come compressed and retained with a fiberglass strap. Do NOT remove the strap until the shock is installed.*

6   Install the wheel and lug nuts, then lower

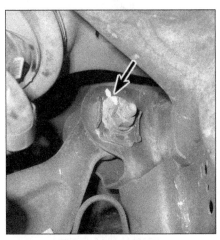

10.4 Mark the relationship of the upper control arm cam adjuster to the subframe

the vehicle. Tighten the lug nuts to the torque listed in the Chapter 1 Specifications.

10   Suspension arms (rear) - removal and installation

Warning: *The manufacturer recommends replacing all fasteners with new ones during installation.*

1   Loosen the wheel lug nuts, raise the vehicle and support it securely on jackstands. Block the front wheels to prevent the vehicle from rolling. Remove the wheel.

Upper control arm

2   Detach the ABS wheel speed sensor harness clips from the control arm.

3   Support the lower control arm with a floor jack, then remove the upper control arm-to-knuckle nut and bolt **(see illustration)**.

4   Scribe or draw an alignment mark on the upper control arm adjuster cam to the subframe **(see illustration)**. This will help restore

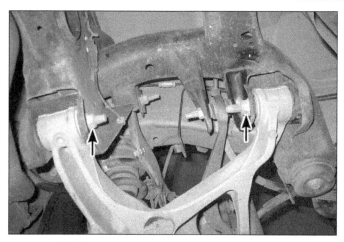

10.10 Lower control arm-to-subframe fasteners

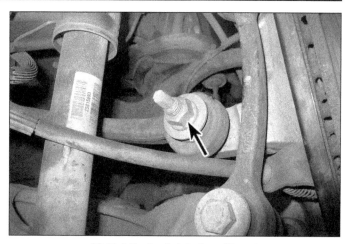

10.14 Adjuster link-to-knuckle nut

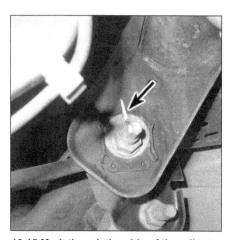

10.15 Mark the relationship of the adjuster link cam to the subframe

alignment during assembly.
5   Hold the adjuster cam bolt with a wrench to prevent it from turning, then unscrew the nut. Remove the nut and cam, then remove the pivot bolt.
6   Remove the upper control arm from the knuckle and subframe.
7   Installation is the reverse of removal. Raise the lower control arm to simulate normal ride height, then align the marks on the adjuster cam and subframe and tighten the inner and outer fasteners to the torque listed in this Chapter's Specifications.
8   Install the wheel and lug nuts. Lower the vehicle and tighten the lug nuts to the torque listed in the Chapter 1 Specifications. Have the wheel alignment check and, if necessary, adjusted.

Lower control arm

9   Remove the coil spring (see Section 11).
10   Remove the control arm-to subframe pivot bolts and detach the arm from the subframe **(see illustration)**.
11   Installation is the reverse of removal. Before tightening the fasteners to the torque

listed in this Chapter's Specifications, raise the outer end of the control arm with a floor jack to simulate normal ride height.
12   Install the wheel and lug nuts. Lower the vehicle and tighten the lug nuts to the torque listed in the Chapter 1 Specifications.

Adjuster link

13   Remove the rear shock absorber (see Section 9).
14   Remove the adjuster link-to-knuckle nut **(see illustration)**.
15   Scribe or draw an alignment mark on the adjuster link adjuster cam to the subframe **(see illustration)**. This will help restore alignment during assembly.
16   Hold the adjuster cam bolt with a wrench to prevent it from turning, then unscrew the nut. Remove the nut and cam, then remove the pivot bolt.
17   Remove the adjuster link from the knuckle and subframe.
18   Installation is the reverse of removal. Before tightening the fasteners to the torque listed in this Chapter's Specifications, raise the outer end of the lower control arm with a floor jack to simulate normal ride height.
19   Install the wheel and lug nuts. Lower the vehicle and tighten the lug nuts to the torque listed in the Chapter 1 Specifications. Have the wheel alignment check and, if necessary, adjusted.

Upper control arm-to-lower control arm link

20   Support the lower control arm with a floor jack.
21   Remove the upper control arm-to knuckle bolt **(see illustration 10.3)**
22   Remove the link-to-knuckle bolt and detach the link **(see illustration 11.5)**.
23   Installation is the reverse of removal. Before tightening the fasteners to the torque listed in this Chapter's Specifications, raise the outer end of the control arm with a floor jack to simulate normal ride height.

24   Install the wheel and lug nuts. Lower the vehicle and tighten the lug nuts to the torque listed in the Chapter 1 Specifications. Have the wheel alignment check and, if necessary, adjusted.

11 Coil spring (rear) - removal and installation

Warning: *Always replace the springs as a set - never replace just one of them.*
1   Loosen the wheel lug nuts, raise the vehicle and support it securely on jackstands. Block the front wheels to prevent the vehicle from rolling. Remove the wheel.
2   Support the lower control arm with a floor jack.
3   Detach the stabilizer bar link from the lower control arm (see Section 14).
4   Detach the shock absorber from the lower control arm (see Section 9).
5   Remove the lower control arm-to-knuckle nut and bolt and the upper control arm-to-lower control arm link bolt **(see illustration)**.

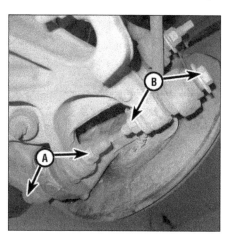

11.5 Lower control arm-to-rear knuckle nut and bolt (A) and upper control arm-to-lower control arm link bolt (B)

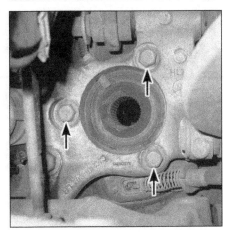

12.4 Rear hub and bearing-to-knuckle bolts

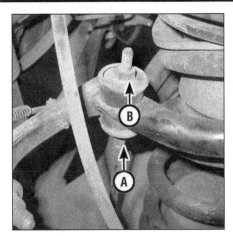

14.3 Hold the hexagonal portion of the link (A) with a wrench while unscrewing the nut (B)

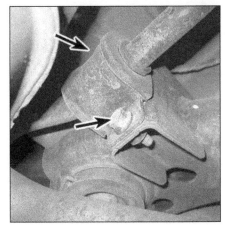

14.4 Remove the stabilizer bar bushing bracket nuts

6   Slowly and carefully lower the floor jack until the coil spring is fully extended, then maneuver the spring and jounce bumper out from the suspension arms. Retrieve the spring seat.

7   Inspect the spring seat and jounce bumper for wear and damage, replacing as necessary. Check the coil spring for nicks and cracks.

8   Installation is the reverse of removal. Raise the outer end of the lower control arm with the floor jack to simulate normal ride height, then tighten the fasteners to the torque values listed in this Chapter's Specifications.

9   Install the wheel and lug nuts, then lower the vehicle. Tighten the lug nuts to the torque listed in the Chapter 1 Specifications.

---

12   Hub and bearing assembly (rear) - removal and installation

---

Caution: *Don't allow any magnets near the hub and bearing assembly, and don't drop the assembly - the encoder for the ABS wheel speed sensor will be damaged.*

Removal

1   Loosen the rear wheel lug nuts. Raise the rear of the vehicle and support it securely on jackstands, then remove the wheel. Block the front wheels to keep the vehicle from rolling off the stands.

2   On AWD models, remove the rear driveaxle/hub nut (see Chapter 8).

3   Remove the brake disc (see Chapter 9).

4   Remove the bolts and detach the hub and bearing assembly from the knuckle **(see illustration)**. On AWD models, guide the driveaxle stub shaft out of the hub splnes as this is done. If the shaft sticks in the hub, use a rubber mallet or brass hammer to tap it out. If it's really stuck, use a puller to push it out.

Caution: *On AWD models, be careful not to overextend the CV joints.*

Installation

5   Installation is the reverse of removal. Tighten the hub and bearing bolts to the torque listed in this Chapter's Specifications. On AWD models, tighten the driveaxle/hub nut to the torque listed in the Chapter 8 Specifications.

6   Install the wheel and lug nuts. Lower the vehicle and tighten the lug nuts to the torque listed in the Chapter 1 Specifications.

---

13   Rear knuckle - removal and installation

---

1   Loosen the rear wheel lug nuts. Raise the rear of the vehicle and support it securely on jackstands, then remove the wheel. Block the front wheels to keep the vehicle from rolling off the stands.

2   Support the lower control arm with a floor jack.

Warning: *The jack must remain in this position throughout the entire procedure.*

3   On AWD models, remove the driveaxle/hub nut (see Chapter 8).

4   Remove the parking brake assembly, actuator and backing plate (see Chapter 9).

5   Remove the wheel speed sensor from the knuckle (see Chapter 9).

6   Remove the hub and bearing assembly (see Section 12).

7   Detach the upper control arm from the knuckle (see Section 10).

8   Detach the adjuster link from the knuckle (see Section 10).

9   Detach the shock absorber and stabilizer bar link from the lower control arm.

10   Remove the lower arm-to-knuckle nut and bolt (see Section 11).

11   Remove the upper contol arm-to-lower control arm link bolt from the knuckle (see Section 11).

12   Installation is the reverse of removal; tighten all suspension fasteners to the torque values listed in this Chapter's Specifications.

Tighten the brake fasteners to the torque values listed in the Chapter 9 Specifications. On all-wheel drive models, tighten the driveaxle/hub nut(s) to the torque listed in the Chapter 8 Specifications.

13   Install the wheel and lug nuts, then lower the vehicle. Tighten the lug nuts to the torque listed in the Chapter 1 Specifications.

---

14   Stabilizer bar and links (rear) - removal and installation

---

Removal

1   Loosen the rear wheel lug nuts. Raise the rear of the vehicle and support it securely on jackstands, then remove the wheels. Block the front wheels to keep the vehicle from rolling off the stands.

2   Remove the spare tire.

3   Remove the left and right side stabilizer bar link-to-stabilizer bar nuts **(see illustration)**. Remove the washers and bushings.

Note: *Hold the link with a wrench to prevent it from turning while loosening the nut*

4   Remove the stabilizer bar bushing brackets and the stabilizer bar **(see illustration)**.

5   If necessary, remove the stabilizer bar links and bushings from the lower control arm.

6   While the stabilizer bar is off the vehicle, inspect the bushings; replace any that are cracked, worn or deteriorated.

7   Clean the bushing area of the stabilizer bar with a stiff wire brush to remove any rust or dirt. Lubricate the inside and outside of the new bushing with vegetable oil (used in cooking) to simplify reassembly.

Caution: *Don't use petroleum or mineral-based lubricants or brake fluid - they will lead to deterioration of the bushings.*

8   Installation is the reverse of removal; make sure the bar is centered. Tighten the fasteners to the torque values listed in this Chapter's Specifications.

9   Install the spare tire.

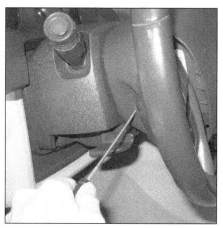

15.3 Depress the retaining spring clips through the holes on each side of the steering wheel to release the airbag (on some models it might be necessary to turn the wheel 90-degrees to access the holes. If so, turn the wheel back to the straight-ahead position before removing it)

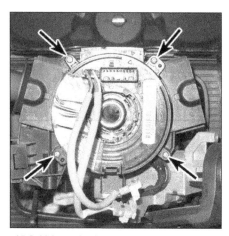

15.8 Airbag clockspring retaining screws

10   Install the wheel and lug nuts, then lower the vehicle. Tighten the lug nuts to the torque listed in the Chapter 1 Specifications.

---

15   Steering wheel - removal and installation

Warning: *These models are equipped with a Supplemental Restraint System (SRS), more commonly known as airbags. Always disable the airbag system before working in the vicinity of any airbag system component to avoid the possibility of accidental deployment of the airbag(s), which could cause personal injury (see Chapter 12).*
Warning: *Do not use a memory saving device to preserve the PCM or radio memory when working on or near airbag system components.*

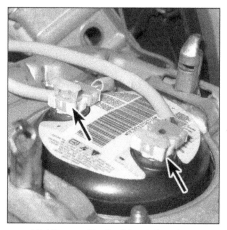

15.4 Pry up the locking tabs, then disconnect the airbag connectors

Removal

1   Turn the ignition key to Off, then disconnect the cable from the negative terminal of the battery. Wait at least two minutes before proceeding.
2   Turn the steering wheel so the wheels are pointing straight ahead.
3   Insert a long, thin screwdriver or other similar object through the holes in the backside of the steering wheel to release the airbag module retaining spring **(see illustration)**. Gently pull the airbag module away from the wheel on the side you have released, then repeat this Step on the other side of the wheel.
4   Pry up the connector locks and disconnect the airbag module electrical connectors **(see illustration)**.
5   Remove the airbag and set it aside in a safe, isolated area.
Warning: *Carry the airbag module with the trim side facing away from you, and set the steering wheel/airbag module down with the trim side facing up. Don't place anything on top of the steering wheel/airbag module. Also, do not drop the airbag module or expose it to heat.*
6   Remove the steering wheel bolt.
7   Remove the steering wheel. This vehicle does not require any special tools to pull the steering wheel off the hub. No special marks are needed as there is a keyway/guide groove that aligns the steering wheel **(see illustration)**.
Warning: *If the steering wheel sticks, try rocking the wheel lightly to remove it from the shaft. Do not hit the steering shaft with a hammer or you will damage the shaft.*
8   If it's necessary to remove the airbag clockspring, remove the steering column covers (see Chapter 11). Remove the four clockspring retaining screws **(see illustration)** and detach it from the steering column, then disconnect the electrical connectors.

Installation

Note: *Steps 10 and 11 can be skipped if you're sure the airbag clockspring hasn't been*

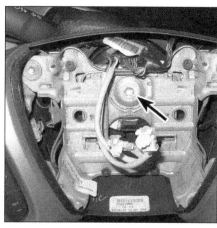

15.7 Steering wheel bolt

rotated after steering wheel removal.
9   Make sure that the front wheels are facing straight ahead.
10   Turn the hub of the clockspring clockwise by hand until it becomes harder to turn the cable (don't apply too much force). Rotate it counterclockwise and count the turns until the same resistance is met.
11   Rotate the clockspring back to half that number of turns to center it. Install the clockspring, making sure its hub doesn't rotate in relation to the housing.
12   The remainder of installation is the reverse of removal.
Caution: *Do not force the steering wheel on. If it does not fit, remove the wheel and look for the blank areas on the splines of the shaft and the hub of the steering wheel. Line up the line up the blank areas.*
13   Tighten the steering wheel bolt to the torque listed in this Chapter's Specifications.
14   Install the airbag module.
Caution: *Make sure the connectors fit correctly on the airbag module (with the safety tabs still UP). When connecting the airbag connectors, match the key on the connector to the corresponding keyway on the module. Do not push the connectors on if the clips are down. After the clips are secured, push the airbag module down onto the steering wheel until the retaining clips lock. When seated properly, the gap between the airbag module trim cover and the steering wheel should be even and consistent.*

---

16   Steering column - removal and installation

Warning: *These models are equipped with a Supplemental Restraint System (SRS), more commonly known as airbags. Always disable the airbag system before working in the vicinity of any airbag system component to avoid the possibility of accidental deployment of the airbag(s), which could cause personal injury (see Chapter 12).*

16.10 Intermediate shaft upper pinch bolt

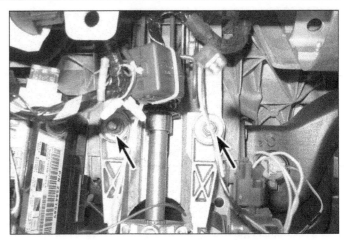

16.12 Steering column lower mounting nuts (upper bolts not visible in this photo)

Warning: *Do not use a memory saving device to preserve the PCM or radio memory when working on or near airbag system components.*

Removal

1 Park the vehicle with the wheels pointing straight ahead. Lock the steering in place. Failure to do so can result in damage to the airbag clockspring. Disconnect the cable from the negative terminal of the battery (see Chapter 5).
2 Remove the steering wheel (see Section 15)
3 Remove the left side under-dash panel (see Chapter 11).
4 Remove the driver's knee bolster (see Chapter 11).
5 Remove the steering column covers (see Chapter 11).
6 Remove the instrument cluster bezel (see Chapter 11).
7 Remove the steering column housing cover (see Chapter 11).
8 Disconnect all electrical connectors from the steering column components.
9 Remove the lower steering column brace bolts and bracket, if so equipped.

10 Remove the intermediate steering shaft upper pinch bolt. Discard the bolt and obtain a new one for reassembly **(see illustration).**
11 Disconnect the intermediate steering shaft from the steering column. Ensure that the intermediate steering shaft lower seal is not pulled away from the body panel.
12 Remove the steering column mounting fasteners, then remove the column out from under the instrument panel **(see illustration).**

Installation

13 Position the steering column in the vehicle. Install the steering column nuts first, then the bolts. Finger tighten them only at this time.
14 Push the steering column flush against the instrument panel carrier and hand-tighten the fasteners, beginning with the nuts, then the bolts.
15 Connect the intermediate shaft to the steering column, install a NEW pinch bolt and tighten it to the torque listed in this Chapter's Specifications. Inspect the shaft seal and ensure that it is seated correctly. If not, disconnect the intermediate shaft and correct the issue.

16 If equipped, re-install the support bracket and bolts and tighten them securely.
17 If your vehicle is equipped with a telescopic/tilt steering column, it may be necessary that you calibrate the steering module. Take the vehicle to a dealer service department or other qualified repair facility.
18 The remainder of installation is the reverse of removal.

17 Tie-rod ends - removal and installation

Removal

1 Loosen the front wheel lug nuts. Apply the parking brake, raise the front of the vehicle and support it securely on jackstands. Remove the front wheel.
2 Hold the tie-rod with a pair of locking pliers or wrench and loosen the jam nut enough to mark the position of the tie-rod end in relation to the threads **(see illustrations).**
3 Loosen the tie-rod end-to-steering knuckle nut about 1/4-inch, but don't remove it.

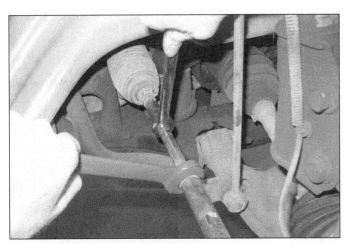

17.2a Hold the tie-rod with a wrench and loosen the jam nut...

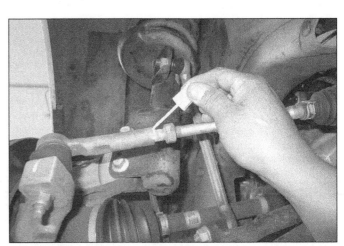

17.2b... then mark the position of the tie-rod end

17.4 Disconnect the tie-rod end from the steering knuckle with a puller

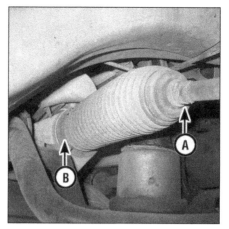

18.3 The outer ends of the steering gear boots are secured by band-type clamps (A); they're easily removed with a pair of pliers. The inner ends are retained by boot clamps (B) which must be cut off and discarded

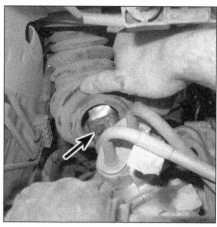

19.7 Push the steering gear input shaft boot up, mark the relationship of the intermediate shaft to the steering gear input shaft, then remove the pinch bolt and disconnect the intermediate shaft from the steering gear

4   Disconnect the tie-rod end from the steering knuckle arm with a tie-rod end remover or small puller **(see illustration)**. Remove the nut and detach the tie-rod end.
5   Unscrew the tie-rod end from the tie-rod.
Installation
6   Thread the tie-rod end on to the marked position. Insert the stud into the steering knuckle arm. Tighten the jam nut securely.
7   Install the nut on the stud and tighten it to the torque listed in this Chapter's Specifications.
8   Install the wheel and lug nuts. Lower the vehicle and tighten the lug nuts to the torque listed in the Chapter 1 Specifications.
9   Have the alignment checked and, if necessary, adjusted.

18   Steering gear boots - replacement

1   Loosen the lug nuts, raise the vehicle and support it securely on jackstands. Remove the wheel.
2   Remove the tie-rod end and jam nut (see Section 17).
3   Remove the outer steering gear boot clamp with a pair of pliers **(see illustration)**. Cut off the inner boot clamp with a pair of diagonal cutters. Slide off the boot.
4   Before installing the new boot, wrap the threads and serrations on the end of the steering rod with a layer of tape so the small end of the new boot isn't damaged.
5   Slide the new boot into position on the steering gear until it seats in the groove in the steering rod and install new clamps.
6   Remove the tape and install the tie-rod end (see Section 17).
7   Install the wheel and lug nuts. Lower the vehicle and tighten the lug nuts to the torque listed in the Chapter 1 Specifications.

19   Steering gear - removal and installation

Warning: *These models are equipped with a Supplemental Restraint System (SRS), more commonly known as airbags. Always disable the airbag system before working in the vicinity of any airbag system component to avoid the possibility of accidental deployment of the airbag(s), which could cause personal injury (see Chapter 12). Make sure the steering column shaft is not turned while the steering gear is removed or you could damage the airbag system clockspring. To prevent the shaft from turning, turn the ignition key to the lock position with the steering wheel in the straight ahead position before beginning work.*
Warning: *Prior to performing this procedure, obtain new subframe mounting bolts.*

Removal

1   Park the vehicle with the front wheels pointing straight ahead. Lock the steering column.
2   Remove as much power steering fluid as possible. Use a large syringe or turkey baster to facilitate this.
Warning: *If a turkey baster is used, never again use it in the preparation of food!*
3   Disconnect the cable from the negative terminal of the battery (see Chapter 5).
4   Loosen the wheel lug nuts, raise the front of the vehicle and support it securely on jackstands. Remove the front wheels.
5   Detach the tie-rod ends from the steering knuckles (see Section 17).
6   Disconnect the stabilizer bar links from the stabilizer bar (see Section 4).
7   Disconnect the intermediate shaft from the steering gear **(see illustration)**.
8   On all-wheel drive models, remove the rear driveshaft (see Chapter 8).
9   Remove the subframe brace.

10   Remove the rear-bank catalytic converter (see Chapter 6).
11   Remove the heat shield protecting the steering gear.
12   Support the rear of the subframe with two floor jacks (one placed on each side).
13   Remove the left and right subframe reinforcement brackets at the rear of the subframe (both sides).
14   Loosen the subframe front reinforcement bolts and loosen the front subframe mounting bolts a few turns.
15   Loosen the subframe rear mounting bolts a few turns.
Warning: *The manufacturer recommends replacing the subframe mounting bolts with new ones whenever they are loosened.*
16   Lower the subframe enough to gain clearance to facilitate steering gear removal.
17   Remove the bolt and retaining plate and detach the fluid lines **(see illustration)**. Also remove the bolt holding the fluid line retainer to the steering gear.

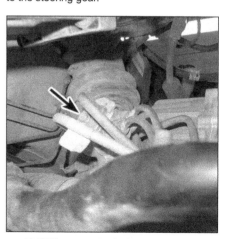

19.17 Remove the bolt and retaining plate, then detach the fluid lines from the steering gear

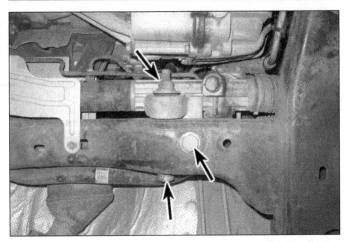

19.18 Power steering gear mounting fasteners - left side (nut on vertical bolt not visible)

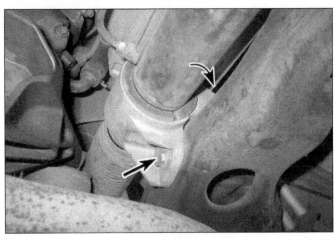

19.19 Steering gear mounting fasteners - right side (vertical bolt not visible)

18   Remove the steering gear nuts and bolts from the left side **(see illustration)**.
19   Remove the steering gear mounting bolts from the right side **(see illustration)**.
20   Maneuver the steering gear out from the left side of the vehicle.

Installation

21   Installation is the reverse of removal, noting the following points:

a) *New subframe mounting bolts should be used. Before tightning the subframe bolts, raise the subframe with the floor jacks. Remove and install one bolt at a time.*

b) *Tighten the steering gear mounting bolts to the torque listed in this Chapter's Specifications.*

c) *When connecting the steering column shaft to the steering gear input shaft, align the matchmarks, and install a new pinch bolt. Tighten the pinch bolt to the torque listed in this Chapter's Specifications.*

d) *Fill the power steering reservoir with the recommended fluid (see Chapter 1). Bleed the power steering system (see Section 21).*

e) *Have the front end alignment checked and, if necessary, adjusted.*

20   Power steering pump - removal and installation

Removal

1   Remove the fuel injection shroud.
2   Disconnect the cable from the negative battery terminal (see Chapter 5).
3   Raise the vehicle and support it securely on jackstands.
4   Using a large syringe or suction gun, remove as much fluid from the power steering fluid reservoir as possible. Place a drain pan under the vehicle to catch any fluid that spills out when the hoses are disconnected.
5   Remove the drivebelt (see Chapter 1).
6   Remove the steering gear heat shield.
7   Remove the rear bank catalytic converter.
8   Disconnect the fluid feed hose from the power steering pump **(see illustration)**.
9   Disconnect the pressure line from the power steering pump.
10   Remove the power steering pump mounting bolts **(see accompanying illustration and illustration 20.8)**.
11   Remove the power steering pump from the vehicle.

Installation

12   Installation is the reverse of removal, noting the following points:

a) *Tighten the mounting bolts to the torque listed in this Chapter's Specifications. Tighten the pressure line fitting securely.*

b) *Fill the power steering reservoir with the recommended fluid (see Chapter 1). Bleed the power steering system (see Section 21).*

21   Power steering system - bleeding

1   The power steering system must be bled whenever a line is disconnected. Bubbles can be seen in power steering fluid that has air in it and the fluid will often have a tan or milky appearance. Low fluid level can cause air to mix with the fluid, resulting in a noisy pump as well as foaming of the fluid.
2   Open the hood and check the fluid level in the reservoir, adding the specified fluid necessary to bring it up to the proper level (see Chapter 1).
3   Start the engine and slowly turn the

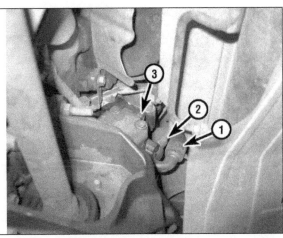

20.8 Power steering pump details (rear view)

1   *Fluid feed hose*
2   *Pressure line fitting*
3   *Pump rear mounting bolt*

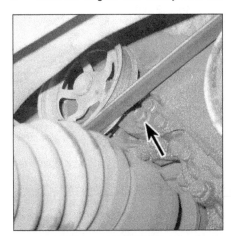

20.10 Power steering pump front mounting bolt

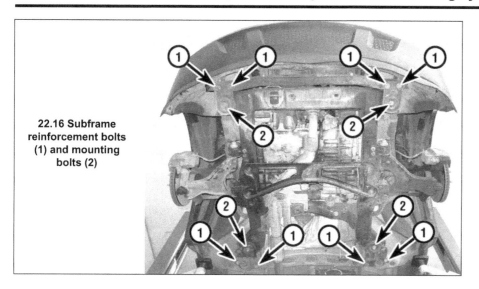

22.16 Subframe reinforcement bolts (1) and mounting bolts (2)

steering wheel several times from left-to-right and back again. Do not turn the wheel completely from lock-to-lock. Check the fluid level, topping it up as necessary until it remains steady and no more bubbles are visible.

22   Subframe (front) - removal and installation

1   Disconnect the cable from the negative battery terminal (see Chapter 5).
2   Support the engine and transaxle assembly from above with an engine support fixture (see Chapter 2B). An engine hoist can be used as an alternative, but will have to be connected after the vehicle is raised and supported on jackstands.
3   Loosen the front wheel lug nuts, raise the front of the vehicle and support it securely on jackstands. Remove both front wheels. **Note:** *The jackstands must be behind the front suspension subframe, not supporting the vehicle by the subframe.*
4   Remove the inner fender lower splash shields (see Chapter 11).
5   Remove the front portion of the exhaust system (see Chapter 4).
6   On all-wheel drive models, remove the driveshaft (see Chapter 8).
7   Separate the control arm balljoints from the steering knuckles (see Section 5).
8   Separate the stabilizer bar links from the stabilizer bar (see Section 4).
9   Remove the heat shields protecting the power steering gear (see Section 20).
10   Remove the pinch bolt and separate the intermediate shaft from the steering gear input shaft (see Section 19).
11   Unbolt the steering gear from the subframe (see Section 19). Hang the steering gear from the exhaust manifold using wire or rope.
12   Disengage the clips and separate the power steering cooler pipe from the left side of the subframe.
13   On the right side, remove the engine mount-to-subframe nut (see Chapter 2A).
14   On the left side, remove the transaxle-to-

subframe nut (see Chapter 7A).
15   Support the subframe with a pair of floor jacks (one on each side).
16   Remove the front and rear subframe reinforcement bolts **(see illustration)**.
17   Remove the subframe mounting bolts and slowly lower the subframe, making sure nothing is still attached.
Warning: *Discard the subframe mounting bolts and obtain new ones for installation.*
18   Installation is the reverse of removal. Tighten all fasteners to the proper torque specifications.

23   Wheels and tires - general information

1   All vehicles covered by this manual are equipped with metric-sized fiberglass or steel belted radial tires **(see illustration)**. Use of other size or type of tires may affect the ride and handling of the vehicle. Don't mix different types of tires, such as radials and bias belted, on the same vehicle as handling may be seriously affected. It's recommended that tires be replaced in pairs on the same axle, but if only one tire is being replaced, be sure it's the same size, structure and tread design as the other.
2   Because tire pressure has a substantial effect on handling and wear, the pressure on all tires should be checked at least once a month or before any extended trips (see Chapter 1).
3   Wheels must be replaced if they are bent, dented, leak air, have elongated bolt holes, are heavily rusted, out of vertical symmetry or if the lug nuts won't stay tight. Wheel repairs that use welding or peening are not recommended.
4   Tire and wheel balance is important in the overall handling, braking and performance of the vehicle. Unbalanced wheels can adversely affect handling and ride characteristics as well as tire life. Whenever a tire is installed on a wheel, the tire and wheel should be balanced by a shop with the proper equipment.

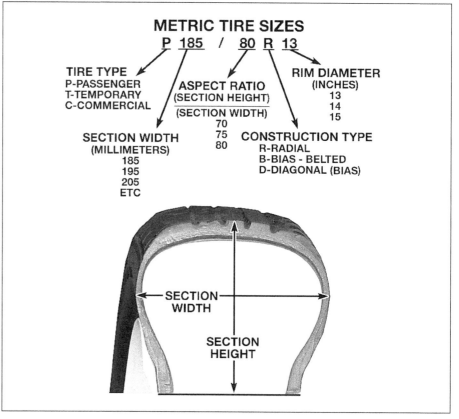

23.1 Metric tire size code

24 Wheel alignment - general information

1   A wheel alignment refers to the adjustments made to the wheels so they are in proper angular relationship to the suspension and the ground. Wheels that are out of proper alignment not only affect vehicle control, but also increase tire wear. The front end angles normally measured are camber, caster and toe-in **(see illustration)**. Toe-in and camber are adjustable on the front and rear; if the caster is not correct, check for bent components.

2   Getting the proper wheel alignment is a very exacting process, one in which complicated and expensive machines are necessary to perform the job properly. Because of this, you should have a technician with the proper equipment perform these tasks. We will, however, use this space to give you a basic idea of what is involved with a wheel alignment so you can better understand the process and deal intelligently with the shop that does the work.

3   Toe-in is the turning in of the wheels. The purpose of a toe specification is to ensure parallel rolling of the wheels. In a vehicle with zero toe-in, the distance between the front edges of the wheels will be the same as the distance between the rear edges of the wheels. The actual amount of toe-in is normally only a fraction of an inch. On the front end, toe-in is controlled by the tie-rod end position on the tie-rod. On the rear end, it's controlled by cam bolts on the inner ends of the adjuster links. Incorrect toe-in will cause the tires to wear improperly by making them scrub against the road surface. Due to the involvement of frame, body, and other factors it is necessary that the vehicle be inspected by a professional who has access to the required alignment tables and tools.

4   Camber is the tilting of the wheels from vertical when viewed from one end of the vehicle. When the wheels tilt out at the top, the camber is said to be positive (+). When the wheels tilt in at the top the camber is negative (-). The amount of tilt is measured in degrees from vertical and this measurement is called the camber angle. This angle affects the amount of tire tread which contacts the road and compensates for changes in the suspension geometry when the vehicle is cornering or traveling over an undulating surface. On the front end it is only adjustable by elongating the strut lower mounting bolt hole

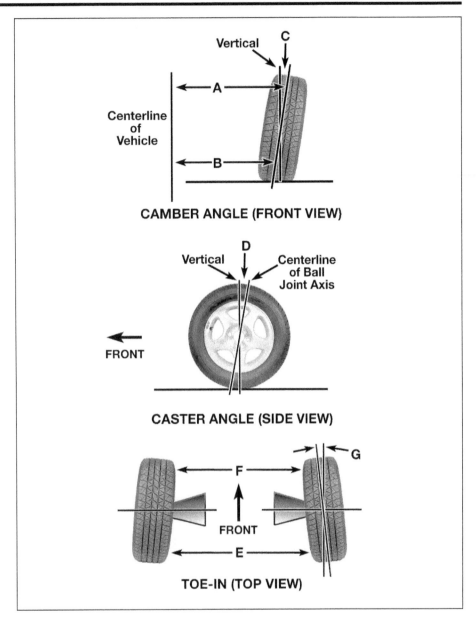

24.1 Front end alignment

A minus B = C (degrees camber)
D = caster (expressed in degrees)

E minus F = toe-in (measured in inches)
G = toe-in (expressed in degrees)

and repositioning the knuckle in the strut. On the rear end it's adjusted by cam bolts on the inner ends of the upper control arms.

5   Caster is the tilting of the front steering axis from the vertical. A tilt toward the rear is positive caster and a tilt toward the front is negative caster. Caster is not adjustable on these vehicles.

Notes

Chapter 11   Body

Contents

	Section
Body repair - major damage	4
Body repair - minor damage	3
Bumper covers - removal and installation	9
Center console - removal and installation	21
Cowl cover - removal and installation	20
Dashboard trim panels - removal and installation	22
Door - removal, installation and adjustment	13
Door latch, lock cylinder and handles - removal and installation	14
Door trim panels - removal and installation	11
Door window glass - removal and installation	15
Door window glass regulator - removal and installation	16
Fastener and trim removal	6

	Section
Front fender - removal and installation	10
General information	1
Hood - removal, installation and adjustment	7
Hood latch and release cable - removal and installation	8
Liftgate - removal, installation and adjustment	18
Liftgate latch and outside handle - removal and installation	19
Mirrors - removal and installation	17
Rear quarter trim panel - removal and installation	12
Repairing minor paint scratches	2
Seats - removal and installation	24
Steering column and housing covers - removal and installation	23
Upholstery, carpets and vinyl trim - maintenance	5

1   General information

Warning: *The models covered by this manual are equipped with Supplemental Restraint Systems (SRS), more commonly known as airbags. Always disable the airbag system before working in the vicinity of any airbag system components to avoid the possibility of accidental deployment of the airbags, which could cause personal injury (see Chapter 12).*

1   Certain body components are particularly vulnerable to accident damage and can be unbolted and repaired or replaced. Among these parts are the hood, doors, tailgate, liftgate, bumpers and front fenders.

2   Only general body maintenance practices and body panel repair procedures within the scope of the do-it-yourselfer are included in this Chapter.

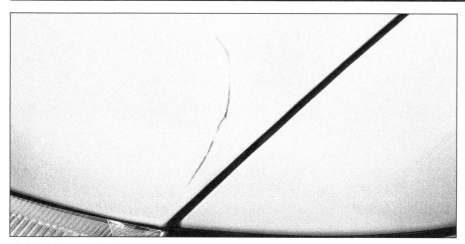

2.1a Make sure the damaged area is perfectly clean and rust free. If the touch-up kit has a wire brush, use it to clean the scratch or chip. Or use fine steel wool wrapped around the end of a pencil. Clean the scratched or chipped surface only, not the good paint surrounding it. Rinse the area with water and allow it to dry thoroughly

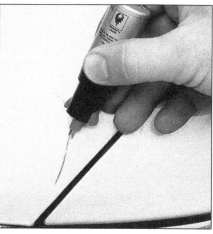

2.1b Thoroughly mix the paint, then apply a small amount with the touch-up kit brush or a very fine artist's brush. Brush in one direction as you fill the scratch area. Do not build up the paint higher than the surrounding paint

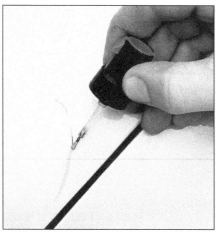

2.1c If the vehicle has a two-coat finish, apply the clear coat after the color coat has dried

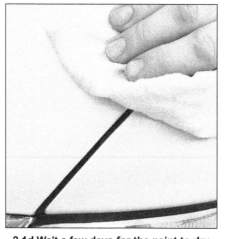

2.1d Wait a few days for the paint to dry thoroughly, then rub out the repainted area with a polishing compound to blend the new paint with the surrounding area. When you're happy with your work, wash and polish the area

2   Repairing minor paint scratches

No matter how hard you try to keep your vehicle looking like new, it will inevitably be scratched, chipped or dented at some point. If the metal is actually dented, seek the advice of a professional. But you can fix minor scratches and chips yourself. Buy a touch-up paint kit from a dealer service department or an auto parts store. To ensure that you get the right color, you'll need to have the specific make, model and year of your vehicle and, ideally, the paint code, which is located on a special metal plate under the hood or in the door jamb.

3   Body repair - minor damage

Plastic body panels

1   The following repair procedures are for minor scratches and gouges. Repair of more serious damage should be left to a dealer ser-

vice department or qualified auto body shop. Below is a list of the equipment and materials necessary to perform the following repair procedures on plastic body panels.

> *Wax, grease and silicone removing solvent*
> *Cloth-backed body tape*
> *Sanding discs*
> *Drill motor with three-inch disc holder*
> *Hand sanding block*
> *Rubber squeegees*
> *Sandpaper*
> *Non-porous mixing palette*
> *Wood paddle or putty knife*
> *Curved-tooth body file*
> *Flexible parts repair material*

Flexible panels (bumper trim)

2   Remove the damaged panel, if necessary or desirable. In most cases, repairs can be carried out with the panel installed.

3   Clean the area(s) to be repaired with a

wax, grease and silicone removing solvent applied with a water-dampened cloth.

4   If the damage is structural, that is, if it extends through the panel, clean the backside of the panel area to be repaired as well. Wipe dry.

5   Sand the rear surface about 1-1/2 inches beyond the break.

6   Cut two pieces of fiberglass cloth large enough to overlap the break by about 1-1/2 inches. Cut only to the required length.

7   Mix the adhesive from the repair kit according to the instructions included with the kit, and apply a layer of the mixture approximately 1/8-inch thick on the backside of the panel. Overlap the break by at least 1-1/2 inches.

8   Apply one piece of fiberglass cloth to the adhesive and cover the cloth with additional adhesive. Apply a second piece of fiberglass cloth to the adhesive and immediately cover the cloth with additional adhesive in sufficient quantity to fill the weave.

9   Allow the repair to cure for 20 to 30 minutes at 60-degrees to 80-degrees F.

10   If necessary, trim the excess repair material at the edge.

11   Remove all of the paint film over and around the area(s) to be repaired. The repair material should not overlap the painted surface.

12   With a drill motor and a sanding disc (or a rotary file), cut a "V" along the break line approximately 1/2-inch wide. Remove all dust and loose particles from the repair area.

13   Mix and apply the repair material. Apply a light coat first over the damaged area; then continue applying material until it reaches a level slightly higher than the surrounding finish.

14   Cure the mixture for 20 to 30 minutes at 60-degrees to 80-degrees F.

15   Roughly establish the contour of the area being repaired with a body file. If low areas or pits remain, mix and apply additional adhesive.

16   Block sand the damaged area with sandpaper to establish the actual contour of the

surrounding surface.

17   If desired, the repaired area can be temporarily protected with several light coats of primer.

18   Because of the special paints and techniques required for flexible body panels, it is recommended that the vehicle be taken to a paint shop for completion of the body repair.

Steel body panels

See photo sequence.

Repair of dents

19   When repairing dents, the first job is to pull the dent out until the affected area is as close as possible to its original shape. There is no point in trying to restore the original shape completely as the metal in the damaged area will have stretched on impact and cannot be restored to its original contours. It is better to bring the level of the dent up to a point that is about 1/8-inch below the level of the surrounding metal. In cases where the dent is very shallow, it is not worth trying to pull it out at all.

20   If the backside of the dent is accessible, it can be hammered out gently from behind using a soft-face hammer. While doing this, hold a block of wood firmly against the opposite side of the metal to absorb the hammer blows and prevent the metal from being stretched.

21   If the dent is in a section of the body which has double layers, or some other factor makes it inaccessible from behind, a different technique is required. Drill several small holes through the metal inside the damaged area, particularly in the deeper sections. Screw long, self-tapping screws into the holes just enough for them to get a good grip in the metal. Now pulling on the protruding heads of the screws with locking pliers can pull out the dent.

22   The next stage of repair is the removal of paint from the damaged area and from an inch or so of the surrounding metal. This is easily done with a wire brush or sanding disk in a drill motor, although it can be done just as effectively by hand with sandpaper. To complete the preparation for filling, score the surface of the bare metal with a screwdriver or the tang of a file or drill small holes in the affected area. This will provide a good grip for the filler material. To complete the repair, see the Section on filling and painting.

Repair of rust holes or gashes

23   Remove all paint from the affected area and from an inch or so of the surrounding metal using a sanding disk or wire brush mounted in a drill motor. If these are not available, a few sheets of sandpaper will do the job just as effectively.

24   With the paint removed, you will be able to determine the severity of the corrosion and decide whether to replace the whole panel, if possible, or repair the affected area. New body panels are not as expensive as most people think and it is often quicker to install a new panel than to repair large areas of rust.

25   Remove all trim pieces from the affected area except those which will act as a guide to the original shape of the damaged body, such as headlight shells, etc. Using metal snips or a hacksaw blade, remove all loose metal and any other metal that is badly affected by rust. Hammer the edges of the hole in to create a slight depression for the filler material.

26   Wire-brush the affected area to remove the powdery rust from the surface of the metal. If the back of the rusted area is accessible, treat it with rust inhibiting paint.

27   Before filling is done, block the hole in some way. This can be done with sheet metal riveted or screwed into place, or by stuffing the hole with wire mesh.

28   Once the hole is blocked off, the affected area can be filled and painted. See the following subsection on filling and painting.

Filling and painting

29   Many types of body fillers are available, but generally speaking, body repair kits which contain filler paste and a tube of resin hardener are best for this type of repair work. A wide, flexible plastic or nylon applicator will be necessary for imparting a smooth and contoured finish to the surface of the filler material. Mix up a small amount of filler on a clean piece of wood or cardboard (use the hardener sparingly). Follow the manufacturer's instructions on the package, otherwise the filler will set incorrectly.

30   Using the applicator, apply the filler paste to the prepared area. Draw the applicator across the surface of the filler to achieve the desired contour and to level the filler surface. As soon as a contour that approximates the original one is achieved, stop working the paste. If you continue, the paste will begin to stick to the applicator. Continue to add thin layers of paste at 20-minute intervals until the level of the filler is just above the surrounding metal.

31   Once the filler has hardened, the excess can be removed with a body file. From then on, progressively finer grades of sandpaper should be used, starting with a 180-grit paper and finishing with 600-grit wet-or-dry paper. Always wrap the sandpaper around a flat rubber or wooden block, otherwise the surface of the filler will not be completely flat. During the sanding of the filler surface, the wet-or-dry paper should be periodically rinsed in water. This will ensure that a very smooth finish is produced in the final stage.

32   At this point, the repair area should be surrounded by a ring of bare metal, which in turn should be encircled by the finely feathered edge of good paint. Rinse the repair area with clean water until all of the dust produced by the sanding operation is gone.

33   Spray the entire area with a light coat of primer. This will reveal any imperfections in the surface of the filler. Repair the imperfections with fresh filler paste or glaze filler and once more smooth the surface with sandpaper. Repeat this spray-and-repair procedure until you are satisfied that the surface of the filler and the feathered edge of the paint are perfect. Rinse the area with clean water and allow it to dry completely.

34   The repair area is now ready for painting. Spray painting must be carried out in a warm, dry, windless and dust free atmosphere. These conditions can be created if you have access to a large indoor work area, but if you are forced to work in the open, you will have to pick the day very carefully. If you are working indoors, dousing the floor in the work area with water will help settle the dust that would otherwise be in the air. If the repair area is confined to one body panel, mask off the surrounding panels. This will help minimize the effects of a slight mismatch in paint color. Trim pieces such as chrome strips, door handles, etc., will also need to be masked off or removed. Use masking tape and several thickness of newspaper for the masking operations.

35   Before spraying, shake the paint can thoroughly, then spray a test area until the spray painting technique is mastered. Cover the repair area with a thick coat of primer. The thickness should be built up using several thin layers of primer rather than one thick one. Using 600-grit wet-or-dry sandpaper, rub down the surface of the primer until it is very smooth. While doing this, the work area should be thoroughly rinsed with water and the wet-or-dry sandpaper periodically rinsed as well. Allow the primer to dry before spraying additional coats.

36   Spray on the top coat, again building up the thickness by using several thin layers of paint. Begin spraying in the center of the repair area and then, using a circular motion, work out until the whole repair area and about two inches of the surrounding original paint is covered. Remove all masking material 10 to 15 minutes after spraying on the final coat of paint. Allow the new paint at least two weeks to harden, then use a very fine rubbing compound to blend the edges of the new paint into the existing paint. Finally, apply a coat of wax

4   Body repair - major damage

1   Major damage must be repaired by an auto body shop specifically equipped to perform body and frame repairs. These shops have the specialized equipment required to do the job properly.

2   If the damage is extensive, the frame must be checked for proper alignment or the vehicle's handling characteristics may be adversely affected and other components may wear at an accelerated rate.

3   Due to the fact that all of the major body components (hood, fenders, etc.) are separate and replaceable units, any seriously damaged components should be replaced rather than repaired. Sometimes the components can be found in a wrecking yard that specializes in used vehicle components, often at considerable savings over the cost of new parts.

These photos illustrate a method of repairing simple dents. They are intended to supplement *Body repair - minor damage* in this Chapter and should not be used as the sole instructions for body repair on these vehicles.

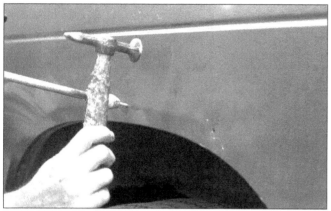

1 If you can't access the backside of the body panel to hammer out the dent, pull it out with a slide-hammer-type dent puller. Tap with a hammer near the edge of the dent to help 'pop' the metal back to its original shape, about 1/8-inch below the surface of the surrounding metal

2 Using coarse-grit sandpaper, remove the paint down to the bare metal. Clean the repair area with wax/silicone remover.

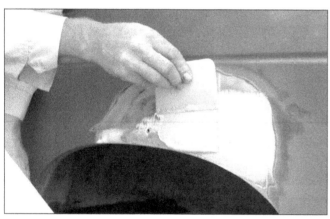

3 Following label instructions, mix up a batch of plastic filler and hardener, then quickly press it into the metal with a plastic applicator. Work the filler until it matches the original contour and is slightly above the surrounding metal

4 Let the filler harden until you can just dent it with your fingernail. File, then sand the filler down until it's smooth and even. Work down to finer grits of sandpaper - always using a board or block - ending up with 360 or 400 grit

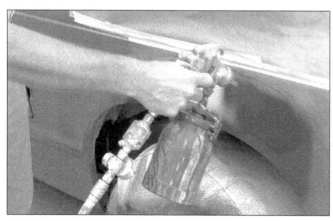

5 When the area is smooth to the touch, clean the area and mask around it. Apply several layers of primer to the area. A professional-type spray gun is being used here, but aerosol spray primer works fine

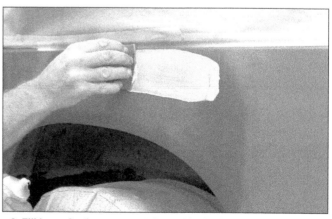

6 Fill imperfections or scratches with glazing compound. Sand with 360 or 400-grit and re-spray. Finish sand the primer with 600 grit, clean thoroughly, then apply the finish coat. Don't attempt to rub out or wax the repair area until the paint has dried completely (at least two weeks)

5   Upholstery, carpets and vinyl trim - maintenance

Upholstery and carpets

1   Every three months remove the floormats and clean the interior of the vehicle (more frequently if necessary). Use a stiff whiskbroom to brush the carpeting and loosen dirt and dust, then vacuum the upholstery and carpets thoroughly, especially along seams and crevices.

2   Dirt and stains can be removed from carpeting with basic household or automotive carpet shampoos available in spray cans. Follow the directions and vacuum again, then use a stiff brush to bring back the nap of the carpet.

3   Most interiors have cloth or vinyl upholstery, either of which can be cleaned and maintained with a number of material-specific cleaners or shampoos available in auto supply stores. Follow the directions on the product for usage, and always spot-test any upholstery cleaner on an inconspicuous area (bottom edge of a backseat cushion) to ensure that it doesn't cause a color shift in the material.

4   After cleaning, vinyl upholstery should be treated with a protectant.

Note: *Make sure the protectant container indicates the product can be used on seats - some products may make a seat too slippery.*

Caution: *Do not use protectant on vinyl-covered steering wheels.*

5   Leather upholstery requires special care. It should be cleaned regularly with saddle-soap or leather cleaner. Never use alcohol, gasoline, nail polish remover or thinner to clean leather upholstery.

6   After cleaning, regularly treat leather upholstery with a leather conditioner, rubbed in with a soft cotton cloth. Never use car wax on leather upholstery.

7   In areas where the interior of the vehicle is subject to bright sunlight, cover leather seating areas of the seats with a sheet if the vehicle is to be left out for any length of time.

Vinyl trim

8   Don't clean vinyl trim with detergents, caustic soap or petroleum-based cleaners. Plain soap and water works just fine, with a soft brush to clean dirt that may be ingrained. Wash the vinyl as frequently as the rest of the vehicle.

9   After cleaning, application of a high-quality rubber and vinyl protectant will help prevent oxidation and cracks. The protectant can also be applied to weather-stripping, vacuum lines and rubber hoses, which often fail as a result of chemical degradation, and to the tires.

6   Fastener and trim removal

1   There is a variety of plastic fasteners used to hold trim panels, splash shields and other parts in place in addition to typical screws, nuts and bolts. Once you are familiar with them, they can usually be removed without too much difficulty.

2   The proper tools and approach can prevent added time and expense to a project by minimizing the number of broken fasteners and/or parts.

3   The following illustration shows various types of fasteners that are typically used on most vehicles and how to remove and install them **(see illustration)**. Replacement fasteners are commonly found at most auto parts stores, if necessary.

Fasteners

This tool is designed to remove special fasteners. A small pry tool used for removing nails will also work well in place of this tool

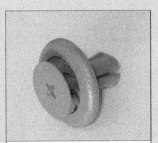
A Phillips head screwdriver can be used to release the center portion, but light pressure must be used because the plastic is easily damaged. Once the center is up, the fastener can easily be pried from its hole

Here is a view with the center portion fully released. Install the fastener as shown, then press the center in to set it

This fastener is used for exterior panels and shields. The center portion must be pried up to release the fastener. Install the fastener with the center up, then press the center in to set it

This type of fastener is used commonly for interior panels. Use a small blunt tool to press the small pin at the center in to release it . . .

. . . the pin will stay with the fastener in the released position

Reset the fastener for installation by moving the pin out. Install the fastener, then press the pin flush with the fastener to set it

This fastener is used for exterior and interior panels. It has no moving parts. Simply pry the fastener from its hole like the claw of a hammer removes a nail. Without a tool that can get under the top of the fastener, it can be very difficult to remove

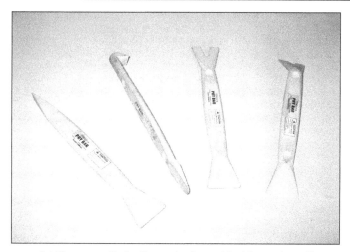

6.4 These small plastic pry tools are ideal for prying off trim panels

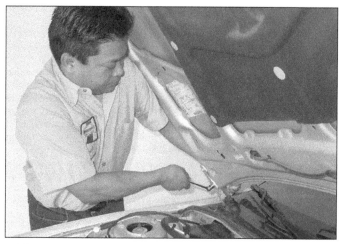

7.4 Support the hood with your shoulder while removing the hood bolts (typical shown)

4   Trim panels are typically made of plastic and their flexibility can help during removal. The key to their removal is to use a tool to pry the panel near its retainers to release it without damaging surrounding areas or breaking-off any retainers. The retainers will usually snap out of their designated slot or hole after force is applied to them. Stiff plastic tools designed for prying on trim panels are available at most auto parts stores **(see illustration)**. Tools that are tapered and wrapped in protective tape, such as a screwdriver or small pry tool, are also very effective when used with care.

---

7   Hood - removal, installation and adjustment

Note: *The hood is awkward to remove and install; at least two people should perform this procedure.*

Removal and installation

1   Open the hood, then place blankets or pads over the fenders and cowl area of the body. This will protect the body and paint as the hood is lifted off.
2   Make marks around the hood hinge to ensure proper alignment during installation.
3   Remove the hood strut and washer hose.
4   Have an assistant support one side of the hood. Take turns removing the hinge-to-hood bolts and lift off the hood **(see illustration)**.
5   Installation is the reverse of removal. Align the hinge bolts with the marks made in Step 2.

Adjustment

6   Fore-and-aft and side-to-side adjustment of the hood is done by moving the hinges after loosening the hinge-to-body bolts.
7   Loosen the bolts and move the hood into correct alignment. Move it only a little at a time. Tighten the hinge bolts and carefully

lower the hood to check the position.
8   Finally, adjust the hood bumpers on the radiator support so the hood, when closed, is flush with the fenders **(see illustration)**.
9   The hood latch assembly, as well as the hinges, should be periodically lubricated with white, lithium-base grease to prevent binding and wear.

---

8   Hood latch and release cable - removal and installation

Latch

1   Open the hood and remove the sight shield. Scribe a line around the latch to aid alignment when installing. Remove the retaining bolts securing the hood latch to the radiator support **(see illustration)** and remove the latch.
2   Disconnect the remote start connector (if equipped).

7.8 To adjust the vertical height of the leading edge of the hood so that it's flush with the fenders, turn each edge cushion clockwise (to lower the hood) or counterclockwise (to raise the hood)

8.1 Hood latch mounting fasteners

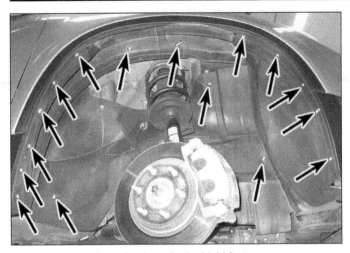

9.2 Inner fender splash shield fasteners

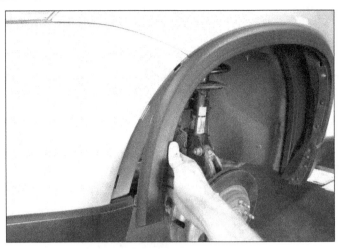

9.3 Carefully separate the fender trim panel from the bumper cover

3   Disconnect the hood release cable by disengaging the cable from the latch.

4   Installation is the reverse of removal.

Cable

5   Open the hood and remove the sight shield. Remove the air filter (see Chapter 4). Remove the cable from the latch and bracket.

6   Remove the driver's side instrument panel cover (see Section 22).

7   Release the cable from the retainer and slide the end out of the handle.

8   Pull the cable through from inside the vehicle.

9   Installation is the reverse of removal.

---

9   Bumper covers - removal and installation

Note: *Refer to Section 6 for fastener and trim removal.*

Front bumper cover

1   Apply the parking brake, raise the vehicle and support it securely on jackstands.

2   Remove the front inner fender splash shields from the bumper cover (**see illustration**).

3   Carefully pry the fender trim panel from the bumper cover (**see illustration**).

4   Remove the fasteners joining the bumper cover and the lower front part of each fender (**see illustration**).

5   Remove the front air dam mounting fasteners (if equipped).

6   Remove the fasteners from the top and the bottom of the bumper cover (**see illustrations**).

7   Carefully pull the bumper cover away from the front and sides of the vehicle and remove it completely. Using a flat screwdriver press the tabs in between the bumper and fenders, and the headlight bracket. Then disconnect all electric connections.

8   Installation is the reverse of removal. An assistant is helpful at this point.

9.4 The bumper cover-to-fender screw location

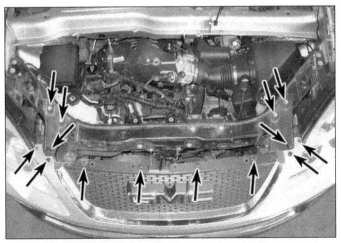

9.6a Bumper cover top mounting fasteners

9.6b Bumper cover bottom mounting fasteners

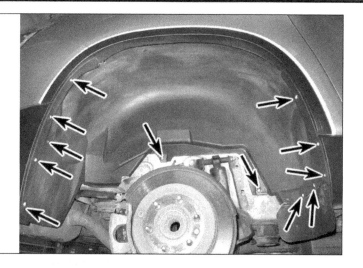

9.9 Inner fender splash shield fasteners (rear)

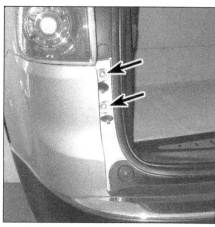

9.10 The rear bumper cover upper mounting bolts are located under these small covers

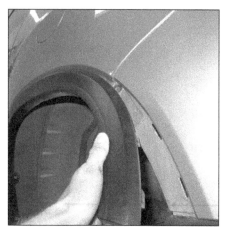

9.11 Carefully separate the fender trim panel from the bumper cover, then remove the fastener from the upper corner of the bumper cover

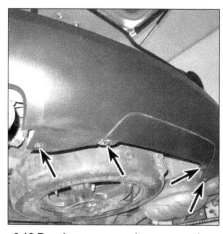

9.12 Rear bumper cover lower mounting fasteners

13   Pull the bumper cover out and away from the vehicle while disengaging the tabs.
14   Disconnect all electrical connections as needed.
15   Installation is the reverse of removal.

10   Front fender - removal and installation

Note: *Refer to Section 6 for fastener and trim removal.*
1   Loosen the front wheel lug nuts. Raise the vehicle, support it securely on jackstands and remove the front wheel.
2   Remove the inner fender splash shield.
3   Pull the rocker molding away enough to remove the fender lower mounting bolt **(see illustration)**.
4   Remove the upper fender mounting bolt inside the doorjamb **(see illustration)**.
5   Inside the fender, remove the fender mounting fastener **(see illustration)**.
6   Remove the fasteners along the top edge of the fender, then lift off the fender **(see**

Rear bumper cover
9   Remove the inner fender splash shield fasteners that attach to the bumper **(see illustration)**.

10   Open the liftgate, pry off the bolt covers from the top of the bumper cover, then remove the fasteners **(see illustration)**.
11   Carefully pry off the fender trim panel from the bumper cover **(see illustration)**.
12   Detach the fasteners securing the bottom of the bumper cover **(see illustration)**.

10.3 Pull the molding away and remove the fender lower mounting bolts

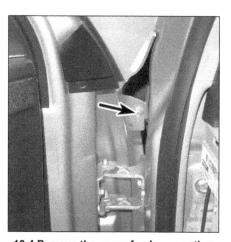

10.4 Remove the upper fender mounting bolt in the doorjamb area

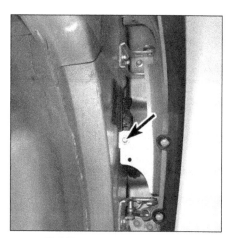

10.5 Remove the inner fender mounting fastener

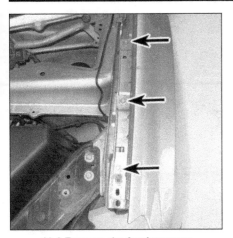

10.6 Remove the fender upper mounting fasteners

11.2a Carefully pry off the trim cover . . .

11.2b . . . then remove the screw

illustration). It's a good idea to have an assistant support the fender while it's being moved away from the vehicle to prevent damage to the surrounding body panels.

7   Installation is the reverse of removal. Check the alignment of the fender to the hood and front edge of the door before final tightening of the fender fasteners.

11   Door trim panels - removal and installation

Caution: *Wear gloves when working inside the door openings to protect against sharp metal edges.*
Note: *Refer to Section 6 for fastener and trim removal.*
1   Disconnect the cable from the negative battery terminal (see Chapter 5).

Front and rear doors
2   Remove the trim cover from the inside handle bezel and remove the screw from the bezel **(see illustrations)**.
3   Pry off the small trim that covers the pull-

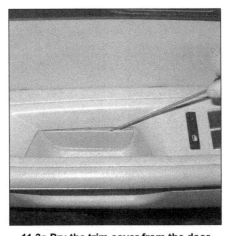

11.3a Pry the trim cover from the door pull handle . . .

handle, then remove the fasteners securing the pull handle **(see illustrations)**.
4   Detach the door trim fasteners by prying the door trim away from the door near the outside edge **(see illustration)**, disconnecting any electrical connectors.

11.3b . . . then remove the fasteners

5   For access to the door outside handle or the door window regulator inside the door, raise the window fully, then carefully peel back the plastic watershield **(see illustration)**.
6   Installation is the reverse of removal. Align the door trim fasteners with the holes in the door, then push the door trim into place.

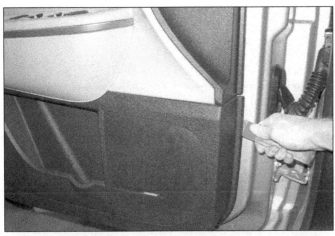

11.4 Pry the door trim away from the door around the edges

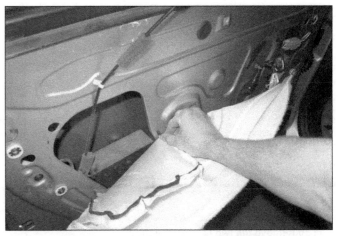

11.5 Carefully peel back the plastic watershield

11.7 Detach the power liftgate strut retaining clip

11.8 Carefully pry off the upper trim panel from the liftgate

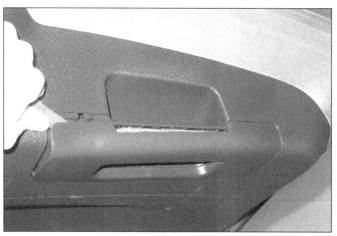

11.9a Pry off the handle cover . . .

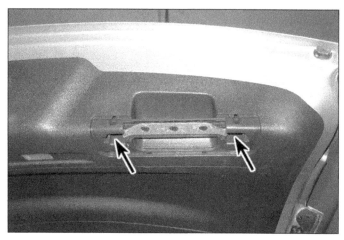

11.9b . . . then remove the two fasteners

Liftgate

7   Support the liftgate in the open position, then disconnect the power liftgate strut, if equipped **(see illustration)**.

8   Carefully pry off the upper trim panel **(see illustration)**.

9   Carefully pry off the pull handle cover, then remove the fasteners below **(see illustrations)**.

10   Remove the fasteners from the bottom of the liftgate **(see illustration)**.

11   Disengage the liftgate trim panel clips using a door panel removal tool **(see illustration),** and remove the lower section of the panel.

12   Disconnect electrical connections. Carefully peel back the plastic watershield as necessary to access other components.

13   Installation is the reverse of removal.

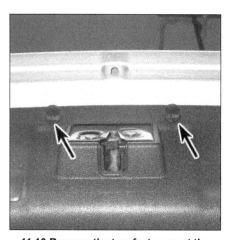

11.10 Remove the two fasteners at the bottom of the liftgate

11.11 Carefully pry the door trim panel loose from its mounting clips

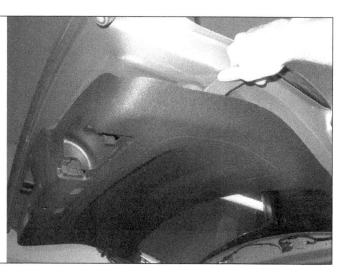

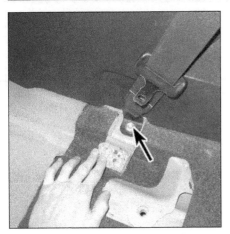

12.3 Remove the seat belt anchor

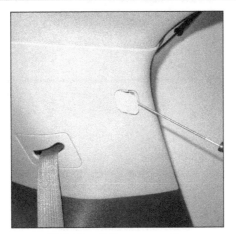

12.4a Carefully pry open the access cover . . .

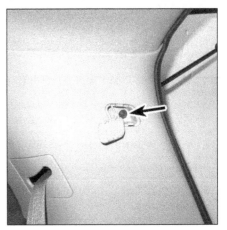

12.4b . . . then remove the fastener

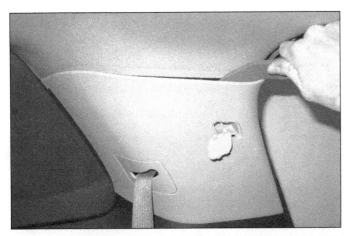

12.5 Use a trim removal tool and carefully pry the upper panel off

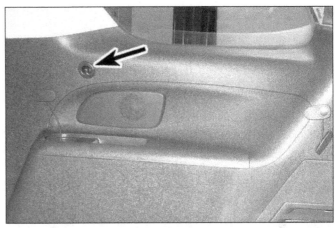

12.6a Remove this cargo hook . . .

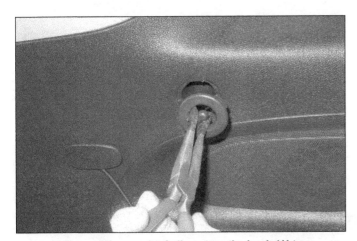

12.6b . . . using a pair of pliers, turn the hook 1/4 turn counterclockwise to remove it . . .

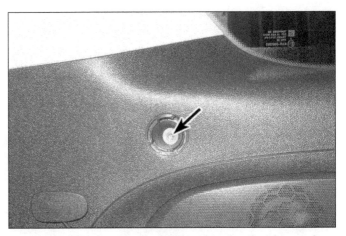

12.6c . . . then remove the fastener securing the trim panel

12   Rear quarter trim panel - removal and installation

Warning: *The models covered by this manual are equipped with Supplemental Restraint Systems (SRS), more commonly known as airbags. Always disable the airbag system* before working in the vicinity of any airbag system components to avoid the possibility of accidental deployment of the airbags, which could cause personal injury (see Chapter 12).

1   Disconnect the cable from the negative battery terminal (see Chapter 5).
2   Remove the third row number two seat (see Section 24).

3   Remove the seat belt anchor fastener **(see illustration)**.
4   Open the access door and remove the upper trim fastener **(see illustrations)**.
5   Carefully pry off the upper trim panel **(see illustration)**.
6   Remove the cargo tie down hook and trim panel fastener **(see illustrations)**.

7 Remove the fasteners securing the rear of the trim panel **(see illustrations).**

8 Pull up the armrest and remove it, then remove the fastener securing the armrest **(see illustrations).**

9 Remove the liftgate door sill cover **(see illustration).**

10 Pull the lower quarter trim outwards carefully to remove it; be sure to disconnect any electrical connections.

11 Installation is the reverse of removal.

---

13 Door - removal, installation and adjustment

Warning: *The models covered by this manual are equipped with Supplemental Restraint Systems (SRS), more commonly known as airbags. Always disable the airbag system before working in the vicinity of any airbag system component to avoid the possibility of accidental deployment of the airbag, which could cause personal injury (see Chapter 12).*

Note: *The door is heavy and somewhat awkward to remove - at least two people should perform this procedure.*

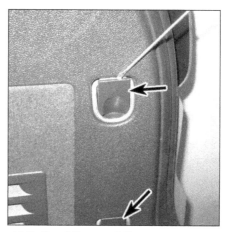

12.7a Carefully pry off the two covers . . .

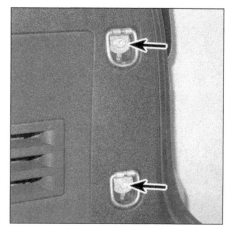

12.7b . . . then remove the fasteners

Removal and installation

1 Raise the window completely in the door.

2 Disconnect the cable from the negative battery terminal (see Chapter 5).

3 Open the door all the way and support it with a jack or blocks covered with rags to prevent damaging the outer surface.

4 Disconnect the electrical connector **(see illustration).**

5 Mark around the door hinges with a pen or a scribe to facilitate realignment during reassembly.

6 With an assistant holding the door, remove the hinge-to-door bolts and lift off the door.

7 Installation is the reverse of removal.

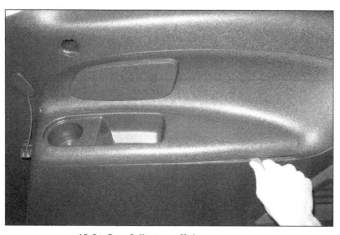

12.8a Carefully pry off the armrest . . .

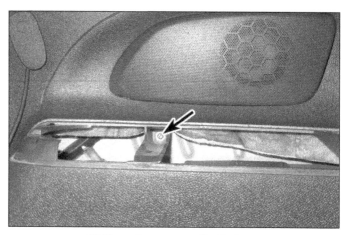

12.8b . . . then remove the fastener

12.9 Carefully pry off the door sill cover

13.4 Lift the latch to disconnect the electrical connecter

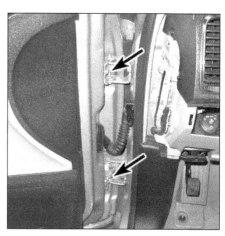

13.6 Remove the hinge fasteners

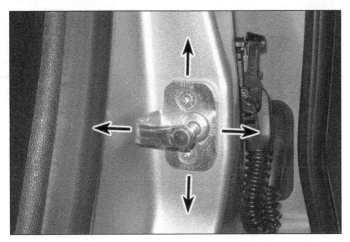

13.11 Adjust the door lock striker by loosening the mounting screws and gently tapping the striker in the desired direction

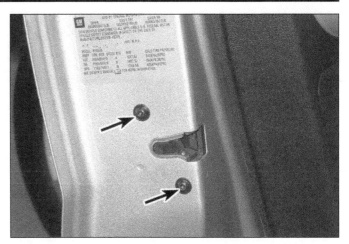

14.5 Door latch mounting fasteners

14.7a Remove the access cover . . .

14.7b . . . then loosen the retainer

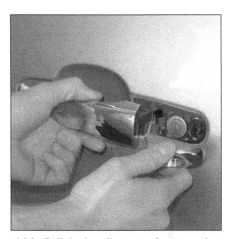

14.8a Pull the handle out and remove the small cover . . .

Adjustment

8   Having proper door-to-body alignment is a critical part of a well-functioning door assembly. First check the door hinge pins for excessive play. Fully open the door and lift up and down on the door without lifting the body. If a door has 1/16-inch or more excessive play, the hinges should be replaced.

9   Door-to-body alignment adjustments are made by loosening the hinge-to-body bolts or hinge-to-door bolts and moving the door. Proper body alignment is achieved when the top of the doors are parallel with the roof section, the front door is flush with the fender, the rear door is flush with the rear quarter panel and the bottom of the doors are aligned with the lower rocker panel. If these goals can't be reached by adjusting the hinge-to-body or hinge-to-door bolts, body alignment shims may have to be purchased and inserted behind the hinges to achieve correct alignment. **Note:** *It is necessary to remove the front fender to access the hinge-to-body bolts (see Section 10).*

10   To adjust the door-closed position, scribe a line or mark around the striker plate to provide a reference point, then check that the door

latch is contacting the center of the latch striker. If not, adjust the up and down position first.
11   Finally adjust the latch striker sideways position, so that the door panel is flush with the center pillar or rear quarter panel and provides positive engagement with the latch mechanism **(see illustration).**

---

14   Door latch, lock cylinder and handles - removal and installation

---

Note: *Based on the modular design of the window regulator, some of the following components are removed as an assembly along with the window regulator module.*

Door latch

1   Remove door trim panel (see Section 11), then remove the watershield.
2   Remove the retainers for the cable.
3   Disconnect the electrical connector from the latch.
4   Detach the rods from the latch.
5   Remove the latch mounting fasteners **(see illustration)** and remove the latch.
6   Installation is the reverse of removal.

14.8b . . . then slide the handle to the left and remove it

Outside handle

7   Open the door and remove the small access cover, then loosen - but don't remove - the retainer **(see illustrations).**
8   Pull the door handle out halfway and remove the small cover, then remove the handle **(see illustrations).**
9   Installation is the reverse of removal.

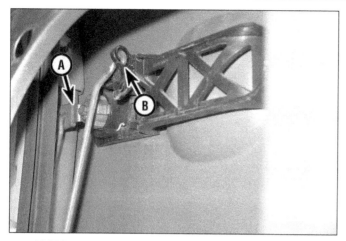

14.12 Disconnect the latch rod (A) and the handle rod (B)

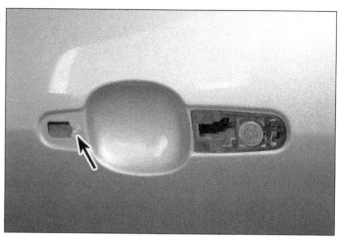

14.13 Remove this fastener securing the door handle housing

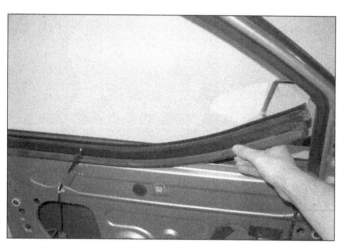

15.2 Pull the seal off of the pinch weld at the window opening

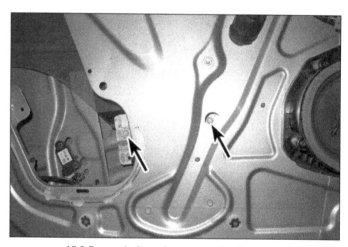

15.3 Door window glass mounting fasteners

Door lock cylinder

10 Remove the door trim panel and plastic watershield (see Section 11).

11 Remove the door handle (see Steps 7 through 9).

12 Disconnect the latch rod from the lock cylinder and the handle rod **(see Illustration).**

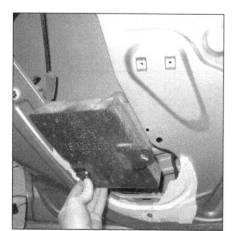

16.2 Remove the fasteners, then remove the energy absorber

13 Remove the fastener securing the door handle housing **(see illustration).**

14 Release the retainer from the rear of the door lock cylinder.

15 Push the lock out of housing.

16 Installation is the reverse of removal.

15　Door window glass - removal and installation

Caution: *Wear gloves when working inside the door openings to protect against cuts from sharp metal edges.*
Note: *Refer to Section 6 for fastener and trim removal.*

Door glass

1 Remove the door trim panel and the plastic watershield (see Section 11).

2 Pull the window seal off of the pinch weld at the window opening **(see illustration).**

3 Lower the window until you can access the fasteners on the window sash. Disconnect the cable from the negative battery terminal (see Chapter 5) to prevent accidental raising or lowering of window, then remove the fas-

teners **(see illustration).**

4 Guide the window up and tilt it forward to remove it through the window opening.

5 Installation is the reverse of removal.

Liftgate, windshield and quarter window glass

6 Replacement of these glass pieces requires the use of special fast-setting urethane adhesive materials, and specialized tools and techniques. These operations should be left to a dealer service department or a shop specializing in glass work.

16　Door window glass regulator - removal and installation

Caution: *Wear gloves when working inside the door openings to protect against cuts from sharp metal edges.*
Note: *Refer to Section 6 for fastener and trim removal.*

1 Remove door trim panel and watershield (see Section 11).

2 Remove the energy absorber **(see illustration).**

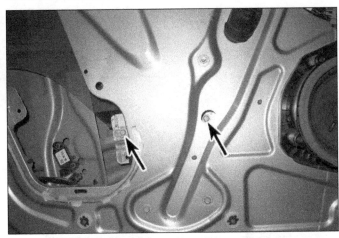

16.3 Window glass clamp mounting fasteners

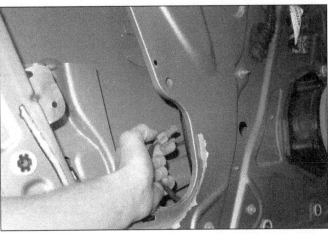

16.4 Disconnect the window regulator motor electrical connector

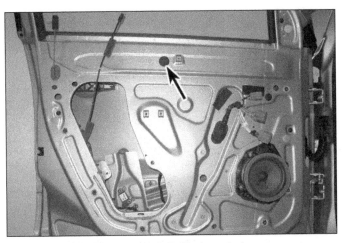

16.5a Remove regulator fastener hole cover

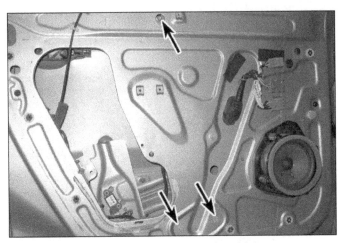

16.5b Window regulator assembly mounting fasteners

3   Lower the window glass until the glass clamp fasteners are accessible through the access holes **(see illustration)**. Loosen (but don't remove) both glass clamp fasteners. Raise the window and secure it with tape so that it won't go down.

4   Disconnect the regulator motor electrical connector **(see illustration).**

5   Remove the regulator fastener hole cover, then remove the regulator assembly mounting fasteners **(see illustrations)**.

6   Remove the window regulator assembly from the door.

7   Installation is the reverse of removal. Lubricate the rollers and wear points on the regulator with white grease before installation.

---

17   Mirrors - removal and installation

Outside mirrors

1   Remove the door trim panel (see Section 11) and molding.

2   Disconnect the electrical connector **(see illustration).**

3   Remove the mirror fasteners **(see illustration 17.2)**.

4   Installation is the reverse of removal.

Inside mirror

5   Remove the mirror cover fastener.

6   Disconnect the electrical connector from the mirror, if equipped.

7   Remove the setscrew located at the base of the mirror stalk.

8   Slide the mirror upwards to remove it from its mount.

9   If the mount plate itself has come off the windshield, adhesive kits are available at auto parts stores to resecure it. Follow the instructions included with the kit. Be sure to position the flanged part of the mount away from the glass and the narrower part pointing up.

17.2 Outside mirror electrical connecter (A) and mounting fasteners (B)

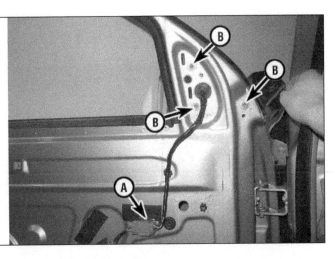

18.4 Disconnect the electrical connections

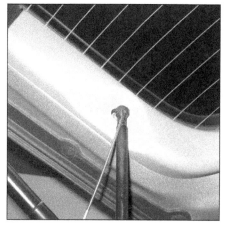

18.5 Detach the liftgate strut retaining clips

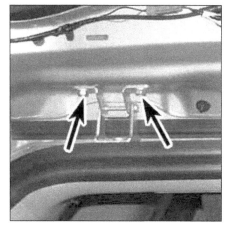

18.6 Have an assistant support the liftgate as you detach the liftgate hinges

18 Liftgate - removal, installation and adjustment

Note: *The liftgate is heavy and awkward to remove - at least two people should perform this procedure.*

Removal and installation

1   Disconnect the cable from the negative battery terminal (see Chapter 5).

2   Open the liftgate and support it securely.
3   Remove the liftgate upper trim panel (see Section 11).
4   Disconnect the electrical connections **(see illustration).**
5   With an assistant holding the liftgate, disconnect the support struts and actuator rod by prying out the small clips, then pulling the struts from their ballstuds **(see illustration)**.
6   With an assistant holding the liftgate, remove the hinge-to-liftgate fasteners and lift off the liftgate **(see illustration)**.
Note: *Draw a reference line around the hinges on the liftgate before removing the fasteners.*
7   Installation is the reverse of removal.

Adjustment

8   Having proper liftgate-to-body alignment is a critical part of a well-functioning liftgate assembly. First check the liftgate hinge pins for excessive play. Fully open the liftgate and lift up and down on the liftgate without lifting the body. If a liftgate has 1/8-inch or more excessive play, the hinges should be replaced.
9   Liftgate-to-body alignment adjustments are made by loosening the hinge-to-body bolts or hinge-to-liftgate bolts and moving the liftgate. Proper body alignment is achieved when the top of the liftgate is parallel with the roof section and the sides of the liftgate are flush with the rear quarter panels and the bottom of the liftgate is aligned with the lower

liftgate sill. If these goals can't be reached by adjusting the hinge-to-body or hinge-to-liftgate bolts, body alignment shims may have to be purchased and inserted behind the hinges to achieve correct alignment.
10   To adjust the liftgate-closed position, scribe a line or mark around the striker plate to provide a reference point, then check that the liftgate latch is contacting the center of the latch striker. If not, adjust the up and down position first.
11   Finally adjust the latch striker sideways position, so that the liftgate panel is flush with the rear quarter panel and provides positive engagement with the latch mechanism.

19 Liftgate latch and outside handle - removal and installation

1   If the liftgate cannot be opened electrically, a manual over-ride is provided. Remove the plug in the liftgate trim and insert a blunt tool in the hole to release the latch and open the liftgate **(see illustrations)**.
2   Open the liftgate and remove the trim panel and watershield (see Section 11).

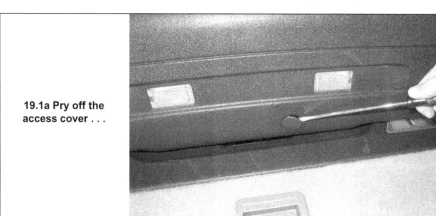

19.1a Pry off the access cover . . .

19.1b . . . then insert a blunt tool in the hole to release the latch

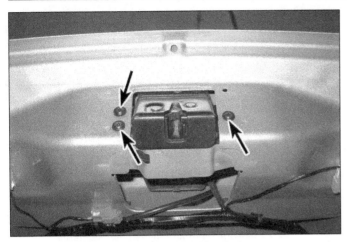

19.4 Liftgate latch mounting fasteners

20.2 Remove the cowl cover mounting fasteners

20.3 Carefully pry off the cowl cover

21.2 Carefully pry off the trim panel

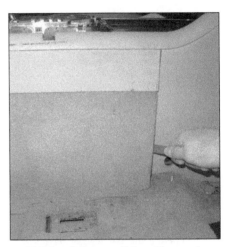

21.3 Carefully pry off the side panels

Liftgate latch

3   Disconnect the electrical connector from the latch.
4   Remove the fasteners securing the latch **(see illustration)**.
5   Installation is the reverse of removal.

Liftgate outside handle

6   Remove the handle fasteners, then disconnect the electrical connectors.
7   Installation is the reverse of removal.

20   Cowl cover - removal and installation

1   Remove the wiper arms (see Chapter 12). Make sure the wipers are in the parked position and note the locations of the blades on the windshield.
2   Remove the fasteners securing the cowl cover **(see illustration)**.
3   Carefully pry up the cowl cover, then remove the cowl **(see illustration)**.

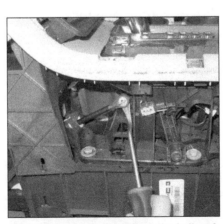

21.4a Pry the cable end from the ballstud on the shifter

21   Center console - removal and installation

1   Remove the front seats (see Section 24).
2   Remove the shifter console trim **(see illustration)**.

21.4b Pinch the tabs while pulling up on the cable to release it from the console

3   Remove the side panels from both sides of the console **(see illustration)**.
4   Disconnect the transmission shift cable from the shifter and disconnect the cable from the center console **(see illustrations)**.

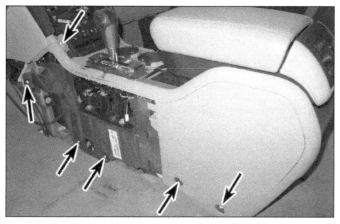

21.5 Center console mounting fasteners (left side shown; right side similar)

22.2 Knee bolster mounting fastener locations

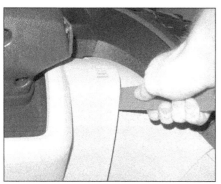

22.3 Carefully pry off the knee bolster

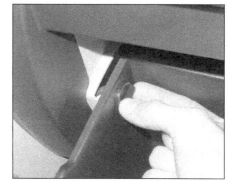

22.5 Twist the door stop pins to remove them

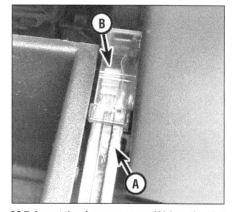

22.7 Insert the dampener arm (A) into the slot on the gear assembly until the teeth on the arm are meshed with the dampener gear (B)

5   Remove the center console mounting fasteners (see illustration).
6   Lift up the console, disconnect any electrical connectors, then remove the console.
7   Installation is the reverse of removal.

22   Dashboard trim panels - removal and installation

Warning: *These models are equipped with a Supplemental Restraint System (SRS), more commonly known as airbags. Always disable the airbag system before working in the vicinity of any airbag system component to avoid the possibility of accidental deployment of the*

airbag(s), which could cause personal injury (see Chapter 12).
Note: *Refer to Section 6 for fastener and trim removal.*

Driver's knee bolster

1   Remove the pushpin fasteners securing the left side under dash panel, then remove the panel.
2   Remove the knee bolster mounting fasteners (see illustration).
3   Pry the panel directly reward to detach the clips (see illustration), then disconnect any electrical connectors from the knee bolster.
4   Installation is the reverse of removal.

Glove box

5   Twist the two door stop pins and remove them (see illustration).
6   Pull the glove box outward, then remove the glovebox door.
7   Align the dampener arm into the dampener gear assembly (see illustration). The remainder of installation is the reverse of removal.
Note: *The teeth on the dampener arm are located on the bottom side of the arm.*

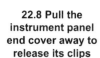

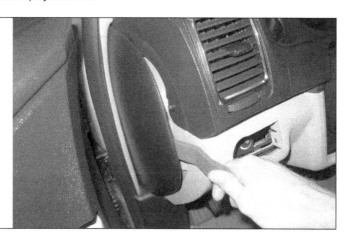

22.8 Pull the instrument panel end cover away to release its clips

22.10 Carefully pry off the center trim panel

22.11 instrument cluster bezel mounting fasteners - 2012 and earlier models

22.12 Pull the cluster bezel rearward to release its clips

22.21 Pry the trim panel out below the instrument cluster bezel

22.22 Remove the four instrument cluster bezel fasteners - 2013 and later Traverse model shown

22.24 Lift the liner out from the center compartment - Traverse model shown

Instrument panel end covers

8 Remove the end cover by pulling or prying it to disengage the clips **(see illustration)**.
9 Installation is the reverse of removal.

Center trim panel

10 Using a suitable tool pry around the outside of the trim panel to release the clips, then remove the trim panel **(see illustration)**.

Instrument cluster bezel

2012 and earlier models

11 Remove the fasteners securing the bezel **(see illustration)**.
12 Pull the bezel rearward to unclip it **(see illustration)**, then remove it from the instrument panel.
13 Installation is the reverse of removal.

2013 and later models

Acadia models

14 Remove the driver's knee bolster (see Steps 1 through 3).
15 Remove the instrument cluster bezel fasteners, then carefully pull the bezel outwards.
16 Disconnect the electrical connectors to the bezel and remove the bezel from the instrument panel.

17 Installation is the reverse of removal.

Enclave models

18 Using a plastic trim tool, pry the cluster trim plate cover from the top of the instrument cluster bezel.
19 Remove the instrument cluster bezel fasteners, then carefully pry the bezel outwards using the plastic trim tool, starting from the center and evenly working around the bezel until it can be removed. Remove the bezel from the instrument panel.
20 Installation is the reverse of removal.

Traverse models

21 Using a plastic trim tool, pry the trim panel out below the instrument cluster bezel **(see illustration)**.
22 Remove the four instrument cluster bezel fasteners **(see illustration)**, then carefully pull the bezel outwards.
23 Installation is the reverse of removal.

Instrument panel center compartment

24 Open the center compartment and remove the liner from the bottom of the compartment **(see illustration)**.
25 On 2012 and earlier Acadia models, remove the center trim panel (see Step 10).

22.26 Remove the center compartment fasteners - Traverse model shown

26 Remove the compartment fasteners **(see illustration)**, then use a plastic trim tool to pry the compartment from the instrument panel.
27 Lift the compartment out then disconnect the electrical connector, if equipped and remove the compartment.
28 Installation is the reverse of removal.

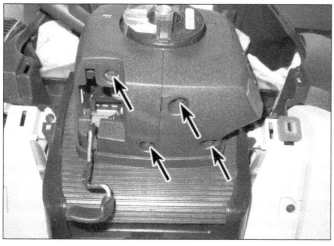

23.2 The steering column cover screws are accessed
from the bottom

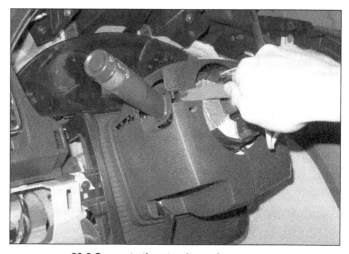

23.3 Separate the steering column covers

23.10 Steering column housing cover
mounting fasteners

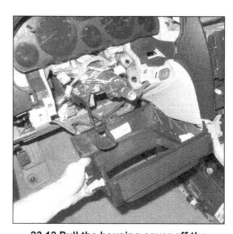

23.12 Pull the housing cover off the
steering column

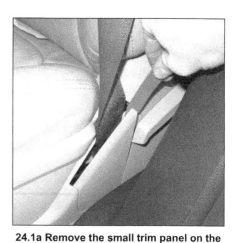

24.1a Remove the small trim panel on the
side of the seat

23   Steering column and housing covers - removal and installation

Warning: *These models are equipped with a Supplemental Restraint System (SRS), more commonly known as airbags. Always disable the airbag system before working in the vicinity of any airbag system component to avoid the possibility of accidental deployment of the airbag(s), which could cause personal injury (see Chapter 12).*

1   Disconnect the cable from the negative terminal of the battery (see Chapter 5).

Column covers

2   Remove the screws from the bottom of the lower cover **(see illustration)**.
3   Pry the upper cover up until it unclips from the lower cover **(see illustration)**.
4   Remove the small trim from around the ignition lock.
5   Remove the covers.
6   Installation is the reverse of removal.

Housing cover

7   Remove the steering wheel (see Chapter 10).
8   Remove the steering column multi-function switches (see Chapter 12).
9   Remove the knee bolster (see Section 22).
10   Remove the fasteners securing the housing cover **(see illustration)**.
11   Remove the ignition switch housing (see Chapter 12).
12   Remove the housing cover **(see illustration)**.
13   Installation is the reverse of removal.

24   Seats - removal and installation

Front seat

Warning: *The models covered by this manual are equipped with Supplemental Restraint Systems (SRS), more commonly known as airbags. Always disable the airbag system before working in the vicinity of any airbag system components to avoid the possibility of accidental deployment of the airbags, which*

could cause personal injury (see Chapter 12).
1   Remove the small trim panel to access the seat belt mounting fastener, then remove the fastener **(see illustrations)**.
2   Remove the mounting bolts at the rear of the seats **(see illustrations)**.
3   Tilt the seat upward and disconnect all electrical connectors, then slide the seat rearward to disengage the front seat mount hooks from the floor and lift out the seat.
4   Installation is the reverse of removal.

Rear seat

5   Remove the floor trim panels **(see illustrations)**.
6   Remove the seat mounting bolts, then remove the seat **(see illustrations)**.
7   Installation is the reverse of removal.

Third seat

8   Remove the cargo area cover as equipped.
9   Remove the seat mounting bolts that are under the cover behind the seat at the floor.
10   Tilt the seat upward and slide the seat rearward to disengage the front seat mount hooks from the floor and lift out the seat.
11   Installation is the reverse of removal.

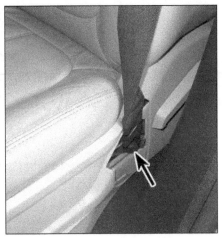

24.1b Remove the seat belt mounting bolt

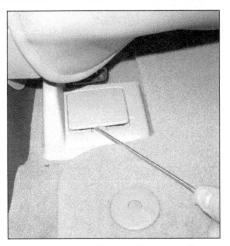

24.2a Carefully pry off the small trim panels . . .

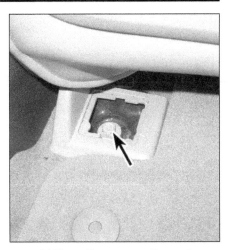

24.2b . . . then remove the seat mounting bolts

24.5a Carefully pry off the trim panels at the front . . .

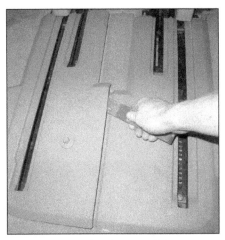

24.5b . . . and at the rear of the seat

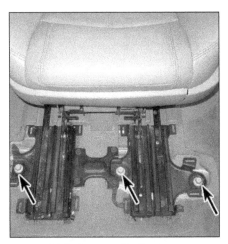

24.6a Remove the front . . .

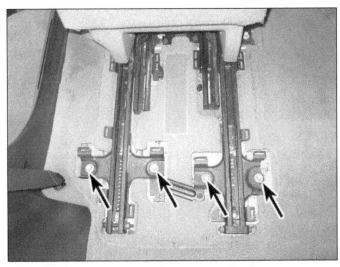

24.6b . . . and rear mounting bolts

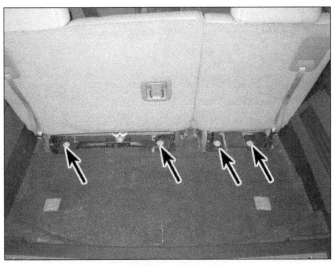

24.9 Third seat mounting bolts

Notes

Chapter 12
Chassis electrical system

Contents

	Section
Airbag system - general information	28
Antenna - removal and installation	13
Bulb replacement	17
Cruise control system - description and check	22
Data Link Communication system - description	27
Daytime Running Lights (DRL) - general information	20
Electric side view mirrors - description	25
Electrical connectors - general information	6
Electrical troubleshooting - general information	2
Fuses, fusible links and circuit breakers - general information	3
General information	1
Headlight bulb - replacement	14
Headlight housing - removal and installation	16
Headlights and fog lights - adjustment	15
Horn - replacement	19

	Section
Ignition switch and key lock cylinder - replacement	9
Instrument cluster - removal and installation	11
Instrument panel switches - replacement	10
Power door lock and keyless entry system - description and check	24
Power seats - description	26
Power window system - description and check	23
Radio and speakers - removal and installation	12
Rear window defogger - check and repair	21
Relays - general information and testing	4
Steering column multi-function switches - replacement	8
Turn signal and hazard flashers - general information	7
Underhood fuse/relay box - removal and installation	5
Wiper motor - replacement	18
Wiring diagrams - general information	29

1   General information

1   The electrical system is a 12-volt, negative ground type. Power for the lights and all electrical accessories is supplied by a lead/acid-type battery that is charged by the alternator.

2   This Chapter covers repair and service procedures for the various electrical components not associated with the engine. Information on the battery, alternator, ignition system and starter motor can be found in Chapter 5.

3   It should be noted that when portions of the electrical system are serviced, the negative cable should be disconnected from the battery to prevent electrical shorts and/or fires.

2   Electrical troubleshooting - general information

1   A typical electrical circuit consists of an electrical component, any switches, relays, motors, fuses, fusible links or circuit breakers related to that component and the wiring and connectors that link the component to both the battery and the chassis. To help you pinpoint an electrical circuit problem, wiring diagrams are included at the end of this Chapter.

2   Before tackling any troublesome electrical circuit, it would be wise to understand the basics of electrical theory and how a circuit in an automobile is connected. Knowing how any system works before attempting repairs will greatly reduced the possibility of replacing unneeded components. A good place to start is to study the appropriate wiring diagrams to get a complete understanding of what makes up that individual circuit. Trouble spots, for instance, can often be narrowed down by noting if other components related to the circuit are operating properly. Taking it step by step and following the guidelines provided will reduce your time in diagnosing electrical issues.

3   Electrical problems usually stem from simple causes, such as loose or corroded connections, worn or chafed wiring, a blown fuse, a melted fusible link, or faulty components. Visually inspect the condition of all fuses, wires and connections in a problem circuit before troubleshooting the circuit. Be sure to check not only the positive signals but the negative signals as well. Faulty grounds, or weak ground connections are a leading factor in system failures.

4   If test equipment and instruments are going to be utilized, be sure you understand how to use the equipment properly before attempting a repair. A bad diagnostic routine can start with bad equipment or the lack of proper use of the equipment. Use the wiring diagrams to plan ahead of time where you will make the necessary connections in order to accurately pinpoint your test connections as well as were the possible trouble could be.

5   The basic tools needed for electrical troubleshooting include a multi-meter that is capable of reading DC and AC voltage, Ohms (resistance), and Amps, a test light, jumper wires with alligator clips at each end, a jumper wire preferably with a circuit breaker incorporated, and a few sharp pins (straight pins work well) which can be used to bypass elec-

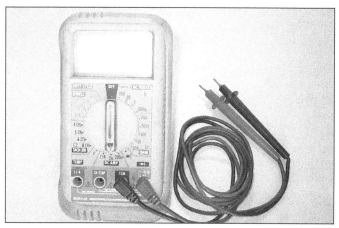

2.5a The most useful tool for electrical troubleshooting is a digital multimeter that can check volts, amps, and test continuity

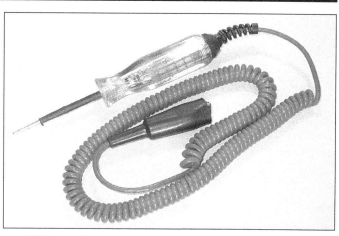

2.5b A test light is a very handy tool for checking voltage

2.6 In use, a test light's lead is clipped to a known ground, then the pointed probe can test connectors, wires or electrical sockets - if the bulb lights, the circuit being tested has battery voltage.

2.9 With the multimeter set to the ohm scale, resistance can be checked across two terminals - when checking for continuity, a low reading indicates continuity; a high reading or infinity indicates diminished or lack of continuity.

trical components **(see illustrations)**. Before attempting to locate a problem with test instruments, use the wiring diagram(s) to decide where to make the connections.

Voltage checks

Note: *Keep in mind that some circuits receive voltage only when the ignition key is in the Accessory or Run position.*

6   Voltage checks should be performed if a circuit is not functioning properly. Connect one lead of a circuit tester to either the negative battery terminal or a known good ground. Always check your test light before checking the actual circuit you're working on to be sure it is making good contact with the negative and the positive leads. Connect the other lead to a connector in the circuit being tested, preferably nearest to the battery or fuse **(see illustration)**. If the bulb of the tester lights, voltage is present, which means that the part of the circuit between the connector and the battery is problem free. Continue checking the rest of the circuit in the same fashion. When you reach a point at which no voltage is present, the prob-

lem lies between that point and the last test point with voltage. Most of the time the problem can be traced to a loose connection.

Finding a short

7   A short occurs when the path of electricity takes a route to ground that it was not designed for. This is usually associated with a blown fuse or melted fusible link. One method of finding shorts in a circuit is to remove the fuse and connect a test light or voltmeter in place of the fuse terminals. A fuse terminal has two connections. One is the supplied voltage to the fuse while the other is the send lead to that circuit. There should be no readable voltage at those two connectors because they should be of the same potential. Moving the wiring harness from side-to-side while watching the test light may also allow you to find any chaffed wiring that might have blown the fuse originally. If the bulb is on, there is a negative and a positive potential at the fuse connection. When the light is on, there is a short to ground somewhere in that area, prob-

ably where the insulation has rubbed through. The same test can be performed on each component in the circuit, or on a switch.

Ground check

8   Perform a ground test to check whether a component is properly grounded. Disconnect the battery and connect one lead of a continuity tester or multimeter (set to the ohms scale), to a known good ground. Connect the other lead to the wire or ground connection being tested. If the resistance is low (less than 5 ohms), the ground is good. If the bulb on a self-powered test light does not go on, the ground is bad. A more accurate method is the voltage drop test. Place the positive side of your multimeter on the positive post of the battery, then place the negative side on the chassis. Note the reading, then move the negative lead to the suspected bad ground area, such as the engine (which should have the same grounded leads to it). Take another reading. The two readings should be exactly the same.

Continuity check

9   A continuity check is done to determine if there are any breaks in a circuit - if it is passing electricity properly. With the circuit off (no power in the circuit), a self-powered continuity tester or multimeter can be used to check the circuit. Connect the test leads to both ends of the circuit (or to the power end and a good ground), and if the test light comes on the circuit is passing current properly **(see illustration)**. If the resistance is low (less than 5 ohms), there is continuity; if the reading is 10,000 ohms or higher, there is a break somewhere in the circuit. The same procedure can be used to test a switch, by connecting the continuity tester to the switch terminals. With the switch turned On, the test light should come on (or low resistance should be indicated on a meter).
Caution: *Always check your multimeter before hooking it up to any circuit so that you know you are on the right scale. If there is voltage present on a lead and you are trying to measure resistance you can do permanent damage to your meter.*

3.1a The main fuse/relay panel is in the engine compartment; disengage the locking tabs and remove the cover for access to the fuses and relays

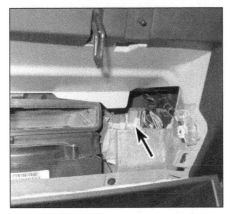

3.1b The interior fuse/relay panel is located under a cover on the right side, just below and behind the glove box area

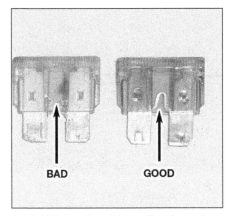

3.4 When a fuse blows, the element between the terminals melts

Finding an open circuit

10   When diagnosing for possible open circuits, it is often difficult to locate them by sight because the connectors hide oxidation or terminal misalignment. Merely wiggling a connector on a sensor or in the wiring harness may correct the open circuit condition. Remember this when an open circuit is indicated when troubleshooting a circuit. Intermittent problems may also be caused by oxidized or loose connections.

11   Electrical troubleshooting is simple if you keep in mind that all electrical circuits are basically electricity running from the battery, through the wires, switches, relays, fuses and fusible links to each electrical component (light bulb, motor, etc.) and to ground, from which it is passed back to the battery. Any electrical problem is an interruption in the flow of electricity to and from the battery.

Finding a battery drain

Note: *Before attempting to find a drain, test the battery to be sure the battery itself is not the cause.*

12   Battery drain is any electrical load that is present when it shouldn't be, which will cause the battery to have insufficient amperage/voltage to restart the vehicle. Today's vehicles have what is referred to as parasitic battery drain. This is a normal process that occurs with all newer vehicles. Each of the different computer based systems in the vehicle have a certain amount of constant current required to maintain enough electricity to restart. The required voltage is very small, so small a standard test light or volt meter will not pick up the signal correctly. An amperage meter in line with the battery negative post and negative clamp is recommended to read the amount of current being passed to the vehicle. A reading of less than 0.02 to 0.04 amps indicates a lack of battery drain. Anything above that would indicate something has been left on, or one of the computer based systems is still activated.

13   Modules all have a sleep mode; this varies with each module or system. Some will

carry out their functions shortly after the last door is closed or when the key is turned off. Delay systems such as dome light entry and exit are a good example of a module cycling through the sleep mode. When the light comes on, it's awake, and when it goes out a few seconds later, it's asleep.

14   Finding a battery drain can be quite challenging. If you are hesitant in trying to locate the drain, take your vehicle to your local dealer or qualified independent shop that specializes in electrical repairs.

Connectors

15   Most electrical connections on these vehicles are made with multiwire plastic connectors. The mating halves of many connectors are secured with locking clips molded into the plastic connector shells. The mating halves of large connectors, such as some of those under the instrument panel, are held together by a bolt through the center of the connector.

16   To separate a connector with locking clips, use a small screwdriver to pry the clips apart carefully, then separate the connector halves. Pull only on the shell - never pull on the wiring harness, as you may damage the individual wires and terminals inside the connectors. Look at the connector closely before trying to separate the halves. Often the locking clips are engaged in a way that is not immediately clear. Additionally, many connectors have more than one set of clips.

17   Each pair of connector terminals has a male half and a female half. When you look at the end view of a connector in a diagram, be sure to understand whether the view shows the harness side or the component side of the connector. Connector halves are mirror images of each other, and a terminal shown on the right side end-view of one half will be on the left side end-view of the other half. It is often necessary to take circuit voltage measurements with a connector connected. Whenever possible, carefully insert a small straight pin (not your meter probe) into the rear of the connector shell to contact the terminal inside, then clip your meter lead to the pin. This kind

of connection is called backprobing. When inserting a test probe into a terminal, be careful not to distort the terminal opening. Doing so can lead to a poor connection and corrosion at that terminal later. Using the small straight pin instead of a meter probe results in less chance of deforming the terminal connector.

3   Fuses, fusible links and circuit breakers - general information

Fuses

1   The electrical circuits of the vehicle are protected by a combination of fuses, circuit breakers and fusible links. The main fuse/relay panel is in the engine compartment **(see illustration)**, while the interior fuse/relay panel is located on the right side, below the glove box area **(see illustration)**. The covers have a legend on the underside to identify the fuses and relays.

2   Each of the fuses is designed to protect a specific circuit, and the various circuits are identified on the fuse panel itself.

3   Several sizes of fuses are employed in the fuse blocks. There are small, medium and large sizes of the same design, all with the same blade terminal design. The medium and large fuses can be removed with your fingers, but the small fuses require the use of pliers or the small plastic fuse-puller tool found in most fuse boxes.

4   If an electrical component fails, always check the fuse first. The best way to check the fuses is with a test light. Check for power at the exposed terminal tips of each fuse. If power is present at one side of the fuse but not the other, the fuse is blown. A blown fuse can also be identified by visually inspecting it.

5   Be sure to replace blown fuses with the correct type. Fuses (of the same physical size) of different ratings may be physically interchangeable, but only fuses of the proper rating should be used. Replacing a fuse with one of a higher or lower value than specified is not recommended. Each electrical circuit needs a specific amount of protection. The

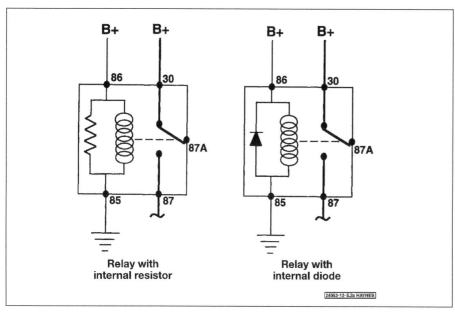

4.2a Typical ISO relay designs, terminal numbering and circuit connections

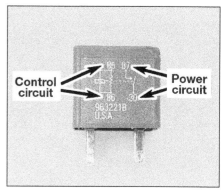

4.2b Most relays are marked on the outside to easily identify the control circuit and power circuit - this one is of the four-terminal type

amperage value of each fuse is molded into the top of the fuse body.

6   If the replacement fuse immediately fails, don't replace it again until the cause of the problem is isolated and corrected. In most cases, this will be a short circuit in the wiring caused by a broken or deteriorated wire.

Fusible links

7   Some circuits are protected by fusible links. The links are used in circuits that are not ordinarily fused, such as the alternator circuit.

8   On this particular vehicle, the alternator circuit (main lead - large red wire) runs from the battery, to the underhood fuse box, then to an inline 200 amp fusible link. Voltage is present at all times on this lead.

Circuit breakers

9   Circuit breakers protect certain circuits, such as the power windows and power seats. Depending on the vehicle's accessories, there may be two 25-amp circuit breakers for the door locks and one 30-amp circuit breaker for the power seats, located in the interior fuse/relay box under the left rear seat.

10   Because the circuit breakers reset automatically, an electrical overload in a circuit-breaker-protected system will cause the circuit to fail momentarily, then come back on. If the circuit does not come back on, check it immediately.

Caution: *If there is a short in the circuit breaker wiring, the metal casing of the breaker may-be extremely hot. Use caution when removing the breaker.*

11   With the voltmeter negative lead on a good chassis ground, touch each end prong of the circuit breaker with the positive meter probe. There should be battery voltage at each end. If there is battery voltage only at one end, the circuit breaker must be replaced.

4   Relays - general information and testing

General information

1   Several electrical accessories in the vehicle, such as the fuel injection system, horns, starter, and fog lamps use relays to transmit the electrical signal to the component. Relays use a low-current circuit (the control circuit) to open and close a high-current circuit (the power circuit). If the relay is defective, that component will not operate properly. Most relays are mounted in the engine compartment and interior fuse/relay boxes. If a faulty relay is suspected, it can be removed and tested using the procedure below or by a dealer service department or a repair shop. Defective relays must be replaced as a unit.

Testing

2   Most of the relays used in these vehicles are of a type often called ISO relays, which refers to the International Standards Organization. The terminals of ISO relays are numbered to indicate their usual circuit connections and functions. There are two basic layouts of terminals on the relays used in these vehicles **(see illustrations)**.

3   Refer to the wiring diagram for the circuit to determine the proper connections for the relay you're testing. If you can't determine the correct connection from the wiring diagrams, however, you may be able to determine the test connections from the information that follows.

4   Two of the terminals are the relay control circuit and connect to the relay coil. The other relay terminals are the power circuit. When the relay is energized, the coil creates a magnetic field that closes the larger contacts of

the power circuit to provide power to the circuit loads.

5   Terminals 85 and 86 are normally the control circuit. If the relay contains a diode, terminal 86 must be connected to battery positive (B+) voltage and terminal 85 to ground. If the relay contains a resistor, terminals 85 and 86 can be connected in either direction with respect to B+ and ground.

6   Terminal 30 is normally connected to the battery voltage (B+) source for the circuit loads. Terminal 87 is connected to the ground side of the circuit, either directly or through a load. If the relay has several alternate terminals for load or ground connections, they usually are numbered 87A, 87B, 87C, and so on.

7   Use an ohmmeter to check continuity through the relay control coil.

a)   *Connect the meter according to the polarity shown in the illustration for one check; then reverse the ohmmeter leads and check continuity in the other direction.*

b)   *If the relay contains a resistor, resistance should be indicated on the meter, and should be the same value with the ohmmeter in either direction.*

c)   *If the relay contains a diode, resistance should be higher with the ohmmeter in the forward polarity direction than with the meter leads reversed.*

d)   *If the ohmmeter shows infinite resistance in both directions, replace the relay.*

8   Remove the relay from the vehicle and use the ohmmeter to check for continuity between the relay power circuit terminals. There should be no continuity between terminal 30 and 87 with the relay de-energized.

9   Connect a fused jumper wire to terminal 86 and the positive battery terminal. Connect another jumper wire between terminal 85 and ground. When the connections are made, the relay should click.

10   With the jumper wires connected, check for continuity between the power circuit terminals. Now, there should be continuity between terminals 30 and 87.

11   If the relay fails any of the above tests, replace it.

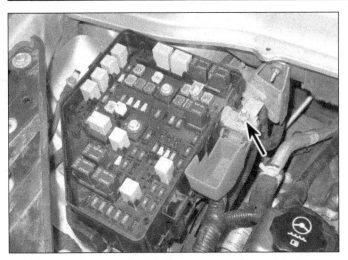

5.3 Remove the nut securing the battery positive cables to the underhood fuse/relay box

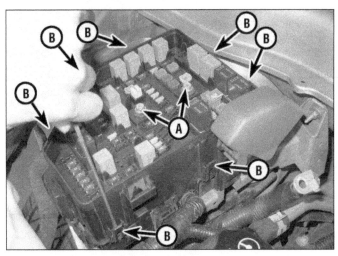

5.4 Completely loosen the captive bolts (A), then use a screwdriver to disengage the tabs (B, not all visible - approximate locations given) retaining the top of the box to the bottom

5.5a Remove the top half of the box . . .

5.5b . . . then pull the junction blocks and wiring harnesses from the bottom half and set them aside

5.6 Ground cable bolt

5   Underhood fuse/relay box - removal and installation

1   Disconnect the cable from the negative terminal of the battery (see Chapter 5).

2   Remove the cover from the fuse/relay box.

3   Flip open the protective cover from the remote positive terminal, remove the nut and detach the battery positive cables from the stud **(see illustration)**.

4   Unscrew the two bolts securing the upper portion of the fuse/relay box to the junction blocks **(see illustration)**. Using a screwdriver, dislodge the tabs securing the perimeter of the upper portion of the box to the lower portion.

5   Lift off the top half of the box, then remove the junction blocks and wiring harnesses from the lower half **(see illustrations)**.

6   Remove the bolt and detach the ground cable from the inner fender panel

7   Release the tabs and detach the lower half of the box from the base **(see illustration)**.

8   Remove the fasteners and detach the

5.7 Use a screwdriver to disengage the tabs

base from the inner fender panel **(see illustration)**.

9   Installation is the reverse of the removal procedure.

5.8 Underhood fuse/relay box mounting base fasteners

6   Electrical connectors - general information

1   Most electrical connections on these vehicles are made with multiwire plastic connectors. The mating halves of many connectors are secured with locking clips molded into the plastic connector shells. The mating halves of some large connectors, such as some of those under the instrument panel, are held together by a bolt through the center of the connector.

2   To separate a connector with locking clips, use a small screwdriver to pry the clips apart carefully, then separate the connector halves. Pull only on the shell, never pull on the wiring harness, as you may damage the individual wires and terminals inside the connectors. Look at the connector closely before trying to separate the halves. Often the locking clips are engaged in a way that is not immediately clear. Additionally, many connectors have more than one set of clips.

3   Each pair of connector terminals has a male half and a female half. When you look at the end view of a connector in a diagram, be sure to understand whether the view shows the harness side or the component side of the connector. Connector halves are mirror images of each other, and a terminal shown on the right side end-view of one half will be on the left side end-view of the other half.

4   It is often necessary to take circuit voltage measurements with a connector connected. Whenever possible, carefully insert a small straight pin (not your meter probe) into the rear of the connector shell to contact the terminal inside, then clip your meter lead to the pin. This kind of connection is called back-probing. When inserting a test probe into a terminal, be careful not to distort the terminal opening. Doing so can lead to a poor connection and corrosion at that terminal later. Using the small straight pin instead of a meter probe results in less chance of deforming the terminal connector.

Electrical connectors

Most electrical connectors have a single release tab that you depress to release the connector

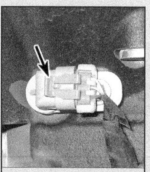

Some electrical connectors have a retaining tab which must be pried up to free the connector

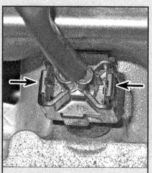

Some connectors have two release tabs that you must squeeze to release the connector

Some connectors use wire retainers that you squeeze to release the connector

Critical connectors often employ a sliding lock (1) that you must pull out before you can depress the release tab (2)

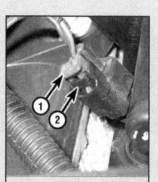

Here's another sliding-lock style connector, with the lock (1) and the release tab (2) on the side of the connector

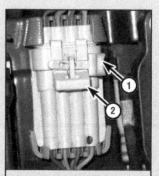

On some connectors the lock (1) must be pulled out to the side and removed before you can lift the release tab (2)

Some critical connectors, like the multi-pin connectors at the Powertrain Control Module employ pivoting locks that must be flipped open

8.5 To remove either multi-function switch, depress the retaining tabs and pull it out of the housing

9.4 Push the retaining pin in to release the lock cylinder, then withdraw the key and cylinder

7 Turn signal and hazard flashers - general information

1   The turn signal and hazard flasher system consists of bulbs, switches, and the Body Control Module (BCM). Each rear taillight housing also consists of a logic module that monitors the bulb condition. Positive voltage is supplied to the BCM from three fuses (LT/ TRN/SIG, RT/TRN/SIG, AND BCK/UP/STOP), while the ground signal is sent to the BCM from either the turn signal switch or the hazard switch. There is no external flasher to replace; the flasher is an integral part of the BCM.
2   When the flasher unit is functioning properly, an audible click can be heard during its operation. If the turn signal indicator on one side of the vehicle flashes much more rapidly than normal, a change in the amount of impedance (resistance) to the flow of current has been detected by the BCM. This can either be a faulty bulb, open wiring, and/or faulty BCM.
3   If the both the front and rear turn signal bulbs on the same side are not flashing, the turn signal and hazard flasher relay function in the BCM is probably defective. Have the BCM replaced by a dealer service department or other repair facility equipped with the proper tools and scanners. This is not a job that you can do at home because the BCM must be programmed with a factory scan tool when it's replaced.

8 Steering column multi-function switches - replacement

Warning: *The models covered by this manual are equipped with Supplemental Restraint Systems (SRS), more commonly known as airbags. Always disable the airbag system before working in the vicinity of any airbag system components to avoid the possibility of accidental deployment of the airbags, which could cause personal injury (see Section 28).*
1   The multi-function switches are located on the left and right sides of the steering col-

umn. Each incorporates multiple functions; the turn signal, headlight, low-to-high beam and flash-to-pass are on the left. The windshield wiper/washer is on the right.
2   Release the tilt wheel lever and place the steering wheel in the center position.
3   Remove the steering column trim covers (see Chapter 11).
4   Depress the retaining tab and disconnect the multi-function switch electrical connector.
5   Depress the multi-function switch retaining tabs, then detach the switch from the steering column housing **(see illustration)**.
6   Installation is the reverse of removal.
Note: *After installing a new multi-function switch assembly, ensure the locking tabs are fully engaged.*

9 Ignition switch and key lock cylinder - replacement

Warning: *The models covered by this manual are equipped with Supplemental Restraint Systems (SRS), more commonly known as airbags. Always disable the airbag system before working in the vicinity of any airbag system components to avoid the possibility of accidental deployment of the airbags, which could cause personal injury (see Section 28).*
1   Disconnect the cable from the negative terminal of the battery (see Chapter 5).
2   Remove the upper and lower steering column covers (see Chapter 11).

Key lock cylinder
3   Insert the ignition key into the lock cylinder and turn it to the RUN position.
4   Insert a pick or other small tool into the hole in the top of the lock cylinder housing and push it in to depress the release button **(see illustration)**. Pull the lock cylinder out of the housing.
5   To install the lock cylinder, make sure the ignition switch is still in the RUN position. If it isn't, rotate it to the RUN position with a screwdriver or needle-nose pliers. The lock

cylinder won't fit all the way into the bore if the switch is in any other position.
6   Insert the lock cylinder into the housing, making sure it engages with the switch and clicks into place.
7   Verify that the ignition switch operates correctly in the Off, ACC, Run and Start positions.

Housing and ignition switch
Removal
8   Remove the steering wheel (see Chapter 10).
9   Remove both multi-function switches (see Section 8). The wiring doesn't have to be disconnected; the switches can be left attached to their respective harnesses while you remove the ignition switch housing.
10   Detach the electrical connectors from the ignition switch.
11   Remove the ignition switch/key lock housing bolt and remove the housing strap.
12   Slide the entire assembly off of the steering column.
13   Remove the ignition switch mounting screws and remove the ignition switch from the housing.

Installation
14   Install the switch into the housing and tighten the screws securely.
15   Slide the housing onto the steering column, then install the strap and tighten the bolt securely.
16   The remainder of installation is the reverse of removal.

10 Instrument panel switches - replacement

Warning: *The models covered by this manual are equipped with Supplemental Restraint Systems (SRS), more commonly known as airbags. Always disable the airbag system before working in the vicinity of any airbag system components to avoid the possibility of accidental deployment of the airbags, which*

10.1 Separate the center trim bezel from the instrument panel

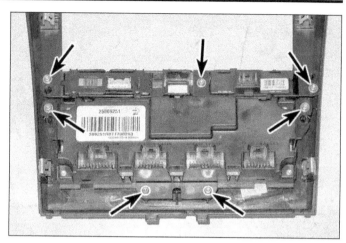

10.3 Center dash trim panel switch fastener locations

11.4 Remove the four instrument cluster mounting screws

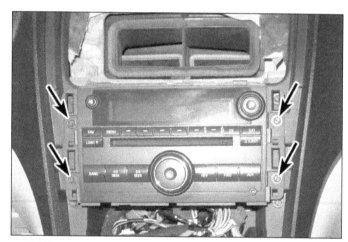

12.3 Remove the mounting screws and pull the radio from the instrument panel

could cause personal injury (see Section 28).
1   Using a plastic flat-bladed trim tool, carefully pry off the center dash trim bezel **(see illustration).**
2   Disconnect the electrical connector from the switches and remove the trim panel.
3   Remove the screws and detach the

12.4 Disconnect the antenna cable and electrical connectors from the back of the radio

switch(es) from the trim panel **(see illustration)**.
4   Installation is the reverse of removal.

11   Instrument cluster - removal and installation

Warning: *The models covered by this manual are equipped with Supplemental Restraint Systems (SRS), more commonly known as airbags. Always disable the airbag system before working in the vicinity of any airbag system components to avoid the possibility of accidental deployment of the airbags, which could cause personal injury (see Section 28).*
1   Disconnect the cable from the negative battery terminal (see Chapter 5).
2   Release the tilt wheel lever and lower the steering wheel to its lowest position.
3   Remove the instrument cluster bezel (see Chapter 11).
4   Remove the instrument cluster mounting screws **(see illustration).**
5   Carefully remove the instrument cluster from the instrument panel.
6   Installation is the reverse of removal.

12   Radio and speakers - removal and installation

Warning: *The models covered by this manual are equipped with Supplemental Restraint Systems (SRS), more commonly known as airbags. Always disable the airbag system before working in the vicinity of any airbag system components to avoid the possibility of accidental deployment of the airbags, which could cause personal injury (see Section 28).*

Radio
1   Disconnect the cable from the negative battery terminal (see Chapter 5).
2   Remove the center trim panel (see Chapter 11).
3   Remove the radio mounting screws and pull the radio from the instrument panel **(see illustration)**.
4   Disconnect the electrical connectors and the antenna cable from the back of the radio **(see illustration)**.
5   Installation is the reverse of removal.

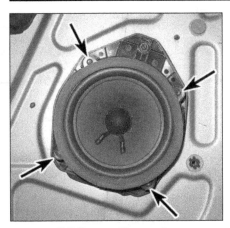

12.7 Remove the speaker mounting screws

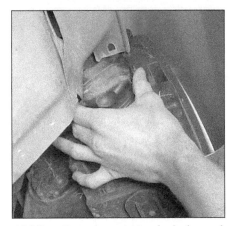

14.2 Turn the cover counterclockwise and remove it from the housing

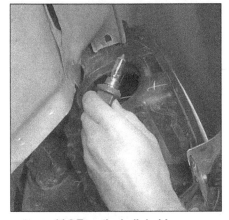

14.3 Turn the bulb holder counterclockwise to remove the bulb from the headlight housing

Speakers

Lower door speakers

6 Remove the door trim panel (see Chapter 11).
7 Remove the speaker mounting screws **(see illustration)**.
8 Pull the speaker from the door, then disconnect the electrical connector.
9 Installation is the reverse of removal.

Upper front door speakers

10 Lift the speaker to slide the lower tabs out of the door panel.
11 Disconnect the electrical connector.
12 Twist the speaker to remove it from the panel.
13 Installation is the reverse of removal.

Rear speaker

14 Remove the carpet retaining trim piece.
15 Remove the pillar trim.
16 Remove the rear upper trim molding.
17 Remove the rear floor storage bin.
18 Remove the cargo hooks.
19 Pull out the quarter trim panel, disengaging its retaining clips (see Chapter 11).
20 Remove the speaker mounting screws, remove the speaker and disconnect the electrical connector.
21 Installation is the reverse of removal.

13 Antenna - removal and installation

Antenna mast

1 The antenna mast simply unscrews from the base. In most cases it can be unscrewed by hand, but if necessary, use a pair of pliers to break it loose.

Antenna base

2 Remove the rear quarter trim panels (see Chapter 11) and the rear garnish molding at the rear of the headliner.
3 Pull the rear of the headliner down for access to the antenna base.

4 Unscrew the antenna base nut, disconnect the cable and detach the antenna base from the roof.
5 Installation is the reverse of removal.

14 Headlight bulb - replacement

Warning: *Halogen bulbs are gas-filled and under pressure and may shatter if the surface is scratched or the bulb is dropped. Wear eye protection and handle the bulbs carefully, grasping only the base whenever possible. Don't touch the surface of the bulb with your fingers because the oil from your skin could cause it to overheat and fail prematurely. If you do touch the bulb surface, clean it with rubbing alcohol.*

Halogen headlights

1 Remove the inner fender splash shield (see Chapter 11).
2 There is a round cover protecting the bulbs. Turn the cover counterclockwise and remove it to expose the bulb(s) **(see illustration)**.
Note: *The outside bulb is the low beam, and the inside bulb is the high beam.*
3 Remove the bulb from the headlight housing **(see illustration)**, then disconnect the electrical connector.
4 Connect the electrical connector to the new bulb holder. Properly align and install the bulb holder into the housing, then twist it clockwise to lock it in place. Check the operation of the bulb.
5 Reinstall the cover, the inner fender splash shield and the wheel and lug nuts. Lower the vehicle and tighten the lug nuts to the torque listed in the Chapter 1 Specifications.

Xenon (HID) headlights

Warning: *Some models use High Intensity Discharge (HID) bulbs instead of conventional halogen bulbs. According to the manufacturer, the high voltages produced by this system can be fatal in the event of shock. Also, the voltage can remain in circuit even after the headlight switch has been turned to OFF and the ignition key*

15.1 Use a Philips screwdriver to turn the adjuster screw

has been removed. Therefore, for your safety, we don't recommend that you try to replace one of these bulbs yourself Instead, have this service performed by a dealer service department or other qualified repair shop.

15 Headlights and fog lights - adjustment

Warning: *The headlights must be aimed correctly. If adjusted incorrectly, they could temporarily blind the driver of an oncoming vehicle and cause an accident or seriously reduce your ability to see the road. The headlights should be checked for proper aim every 12 months and any time a new headlight is installed or front-end bodywork is performed. The following procedure is only an interim step to provide temporary adjustment until the headlights can be adjusted by a properly equipped shop.*

Headlights

1 These models are equipped with composite headlights with adjustment screws that control up-and-down movement **(see illustration)**. Left-and-right movement is not adjustable.

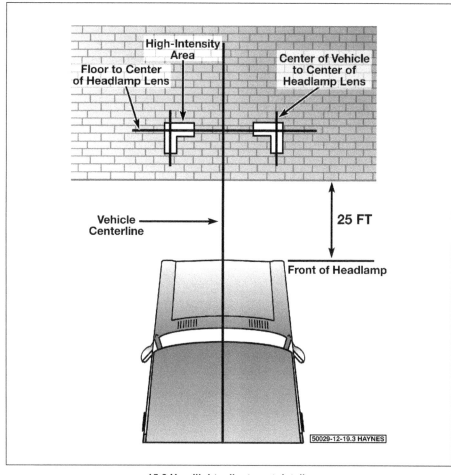

15.2 Headlight adjustment details

vehicle sitting level, the gas tank half-full and no unusually heavy load in the vehicle.

6   Turn on the low beams. Turn the adjusting screw to position the high intensity zone so it is eight inches below the horizontal line.

7   To ensure the headlamps are adjusted properly, have the headlights adjusted by a qualified independent repair shop or a dealer service department at the earliest opportunity.

Fog lights

8   Some models have optional fog lights that can be aimed just like headlights. As with the headlights, there are no left-and-right adjustments.

9   Position tape on a wall 25 feet in front of the vehicle. Tape a horizontal line on the wall that represents the height of the fog light centers, and another tape line four inches below that line.

10   Remove the fasteners from the front portion of the inner fender splash shield to gain access to the adjustment screws for the fog lights **(see illustration)**. Adjust the pattern on the wall so that the top of the fog light beam meets the lower line on the wall.

---

16   Headlight housing - removal and installation

1   Disconnect the cable from the negative battery terminal (see Chapter 5).

2   Remove front bumper cover (see Chapter 11).

3   Remove the headlight housing mounting bolts **(see illustrations)**.

4   Gently pull the headlight housing from the radiator support, then disconnect the electrical connector.

5   Installation is the reverse of removal. If you are reusing the original headlight assemblies, there is no need to readjust the headlight aim. But if you are replacing the assemblies, set the adjustment cam to it's base setting; about six turns from completely seated. This will aid in aligning the bumper cover.

---

2   There are several methods of adjusting the headlights. The simplest method requires a blank wall 25 feet in front of the vehicle and a level floor **(see illustration)**.

3   Position masking tape on the wall in reference to the vehicle centerline and the centerlines of both headlights.

4   Measure the height of the headlight reference marks (in the centers of the headlight lenses) from the ground. Position a horizontal tape line on the wall at the same height as the headlight reference marks.

Note: *It may be easier to position the tape on the wall with the vehicle parked only a few inches away.*

5   Adjustment should be made with the

15.10 Insert a Philips screwdriver into the hole in the bottom of the fog light housing to engage the teeth of the adjuster gear

16.3a Headlight housing upper mounting bolts . . .

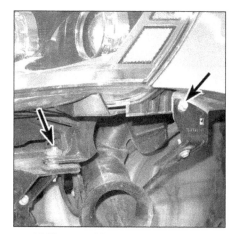

16.3b . . . and lower mounting bolts

Bulb removal

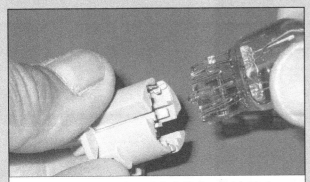

To remove many modern exterior bulbs from their holders, simply pull them out

On bulbs with a cylindrical base ("bayonet" bulbs), the socket is spring-loaded; a pair of small posts on the side of the base hold the bulb in place against spring pressure. To remove this type of bulb, push it into the holder, rotate it 1/4-turn counterclockwise, then pull it out

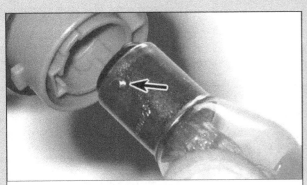

If a bayonet bulb has dual filaments, the posts are staggered, so the bulb can only be installed one way

To remove most overhead interior light bulbs, simply unclip them

17 Bulb replacement

Bulb removal

Exterior light bulbs

Front side marker/turn signal/parking light bulbs

1   Remove the inner fender splash shield (see Chapter 11) or the radiator support cover.
2   Turn the bulb holder counterclockwise and remove it from the headlight housing **(see illustration)**.
3   Pull the bulb straight out of the holder to remove it.
4   Installation is the reverse of removal.

Fog light bulbs

Warning: *Halogen bulbs are gas-filled and under pressure and may shatter if the surface is scratched or the bulb is dropped. Wear eye protection and handle the bulbs carefully, grasping only the base whenever possible. Don't touch the surface of the bulb with your fingers because the oil from your skin could cause it to*
overheat and fail prematurely. If you do touch the bulb surface, clean it with rubbing alcohol.
5   Remove the inner fender splash shield (see Chapter 11).

6   Disconnect the electrical connector from fog lamp bulb. Turn bulb socket counterclockwise and remove it **(see illustration)**.
7   Installation is the reverse of removal.

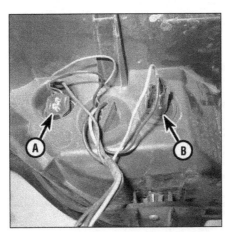

17.2 Front side marker (A, outer) and turn signal/parking light (B, inner) bulbs (left side shown)

17.6 Turn the fog light bulb holder counterclockwise to remove it from the housing

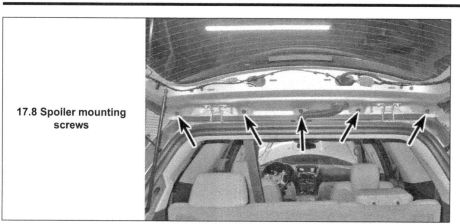

17.8 Spoiler mounting screws

17.13 Pry off the trim caps, then remove the taillight housing screws

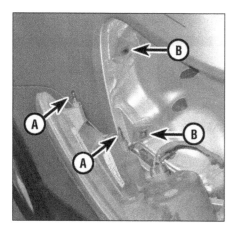

17.14 Pull sharply on the housing to dislodge the ballstuds (A) from their sockets (B)

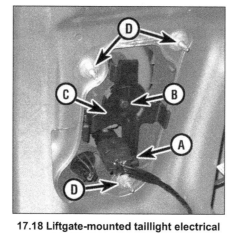

17.18 Liftgate-mounted taillight electrical connector (A), bracket screw (B), bracket (C) and mounting nuts (D)

Center high-mounted stop light

8   Remove the upper interior trim panel from the liftgate (see Chapter 11). Remove the spoiler mounting screws **(see illustration)** and tilt the spoiler upward.

9   Remove the two screws from the bulb assembly and pull it away from the spoiler.

10   Disconnect the electrical connector from the assembly and remove it.

11   Installation is the reverse of removal.

Rear brake/tail, turn signal, brake and back-up light bulbs

Note: *The lights in the taillight housings use*

LED bulbs and are not replaceable; the entire housing must be replaced if defective.

12   Open the liftgate.

Fender-mounted taillight housings

13   Remove the trim caps over the screw access points. Remove the taillight housing mounting screws **(see illustration)**.

14   Pull the light assembly outward to disengage the outer retaining pins **(see illustration)**.

15   Disconnect the electrical connector and remove the housing.

16   Installation is the reverse of removal.

Liftgate-mounted taillight housings

17   Remove the trim panel from the liftgate (see Chapter 11).

18   Disconnect the electrical connector, then remove the screw and bracket. Unscrew the nuts and detach the housing from the liftgate **(see illustration)**.

19   Installation is the reverse of removal.

License plate light bulbs

20   Open the liftgate. Lift to a point where you can easily reach the retaining pins holding the license plate light lens in place.

21   Remove the license plate light housing retaining pins and remove the housing from the liftgate **(see illustration)**.

Note: *You might have to pry the housing from the liftgate.*

22   Turn the bulb holder 90-degrees and detach the housing from the holder **(see illustration)**. Pull the bulb from its socket.

23   Installation is the reverse of removal.

Interior light bulbs

Reading light bulbs

Note: *This applies to the front and rear reading light bulbs.*

17.21 The license plate light housings are held in place by these retaining pins (they look like screws and have to be unscrewed to free them, but can be pushed back into place)

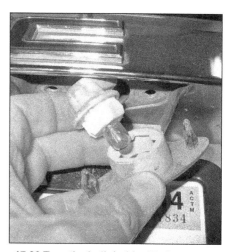

17.22 Turn the bulb holder 90-degrees to free it from the housing

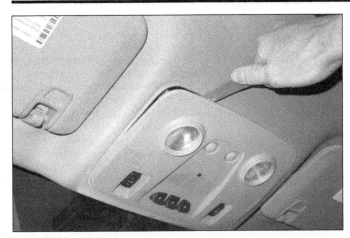

17.24 Carefully pry the overhead console from the headliner

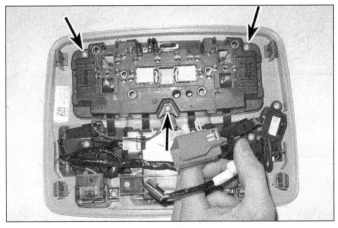

17.26 Pull the electrical connector from the housing, then remove the housing screws

24 Pry the overhead console or reading light housing from the headliner **(see illustration)**.

25 Disconnect the electrical connector and set the overhead console or reading light housing face-down on a clean work surface.

26 Detach the electrical connector from the housing, then remove the screws and separate the housing from the overhead console or reading light housing **(see illustration)**.

27 Pull the bulb straight out of its socket.

28 Installation is the reverse of removal.

Instrument cluster light bulbs

29 Instrument cluster illumination is an integral part of the instrument cluster. Bulbs are not replaceable.

Head Up Display (HUD)

30 The Head Up Display (HUD) is an optional secondary optical display that is projected onto the windshield. The HUD position on the windshield can be controlled by the HUD vertical direction switch. The HUD brightness can be adjusted as well to your personal preference. The HUD can also automatically adjust the brightness based on outside lighting using inputs from a photocell.

Head Up Display switch - replacement

31 Remove the instrument cluster bezel (see Chapter 11).

32 Unsnap the switch from the bezel.

33 Installation is the reverse of removal.

Head Up Display cluster - replacement

34 Remove the upper instrument panel upper trim.

35 Remove the screws and lift the display cluster out of the dash.

36 Disconnect the electrical connectors.

37 Installation is the reverse of removal. No programming is necessary for replacement of the HUD display cluster.

---

18 Wiper motor - replacement

Windshield wiper motor

1 Remove the wiper arm nuts and mark the relationship of the wiper arms to their shafts. Remove both wiper arms.

2 Remove the plastic cowl cover (see Chapter 11).

3 Disconnect the electrical connector from the wiper motor.

4 Remove the two windshield wiper motor and linkage/transmission assembly mounting bolts **(see illustration)** and remove the assembly.

Note: *When removing, tilt the back end up and maneuver the front mounting pin out of its bracket on the cowl.*

5 Using a screwdriver or forked release tool, carefully pry the link rod from the pivot pin on the motor's crank arm.

6 Mark the relationship of the crank arm to the motor shaft and remove the crank arm.

7 Remove the motor mounting bolts and remove the motor from its mounting bracket.

8 Installation is the reverse of removal.

Rear wiper motor

9 Remove the trim cap, then unscrew the wiper arm nut.

10 Mark the relationship of the rear wiper arm to the motor shaft, then remove the arm.

11 Remove the trim panel from the litgate (see Chapter 11).

12 Disconnect the electrical connector from the motor **(see illustration)**.

18.4 Remove the wiper motor and linkage/transmission mounting bolts

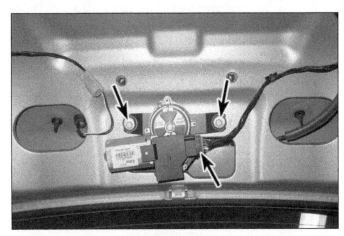

18.12 Rear wiper motor electrical connector and mounting bolts

19.2 Horn bracket mounting bolt (left side shown)

13 Remove the rear wiper motor mounting bolts and remove the motor and bracket.
14 When installing, make sure the pivot shaft gasket is in place.
Note: *The wiper motor pivot shaft gasket has an alignment tab located on the inside of the gasket. Align the tab with the pivot shaft/wiper motor prior to final tightening of the retaining nut.*
15 Installation is the reverse of removal.

19 Horn - replacement

Note: *These models are equipped with a Body Control Module (BCM). Several systems are linked to a centralized control module that allows simple and accurate troubleshooting, but only with a professional-grade scan tool. The Body Control Module governs the door locks, the power windows, the ignition lock and security system, the interior lights, the Daytime Running Lights system, the horn, the windshield wipers, the heating/air conditioning system and the power mirrors. In the event of a malfunction with this system, have the vehicle diagnosed by a dealership service department*
or other qualified automotive repair facility.
1 Remove the inner fender splash shield (see Chapter 11).
Note: *There is a horn on each side of the vehicle.*
2 Remove the mounting bolt, then detach the horn and disconnect the electrical connector.
3 If you're installing a new horn to the bracket, point the horn bell (opening) downward to prevent water from collecting inside the horn.
4 Installation is the reverse of removal.

20 Daytime Running Lights (DRL) - general information

1 The Daytime Running Lights (DRL) system used on all models illuminates the low beam headlights at reduced intensity whenever the ignition is On. The only exception is with the engine running and the shift lever is in Park. Once the parking brake is released or the shift lever is moved, the lights will remain on as

long as the ignition switch is on. When the BCM receives a signal from the ambient light sensor, it sends a ground signal to the DRL relay. Anytime the headlights are on, the DRL are off. Park lamps do not effect the DRL operation.

Ambient light sensor removal

Note: *The sensor is mounted in the grill area of the trim panel.*
2 Remove the upper instrument panel upper trim panel (see Chapter 11).
3 Disconnect the electrical connector and retaining clip.
4 Installation is the reverse of removal.

21 Rear window defogger - check and repair

1 The rear window defogger consists of a number of horizontal elements baked onto the glass surface.
2 Small breaks in the element can be repaired without removing the rear window.

Check

3 Turn the ignition switch and defogger system switches to the ON position. Using a voltmeter, place the positive probe against the defogger grid positive terminal and the negative probe against the ground terminal. If battery voltage is not indicated, check the fuse, defogger switch and related wiring. If voltage is indicated, but all or part of the defogger doesn't heat, proceed with the following tests.
4 When measuring voltage during the next two tests, wrap a piece of aluminum foil around the tip of the voltmeter positive probe and press the foil against the heating element with your finger **(see illustration)**. Place the negative probe on the defogger grid ground terminal.
5 Check the voltage at the center of each heating element **(see illustration)**. If the

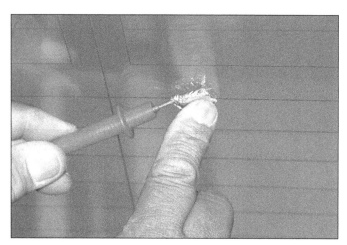

21.4 When measuring the voltage at the rear window defogger grid, wrap a piece of aluminum foil around the positive probe of the voltmeter and press the foil against the wire with your finger

21.5 To determine if a heating element has broken, check the voltage at the center of each element - if the voltage is 5 or 6-volts, the element is unbroken; if the voltage is 10 or 12-volts, the element is broken between the center and the ground side; if there is no voltage, the element is broken between the center and the positive side

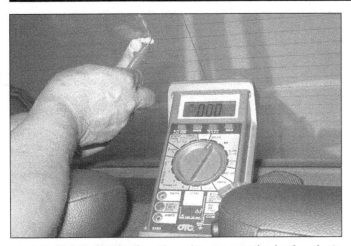

21.7 To find the break, place the voltmeter negative lead against the defogger ground terminal, place the voltmeter positive lead with the foil strip against the heating element at the positive terminal end and slide it toward the negative terminal end - the point at which the voltmeter reading changes abruptly is the point at which the element is broken

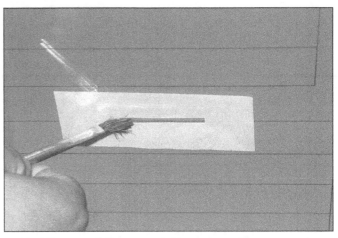

21.13 To use a defogger repair kit, apply masking tape to the inside of the window at the damaged area, then brush on the special conductive coating

voltage is 5 or 6-volts, the element is okay (there is no break). If the voltage is zero, the element is broken between the center of the element and the positive end. If the voltage is 10 to 12-volts, the element is broken between the center of the element and ground. Check each heating element.

6   Connect the negative lead to a good body ground. The reading should stay the same. If it doesn't, the ground connection is bad.

7   To find the break, place the voltmeter negative probe against the defogger ground terminal. Place the voltmeter positive probe with the foil strip against the heating element at the positive terminal end and slide it toward the negative terminal end. The point at which the voltmeter deflects from several volts to zero is the point at which the heating element is broken **(see illustration)**.

Repair

8   Repair the break in the element using a repair kit specifically recommended for this purpose, available at most auto parts stores. Included in this kit is plastic conductive epoxy.

9   Prior to repairing a break, turn off the system and allow it to cool off for a few minutes.

10   Lightly buff the element area with fine steel wool, then clean it thoroughly with rubbing alcohol.

11   Use masking tape to mask off the area being repaired.

12   Thoroughly mix the epoxy, following the instructions provided with the repair kit.

13   Apply the epoxy material to the slit in the masking tape, overlapping the undamaged area about 3/4-inch on either end **(see illustration)**.

14   Allow the repair to cure for 24 hours before removing the tape and using the system.

22   Cruise control system - description and check

1   The cruise control system maintains vehicle speed with the Powertrain Control Module (PCM), throttle actuator control motor, brake switch, control switches and associated wiring. There is no mechanical connection, such as a vacuum servo or cable. Some features of the system require special testers and diagnostic procedures that are beyond the scope of the home mechanic. Listed below are some general procedures that may be used to locate common problems.

2   Check the fuses (see Section 3).

3   The Brake Pedal Position (BPP) switch (or brake light switch) deactivates the cruise control system. Have an assistant press the brake pedal while you check the brake light operation.

4   If the brake lights do not operate properly, correct the problem and retest the cruise control.

5   Check the wiring between the PCM and throttle actuator motor for opens or shorts and repair as necessary.

6   The cruise control system uses information from the BCM, then relays this information to the PCM. The BCM monitors the cruise control switches and their respective resistance values to determine what state the cruise control is in. It in turn sends a signal on the serial data line to the PCM. Vehicle speed, brake pedal position, throttle position, transmission and engine condition are all factors in the cruise control operation.

7   Test drive the vehicle to determine if the cruise control is now working. If it isn't, take it to a dealer service department or other qualified repair shop for further diagnosis.

Note: *A lot of cruise control failures can be attributed to problems that may not seem related to the cruise control itself. Check to make sure the speedometer, brake lights, and the transaxle are operating correctly before assuming the problem is with the cruise control.*

23   Power window system - description and check

Note: *These models are equipped with a Body Control Module (BCM). Several systems are linked to a centralized control module that allows simple and accurate troubleshooting, but only with a professional-grade scan tool. The Body Control Module governs the door locks, the power windows, the ignition lock and security system, the interior lights, the Daytime Running Lights system, the horn, the windshield wipers, the heating/air conditioning system and the power mirrors. In the event of malfunction with this system, have the vehicle diagnosed by a dealership service department or other qualified automotive repair facility.*

1   The power window system operates electric motors, mounted in the doors, which lower and raise the windows. The system consists of the control switches, the motors, regulators, glass mechanisms, the Body Control Module (BCM) and associated wiring.

2   The power windows can be lowered and raised from the master control switch by the driver or by remote switches located at the individual windows. Each window has a separate motor that is reversible. The position of the control switch determines the polarity and therefore the direction of operation.

3   The window motor circuits are protected by a fuse. Each motor is also equipped with

an internal circuit breaker; this prevents one stuck window from disabling the whole system.

4   The power window system will only operate when the ignition switch has activated the Retained Accessory Power (RAP) relay. In addition, many models have a window lockout switch at the master control switch which, when activated, disables the switches at the rear windows and, sometimes, the switch at the passenger's window also. Always check these items before troubleshooting a window problem related to any of the other windows besides the drivers window. Window lock-out does not affect the driver's window.

5   These procedures are general in nature, so if you can't find the problem using them, take the vehicle to a dealer service department or other properly equipped repair facility.

6   If the power windows won't operate, always check the fuse and circuit breaker first.

7   If only the rear windows are inoperative, or if the windows only operate from the master control switch, check the rear window lockout switch for continuity in the unlocked position. Replace it if it doesn't have continuity.

8   Check the wiring between the switches and fuse panel for continuity. Repair the wiring, if necessary.

9   If only one window is inoperative from the master control switch, try the other control switch at the window.

Note: *This doesn't apply to the driver's door window.*

10   If the same window works from one switch, but not the other, check the switch for continuity.

11   If the switch tests OK, check for a short or open in the circuit between the affected switch and the window motor.

12   If one window is inoperative from both switches, remove the switch panel from the affected door. Check for voltage at the motor (refer to Chapter 11 for door panel removal) while the switch is operated.

Note: *Voltage at the switch is not measurable; this is a data line to the BCM and not true positive/negative signals as with older models.*

13   If voltage is reaching the motor, disconnect the glass from the regulator (see Chapter 11). Move the window up and down by hand while checking for binding and damage. Also check for binding and damage to the regulator. If the regulator is not damaged and the window moves up and down smoothly, replace the motor. If there's binding or damage, lubricate, repair or replace parts, as necessary.

14   If voltage isn't reaching the motor, check the wiring in the circuit for continuity between the switches and the body control module, and between the body control module and the motors. You'll need to consult the wiring diagram at the end of this Chapter. If the circuit is equipped with a relay, check that the relay is grounded properly and receiving voltage.

15   Test the windows after you are done to confirm proper repairs. To verify if the motor is getting the needed voltage or ground, a simple but effective test is to sit in the car with the ignition on, open a door and look at the dome light. Then operate the switch to the faulty window. If you see the dome light dimming slightly this is a good indication that the motor is getting power and is probably a stuck motor or faulty wiring. Do not hold the switch on for very long when a motor is stuck or wiring is in question or more damage may occur. Finally, in some cases, a good rap on the door panel in the general area of the window motor - while the key is on and the window switch is depressed in the direction the window needs to move - will free up a stuck motor temporarily. You can damage the door panel or more internal components if you hit it too hard or are too aggressive.

Window Express Down Programming

16   Anytime the battery is disconnected, power is disrupted to the window circuit or the window motor is replaced, the "Express Down" feature will need relearned.

17   Lower the window to its lowest position. Hold the window switch in the down position for an additional five seconds.

18   Raise the window to its highest position. Hold the window switch in the up position for an additional five seconds.

19   Verify proper operation. Repeat the procedure for any additional windows.

---

24   Power door lock and keyless entry system - description and check

---

Note: *These models are equipped with a Body Control Module (BCM). Several systems are linked to a centralized control module that allows simple and accurate troubleshooting, but only with a professional-grade scan tool. The Body Control Module governs the door locks, the power windows, the ignition lock and security system, the interior lights, the Daytime Running Lights system, the horn, the windshield wipers, the heating/air conditioning system and the power mirrors. In the event of malfunction with this system, have the vehicle diagnosed by a dealership service department or other qualified automotive repair facility.*

1   The power door lock system operates the door lock actuators mounted in each door. The system consists of the switches, actuators, lock and unlock relays, Body Control Module (BCM) and associated wiring. Diagnosis can usually be limited to simple checks of the wiring connections and actuators for minor faults that can be easily repaired.

2   Power door lock systems are operated by bi-directional solenoids located in the doors. The lock actuators are mounted as part of the door latch. Remove the door latch for access to the door lock actuator. The lock switches have two operating positions: Lock and Unlock. These switches send a signal to the BCM, which in turn sends a signal to the door lock relays, the relays then send the needed voltage to each of the door lock solenoids.

3   If you are unable to locate the trouble using the following general steps, consult your dealer service department or qualified independent repair shop.

4   Always check the circuit protection first. Some vehicles use a combination of circuit breakers and fuses. Refer to the wiring diagrams at the end of this Chapter.

5   Check for voltage at the switches. If no voltage is present, check the fuse first. If the fuse is good then check the wiring between the fuse panel and the switches for an open lead.

6   If voltage is present, test the switch for continuity. Replace it if there's not continuity in both switch positions. There should be a voltage input and, when switch is depressed, voltage should be going out on the appropriate lead. (Follow the wiring diagram for the actual wire and position on the switch.) To remove the switch, use a flat-bladed trim tool to pry out the door/window switch assembly.

7   If the switch has continuity, check the wiring between the switch, BCM, door lock relay and the door lock solenoid.

8   If all but one lock solenoids operate, remove the trim panel from the affected door (see Chapter 11) and check for voltage at the solenoid while the lock switch is operated. One of the wires should have positive voltage in the Lock position; the other lead should have positive voltage in the Unlock position.

9   If the inoperative solenoid is receiving positive voltage on one lead and negative on the other, the solenoid is most likely defective. Check the connections for good contact; if the connection is good, replace the solenoid.

10   If the inoperative solenoid isn't receiving voltage or ground, check for an open or short in the wire between the lock solenoid and the relay. A good method of non-destructive testing is to squeeze the rubber corrugated tubing and search with your fingers for an individual wire. Follow the wire as far as possible and feel for any breaks in the leads.

Note: *It's not uncommon for wires to break in the portion of the harness between the body and door (opening and closing the door fatigues and eventually breaks the wires).*

11   On the models covered by this manual, power door lock system communication goes through the Body Control Module. If the above tests do not pinpoint a problem, take the vehicle to a dealer or qualified shop with the proper scan tool to retrieve trouble codes from the BCM. Replacing of some components may result in programming issues. To avoid replacing of good components always test thoroughly before any parts are deemed faulty.

Keyless entry system

12   The keyless entry system consists of a remote control transmitter that sends a coded infrared signal to a receiver, which then operates the door lock system.

13   Replace the battery when the transmitter doesn't operate the locks at a distance of ten feet. Normal range should be about 30 feet.

Key remote control battery replacement

14   Use a coin to carefully separate the case halves **(see illustration)**.

15   Replace the battery **(see illustration)**.

16   Snap the case halves together.

Transmitter programming

17   Programming replacement transmitters requires the use of a specialized scan tool. Take the vehicle and the transmitter(s) to a dealer service department or other qualified repair shop equipped with the necessary tool to have the transmitter(s) programmed to the vehicle. Up to eight key fobs can be programmed to one car. All fobs for the vehicle have to be present when reprogramming.

25   Electric side view mirrors - description

Note: *These models are equipped with a Body Control Module (BCM). Several systems are linked to a centralized control module that allows simple and accurate troubleshooting, but only with a professional-grade scan tool. The Body Control Module governs the door locks, the power windows, the ignition lock and security system, the interior lights, the Daytime Running Lights system, the horn, the windshield wipers, the heating/air conditioning system and the power mirrors. In the event of malfunction with this system, have the vehicle diagnosed by a dealership service department or other qualified automotive repair facility.*

1   The electric rear view mirrors use two motors to move the glass; one for up and down adjustments and one for left-right adjustments. Some vehicles are equipped with memory mirrors as well. These mirrors are an integral part of the power seat memory unit as well as incorporated with the Body Control Module (BCM). If there is a problem with these systems it is advised to seek out a qualified independent repair facility or your local dealer.

2   The control switch has a selector portion which sends voltage to the left or right side mirror. With the ignition in the ACC position and the engine OFF, roll down the windows and operate the mirror control switch through all functions (left-right and up-down) for both the left and right side mirrors.

3   Listen carefully for the sound of the electric motors running in the mirrors.

4   If the motors can be heard but the mirror glass doesn't move, there's probably a problem with the drive mechanism inside the

24.14 Carefully separate the case halves . . .

mirror. Power mirrors have no user-serviceable parts inside - a defective mirror must be replaced as a unit (see Chapter 11).

5   If the mirrors don't operate and no sound comes from the mirrors, check the fuses (see Section 3).

6   If the fuses are OK, remove the mirror control switch. Have the switch continuity checked by a dealer service department or other qualified shop.

7   Check the ground connections.

8   If the mirror still doesn't work, remove the mirror and check the wires at the mirror for voltage.

9   If there's not voltage in each switch position, check the circuit between the mirror and control switch for opens and shorts.

Note: *If the mirror is inoperative try holding the switch in one of the directions and swing the door open and closed. If there is a break in the door jamb you may occasionally make contact long enough to avoid removing the mirror from the door for further testing. This will also give you some clue as to where the break is rather than testing things in one given position.*

10   If there's voltage, remove the mirror and test it off the vehicle with jumper wires. Before testing check the wiring diagram for the correct leads that have to be used for each position. Applying voltage to the wrong leads can damage the mirror drive motors. Replace the mirror if it fails this test.

26   Power seats - description

1   These models feature a six-way or an eight-way seat option. The six-way seat is without memory, while the eight-way seat has a four-way lumbar support, memory, heat/air conditioning functions, as well as a microprocessor for the seat memory system. The seats are powered by three reversible motors, mounted in one housing, that are controlled by switches on the side of the seat. Each switch changes the direction of seat travel by reversing polarity to the drive motor. The

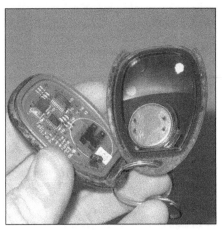

24.15 . . . then pry out the battery

6-way seat uses direct voltage from the seat switch to operate while the 8-way seat voltage comes from the memory processor.

2   Diagnosis is usually a simple matter, using the following procedures.

3   Look under the seat for any object which may be preventing the seat from moving.

4   If the seat won't work at all, check the circuit breaker (see Section 3).

5   With the engine off to reduce the noise level, operate the seat controls in all directions and listen for sound coming from the seat motors.

Note: *A grinding sound when the motor is on indicates either a broken gear (usually plastic gears) or the cable between the motor and the seat transmission has stripped. If you do not hear any noise or motor movement this can be an indication of a frozen motor or seat transmission. Leave a door open and watch the dome light for any signs of dimming while operating the seat switch. The dimming is a good indicator for a stuck or frozen motor, seat transmission, or shorted wiring.*

6   If the motor is getting voltage but doesn't run, test it off the vehicle with jumper wires. If it still doesn't work, replace it.

6-way Seat Diagnosis

7   If the motor isn't getting voltage, remove the switch and check for voltage. If there's no voltage to the switch, check the wiring between the fuse block and the switch. If there's battery voltage at the switch, check the other terminals for voltage while moving the switch around. If the switch is OK, check for an open in the wiring between the switch and motor assembly.

8-way Memory Seat Diagnosis

8   The MSM (Memory Seat Module) sends the necessary voltage to the seat motors by way of a circuit breaker in the instrument panel fuse box. A separate fuse (Marked DSM 10A) sends the needed voltage for the logic circuits of the MSM. The seat switch is connected directly to the MSM. With the MSM system the seat switch is generally referred to as the "door module" (mounted in each door). The signal to and from the switch is on the GMLAN CAN dataline and is not a readable signal with the standard mulitmeter with any real accuracy. Proper equipment and scanning capabilities is needed to diagnose these systems efficiently. See your local independent repair shop or the nearest dealer for diagnostics.

27 Data Link Communication system - description

1   The vehicles covered by this manual have a complex electrical system, encompassing many power accessories, and a number of separate electronic modules.
2   The Powertrain Control Module (PCM) is mainly responsible for engine and transaxle control, but also communicates with other modules around the vehicle through a Data Link Communication system, which sends serial port data very quickly between the various modules. Many of the computer functions involved in the operation of body systems are routed through the Body Control Module (BCM), which communicates with the PCM.
3   Among the modules in the Data Link system besides the BCM and PCM are the Sensing Diagnostic Module (airbag system), the Electronic Brake Control Module, stereo system, GPS, and the instrument panel cluster. The BCM further communicates with various body subsystems.
4   All of the modules in the vehicle have associated trouble codes. When other troubleshooting procedures fail to pinpoint the problem, check the wiring diagrams at the end of this Chapter to see if the BCM or PCM are involved in the circuit. If so, bring your vehicle to a dealer service department or other qualified repair shop with the proper diagnostic tools to extract the trouble codes.
Note: *A scan tool is required to extract any stored trouble codes. Lower quality scanners have limited access to various systems and their related codes. Taking your vehicle to the appropriate repair facility will also give you more information than some of the cheaper solutions in regards of retrieving codes from the vehicle.*

28 Airbag system - general information

1   All models are equipped with a Supplemental Restraint System (SRS), more commonly known as the airbag system. The airbag system is designed to protect the driver and the front seat passenger from serious injury in the event of a head-on or side impact collisions. Airbag systems consist of the impact sensors, airbag modules (including: driver's airbag, passenger's airbag, side curtain/roof rail airbags and side impact airbags), seat belt tensioners and a sensing/diagnostic module mounted under the center console. All of the related airbag wiring harnesses are wrapped

with yellow or orange tape and have double locking tabs at their connectors. Read through this entire Section before removing and/or replacing any airbag component.

Airbag modules
Driver's airbag
2   The airbag inflator module contains a housing incorporating the cushion (airbag) and inflator unit, mounted in the center of the steering wheel. The inflator assembly is mounted on the back of the housing over a hole through which gas is expelled, inflating the bag almost instantaneously when an electrical signal is sent from the system. A spiral cable (or clockspring) assembly on the steering column under the steering wheel carries this signal to the module. This clockspring can transmit an electrical signal regardless of steering wheel position.

Passenger's airbag
3   The airbag is mounted inside the right side of the instrument panel, in the area above the glove box. It's similar in design to the driver's airbag, except that it's larger than the steering wheel unit. The trim cover (on the side of the instrument panel that faces toward the passenger) is textured and colored to match the instrument panel and has a molded seam that splits open when the bag inflates. The passenger airbag is designed to spread not only out towards the passenger but also to encompass the dash and windshield area to prevent the passenger from accelerating into the glass.

Side curtain (roof rail) airbags
4   In addition to the side-impact airbags, extra side-impact protection is also provided by side-curtain airbags on some models. These are long airbags that, in the event of a side impact, come out of the headliner at each side of the car and come down between the side windows and the seats. They are designed to protect the heads of both front seat and rear seat passengers.

Side impact airbags
5   These airbags are located in the outboard sides of the front seats. They are deployed during the same conditions that would trigger a roof rail airbag deployment.

Knee bolsters
6   The knee bolsters are below the steering column and the front passenger knee area. These are designed to cushion the forward motion of the front seat occupants when in a frontal collision.

Sensing and diagnostic module
7   The sensing and diagnostic module supplies the current to the airbag system in the event of the collision, even if battery power is cut off. It checks this system every time the vehicle is started, causing the "AIR BAG" light to go on then off, if the system is operating properly. If there is a fault in the system, the light will go on and stay on, flash, or the dash

will make a beeping sound. If this happens, the vehicle should be taken to your dealer or qualified independent shop immediately for service. This module is mounted under the center console. There is also a roll-over sensor located directly behind it on later models.
Warning: *The diagnostic module maintains a 23 volt reserve power in the unit after the key is turned off. Which means the airbag system is still active even if the key is off. This built in safety feature is active for up to one minute after the key is turned off. Be sure to keep this in mind when servicing the airbag system.*

Seat belt pre-tensioners
8   Some models are equipped with pyrotechnic (explosive) units in the front seat belt retracting mechanisms. During an impact that would trigger the airbag system, the airbag control unit also triggers the seat belt retractors. When the pyrotechnic charges go off, they accelerate the retractors to instantly take up any slack in the seat belt system to more fully prepare the driver and front seat passenger for impact.
9   The airbag system should be disabled any time work is done to or around the seats.
Warning: *Never strike the pillars or floorpan with a hammer or use an impact-driver tool in these areas unless the system is disabled.*

Disarming the system and other precautions
Warning: *Failure to follow these precautions could result in accidental deployment of the airbag and personal injury.*
10   Whenever working in the vicinity of the steering wheel, steering column or any of the other SRS system components, the system must be disarmed. To disarm the airbag system:

a)   *Point the wheels straight ahead and turn the key to the Lock position.*
b)   *Disconnect the cable from the negative battery terminal.*
c)   *Wait at least two minutes for the back-up power supply to be depleted.*

Whenever handling an airbag module:
11   Always keep the airbag opening (the trim side) pointed away from your body. Never place the airbag module on a bench or other surface with the airbag opening facing the surface. Always place the airbag module in a safe location with the airbag opening facing up.
12   Never measure the resistance of any SRS component. An ohmmeter has a built-in battery supply that could accidentally deploy the airbag.
13   Never use electrical welding equipment on a vehicle equipped with an airbag without first disconnecting the electrical connector for each airbag.
14   Never dispose of a live airbag module. Return it to a dealer service department or other qualified repair shop for safe deployment and disposal.

Component removal and installation

Driver's side airbag module and clockspring

15 Refer to Chapter 10 for the driver's side airbag module and clockspring removal and installation procedures.

Other airbag modules

16 We don't recommend removing any of the other airbag modules. These jobs are best left to a professional.

29 Wiring diagrams - general information

1 Since it isn't possible to include all wiring diagrams for every year and model covered by this manual, the following diagrams are those that are typical and most commonly needed.

2 Prior to troubleshooting any circuits, check the fuse and circuit breakers (if equipped) to make sure they're in good condition. Make sure the battery is properly charged and check the cable connections (see Chapter 5).

3 When checking a circuit, make sure that all connectors are clean, no signs of burnt or melted connections, with no broken or loose terminals. When disconnecting a connector, do not pull on the wires. Pull only on the connector housings themselves.

Notes

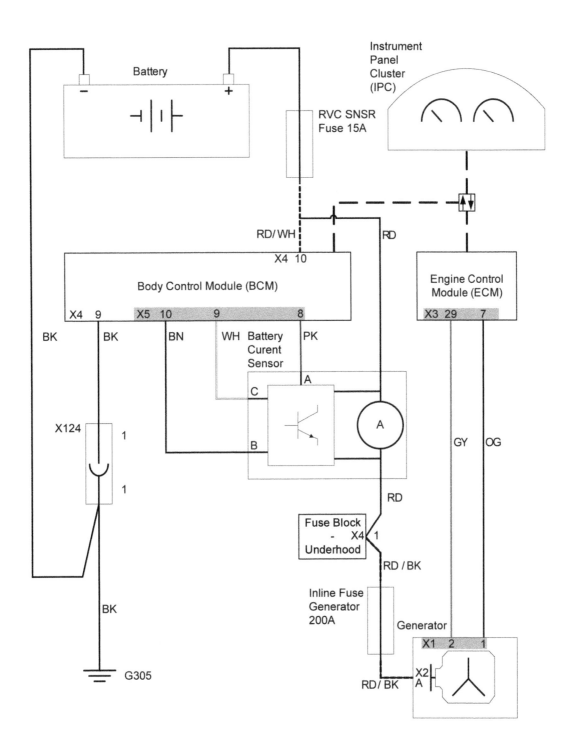

Charging system

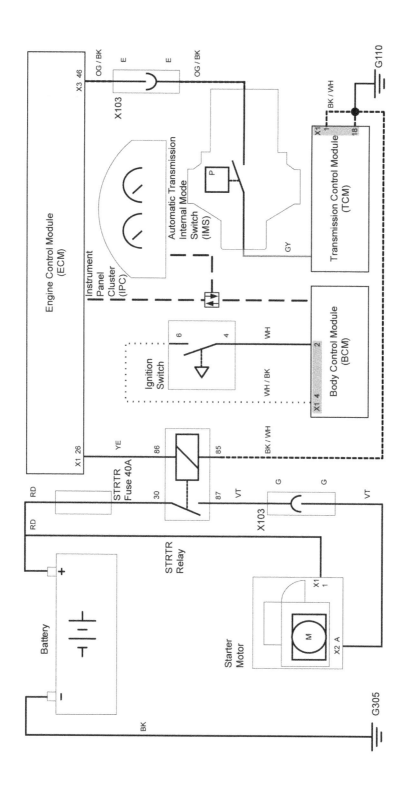

Starting system

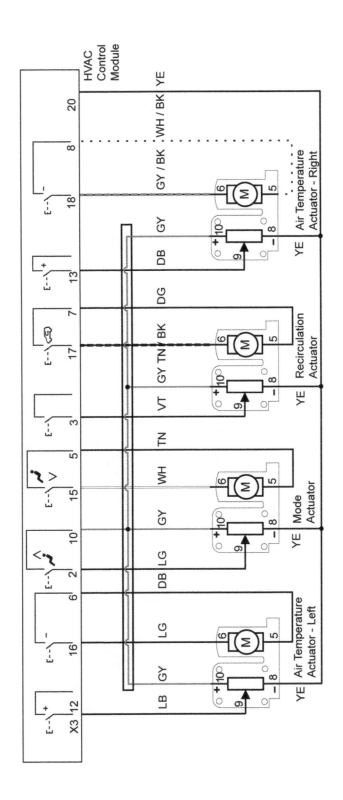

Front air conditioning and heating system actuators (automatic)

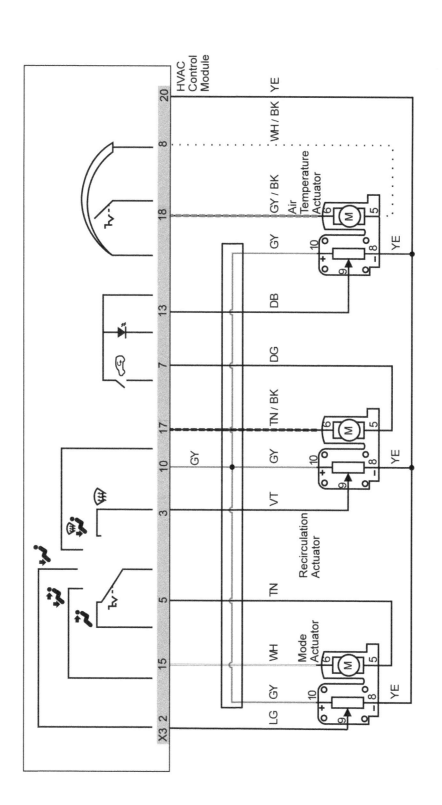

Front air conditioning and heating system actuators (manual)

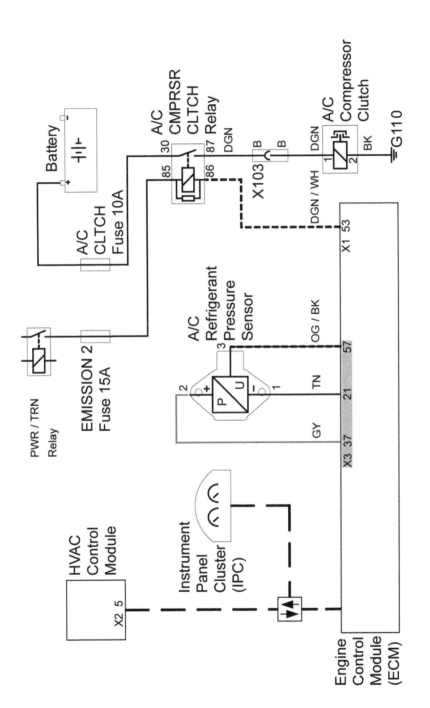

Front air conditioning and heating system compressor controls (automatic)

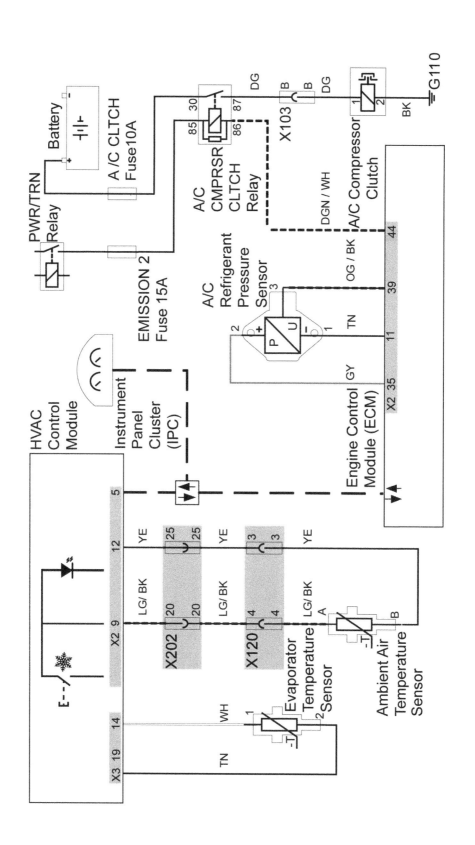

Front air conditioning and heating system compressor controls (manual)

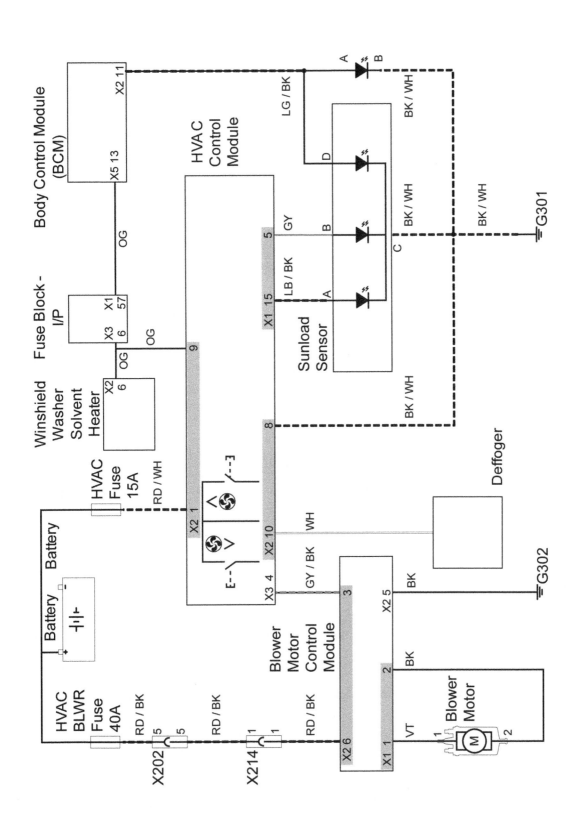

Front air conditioning and heating system - Module power and ground, blower control, data communication and sunload

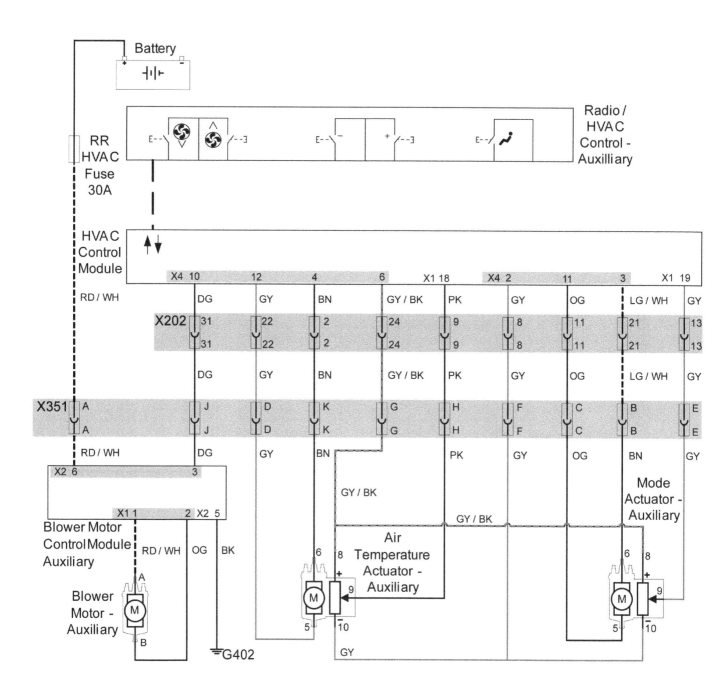

Rear air conditioning and heating system (with UK6)

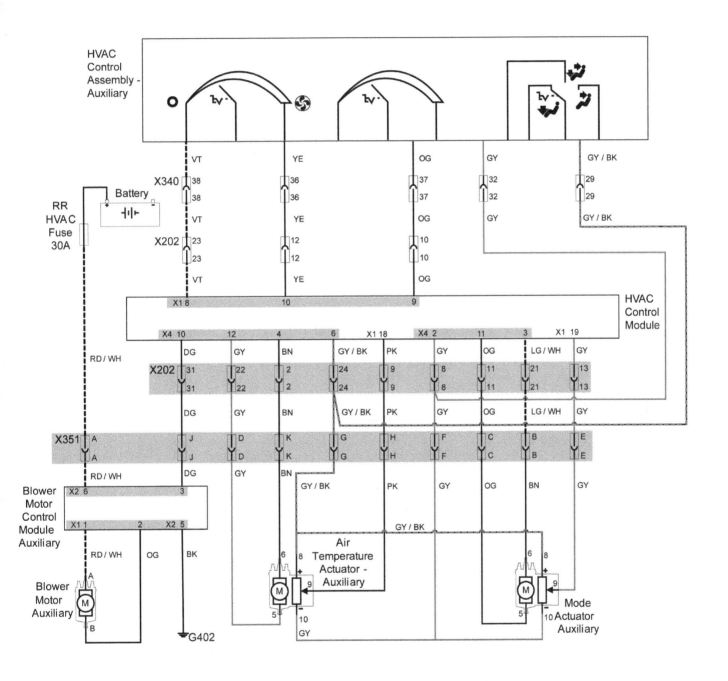

Rear air conditioning and heating system (without UK6)

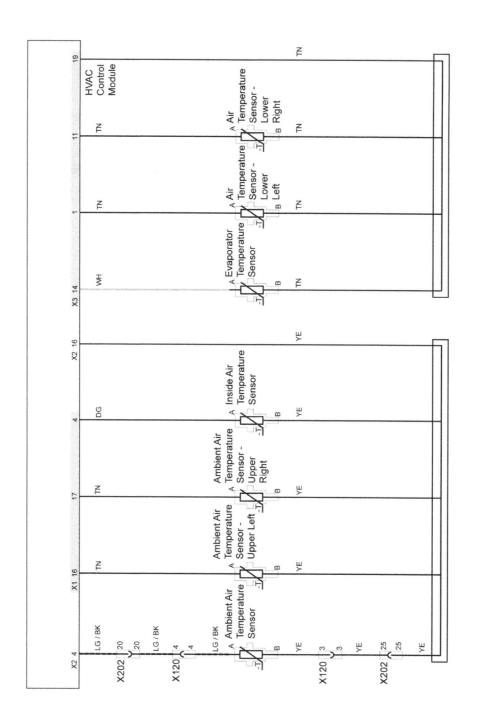

Air conditioning and heating system sensors

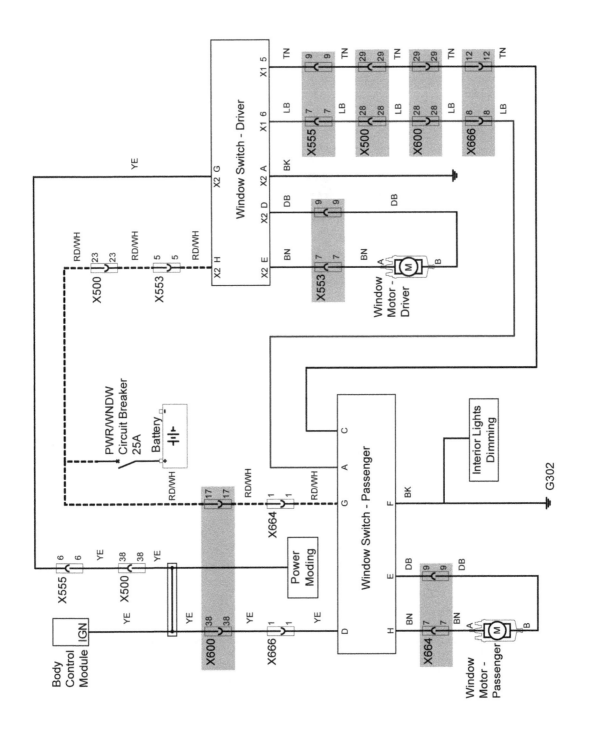

Front power window system (except AXC or AXE)

Power Windows Rear

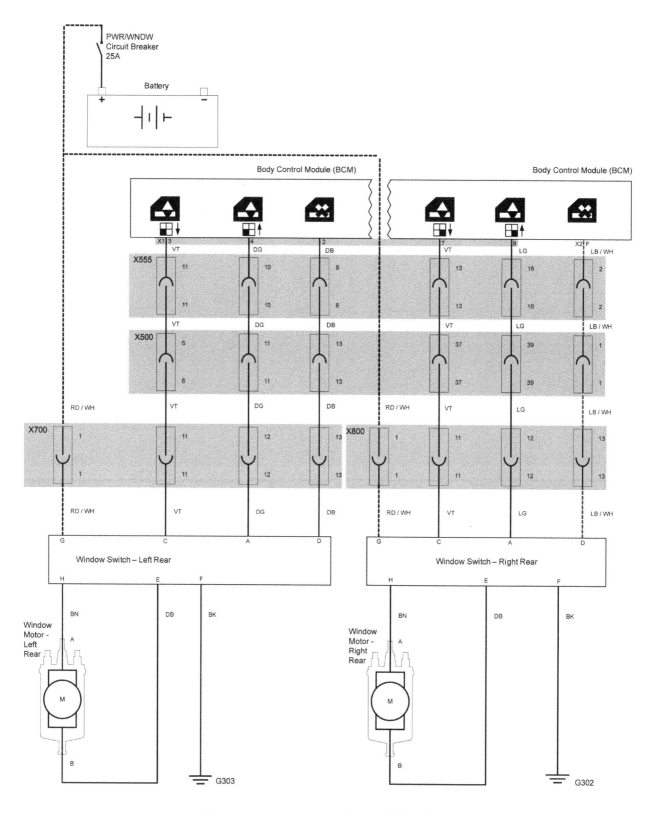

Rear power window system (except AXC or AXE)

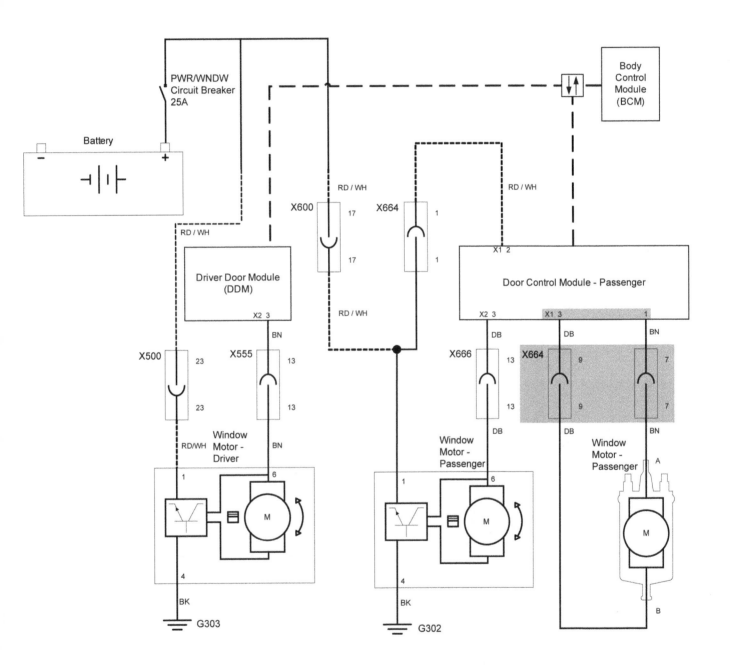

Front power window system (with AXC or AXE)

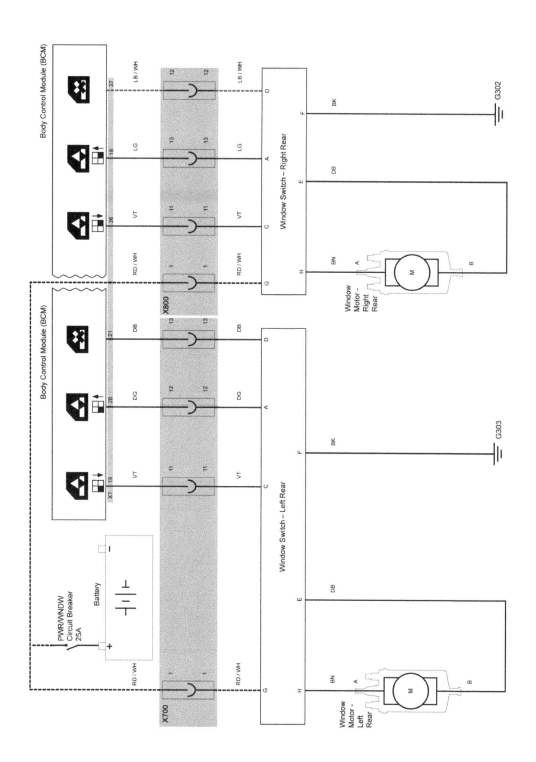

Rear power window system (with AXC or AXE)

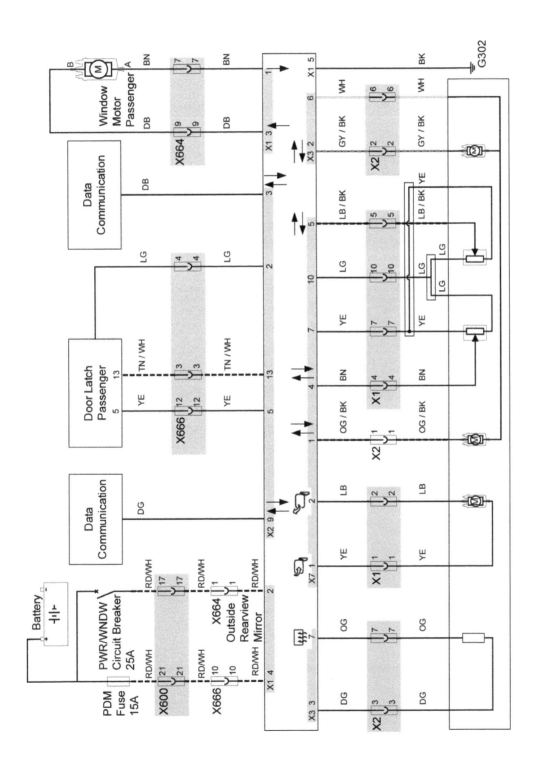

Power door lock system

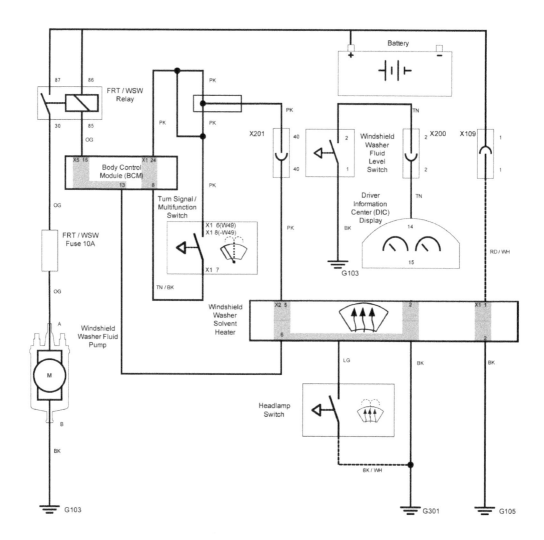

Windshield wiper/washer system

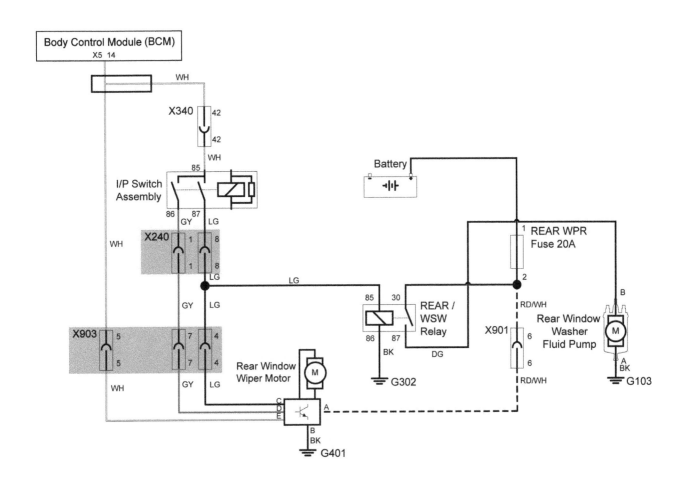

Rear window wiper/washer system

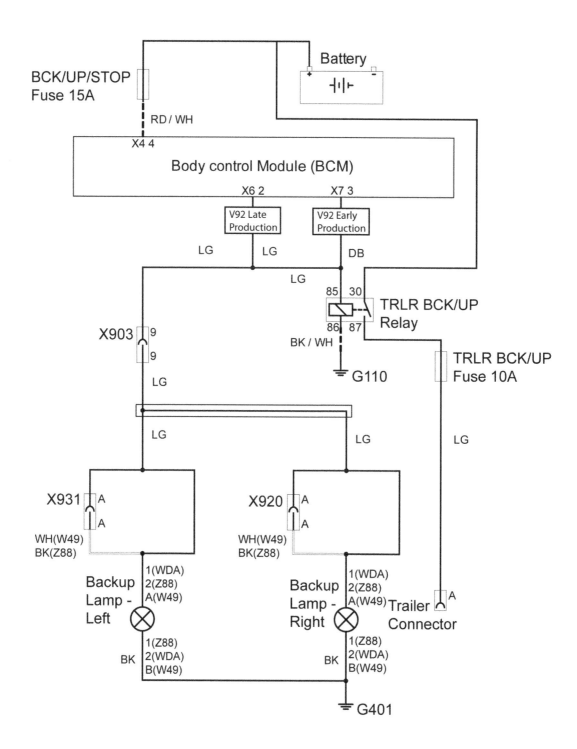

Back-up lights system

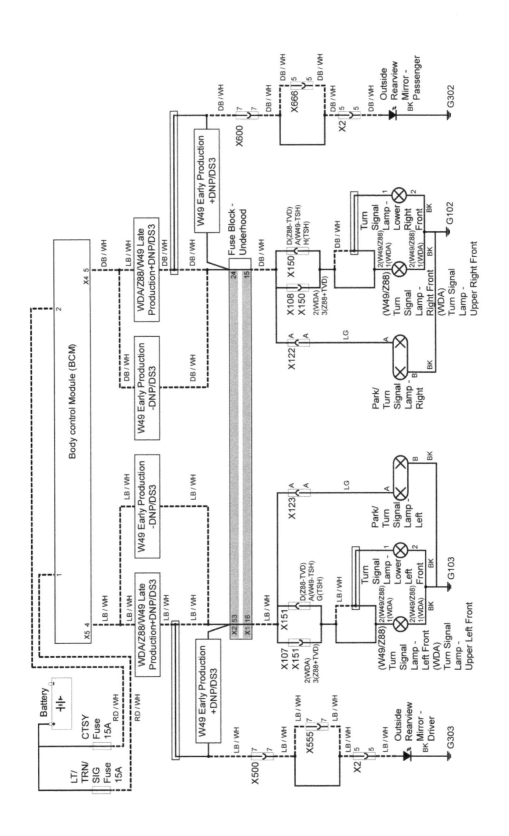

Exterior lighting system (front turn signals and outside mirrors)

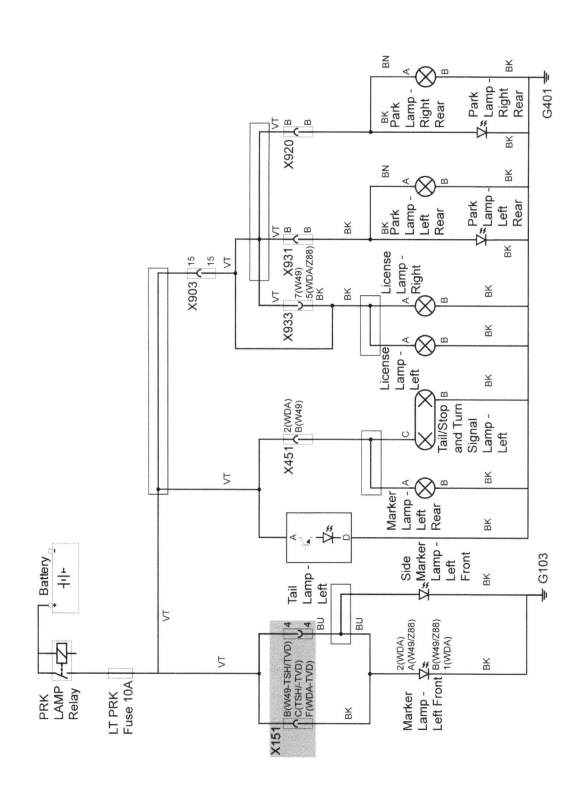

Exterior lighting system (left parking lights) - 2008 and earlier models

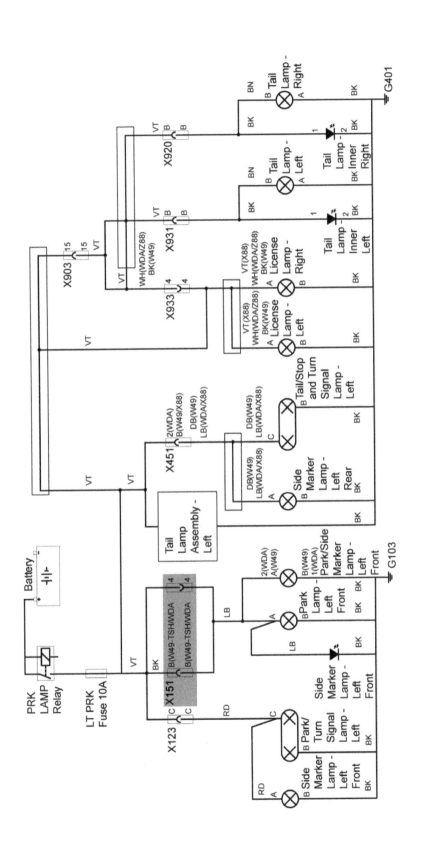

Exterior lighting system (left parking lights) - 2009 and later models

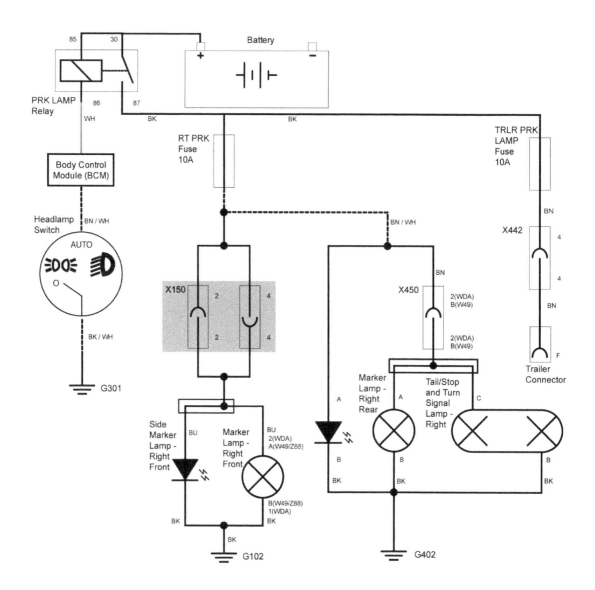

Exterior lighting system (right parking lights and control) - 2008 and earlier models

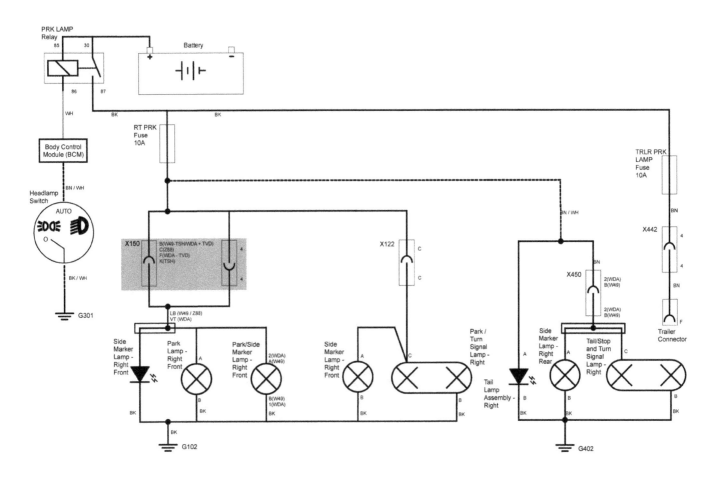

Exterior lighting system (right parking lights and control) - 2009 and later models

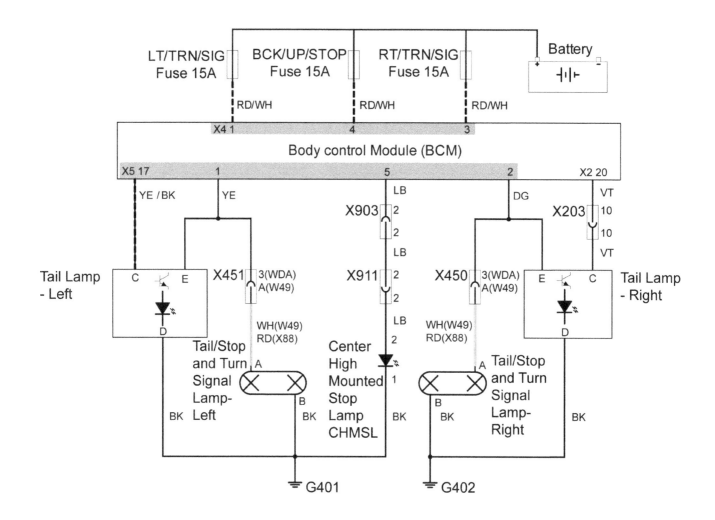

Exterior lighting system (brake and tail lights)

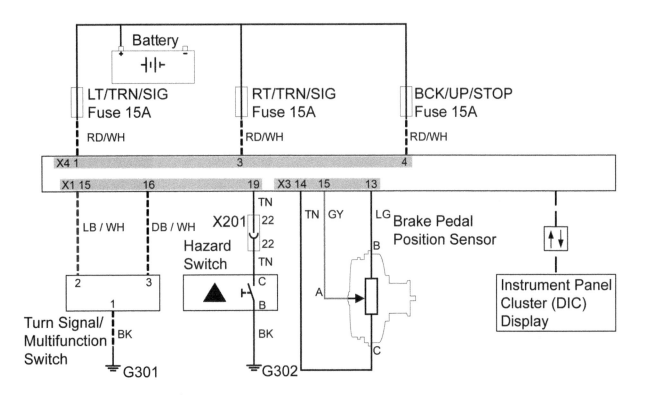

Exterior lighting system (brake and turn signal light controls and indicators) - 2008 and earlier models

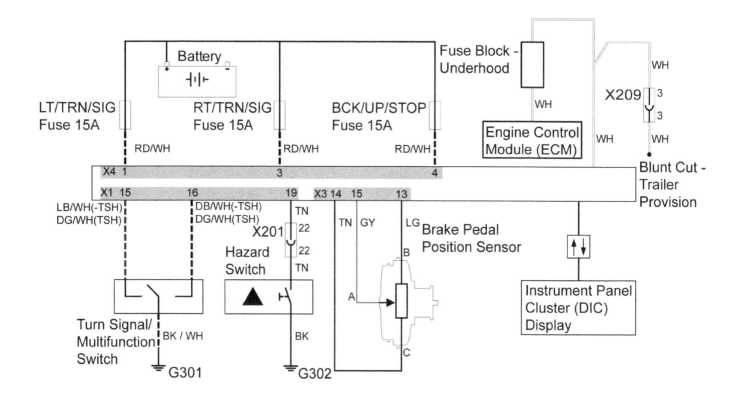

Exterior lighting system (brake and turn signal light controls and indicators) - 2009 and later models

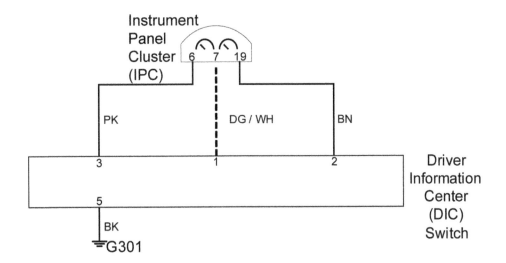

Instrument cluster (driver information center with UH9)

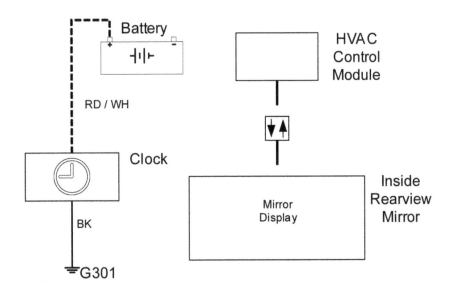

Instrument cluster (driver information system)

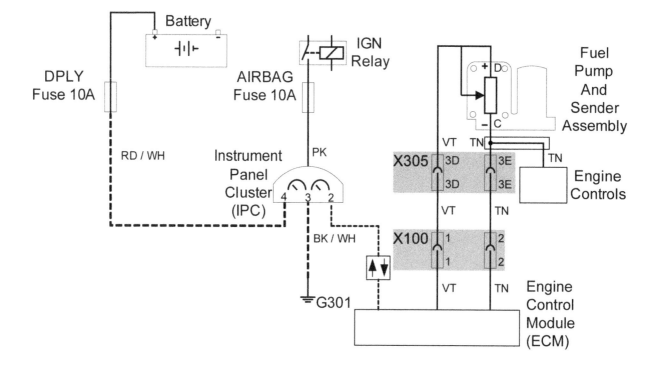

Instrument cluster (gages)

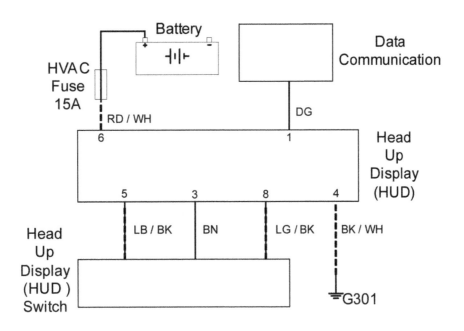

Instrument cluster (head-up display)

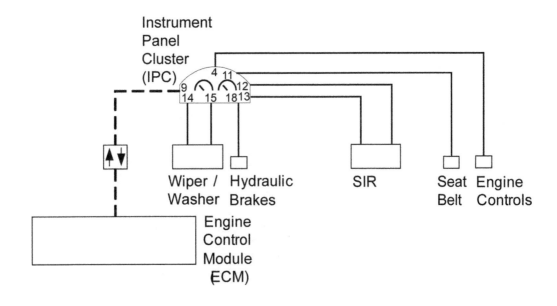

Instrument cluster (lighting and indicators)

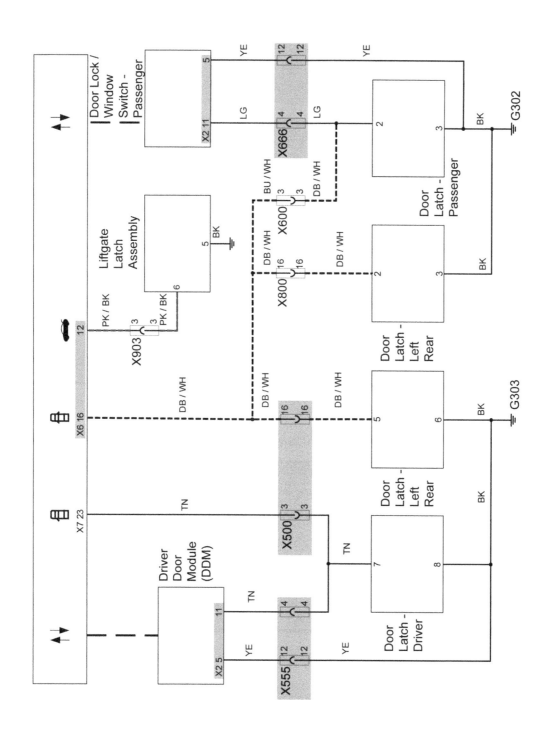

Interior lighting system (door open switches)

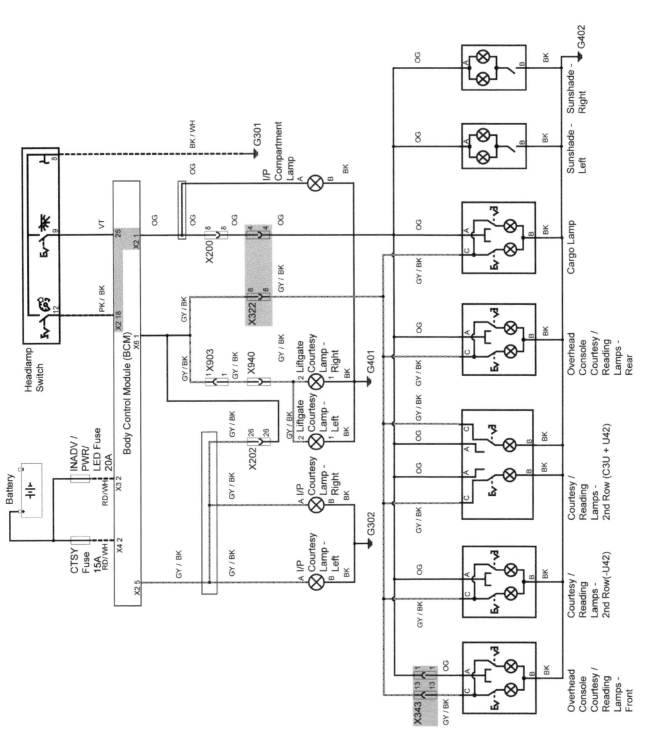

Interior lighting system (headlight switch and power, instrument panel, footwell and liftgate lights)

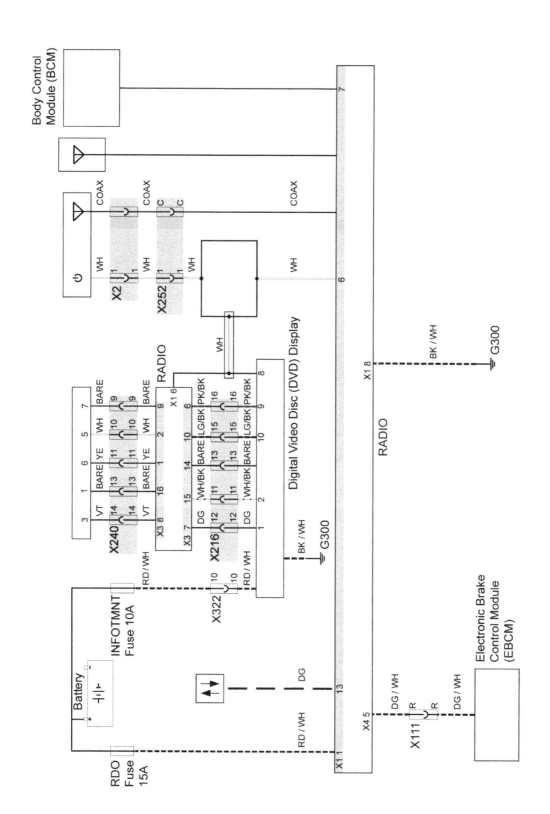

Audio/Navigation system (radio and DVD player)

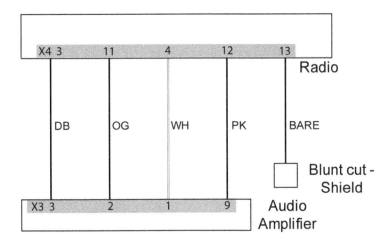

Audio/Navigation system (amplifier subwoofer audio inputs)

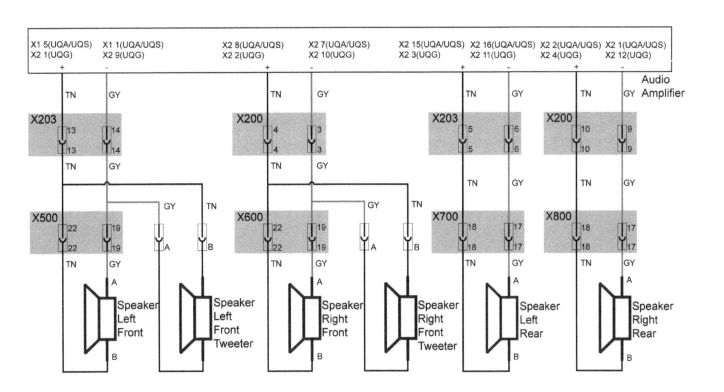

Audio/Navigation system (door and A-pillar speakers)

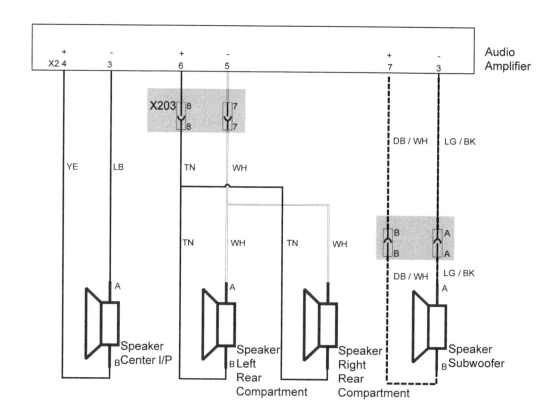

Audio Amplifier

X2 4 (+)   3 (-)   6 (+)   5 (-)   7 (+)   3 (-)

X203   8 | 7
       8 | 7

YE   LB   TN   WH   DB / WH   LG / BK

TN   WH   TN   WH

B   A
B   A

DB / WH   LG / BK

A   A   A   A

Speaker Center I/P

Speaker Left Rear Compartment

Speaker Right Rear Compartment

Speaker Subwoofer

Audio/Navigation system (speakers, Bose audio system)

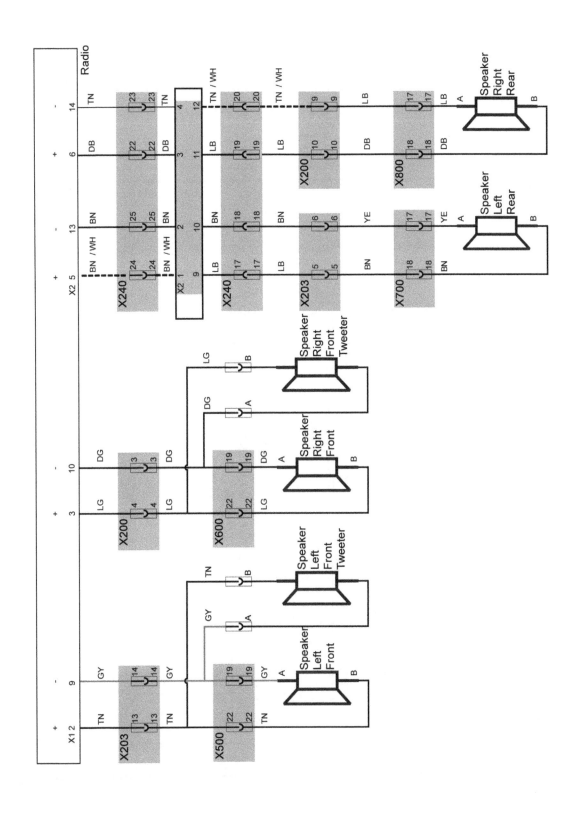

Audio/Navigation system (speakers, standard audio system)

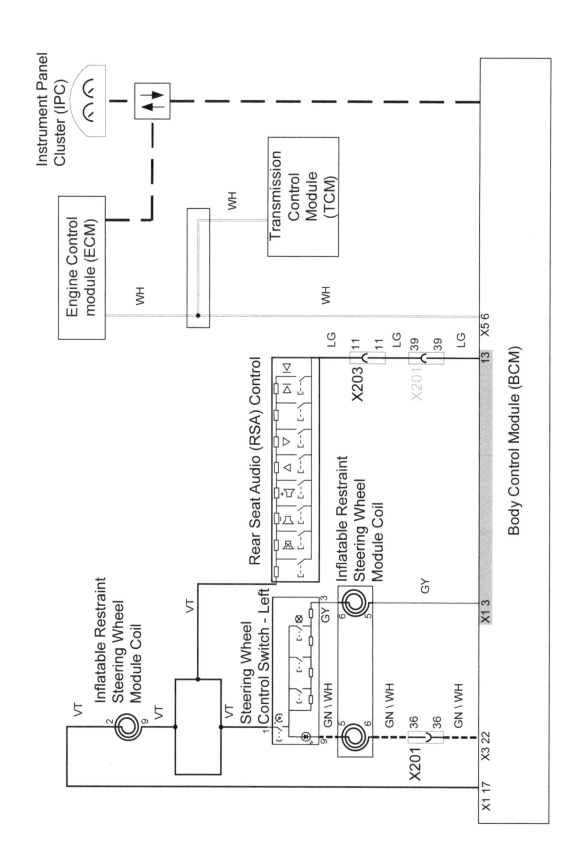

Cruise control system

Notes

Notes

Index

A

About this manual, 0-5
Accelerator Pedal Position (APP) sensor,
    replacement, 6-13
Acknowledgements, 0-2
Air conditioning
    and heating system, check and maintenance, 3-5
    compressor, removal and installation, 3-13
    condenser, removal and installation, 3-14
    receiver-drier, removal and installation, 3-14
    refrigerant pressure sensor, replacement, 3-15
    Thermostatic Expansion valve (TXV), general
        information, 3-15
Air filter
    element, replacement, 1-19
    housing, removal and installation, 4-7
Airbag system, general information, 12-18
Alternator, removal and installation, 5-5
Antenna, removal and installation, 12-9
Anti-lock Brake System (ABS), general information, 9-7
Automatic transaxle, 7A-1
    Brake Transmission Shift Interlock (BTSI) system,
        description, component replacement and
        adjustment, 7A-4
    diagnosis, general, 7A-2
    driveaxle oil seals, replacement, 7A-4
    fluid
        change, 1-21
        level check, 1-11
        type, 1-1
    mount, replacement, 7A-5
    overhaul, general information, 7A-5
    removal and installation, 7A-5
    shift cable, replacement and adjustment, 7A-3
    shift lever, replacement, 7A-2
    shift lever knob, replacement, 7A-2
Automotive chemicals and lubricants, 0-18

B

Balljoints, replacement, 10-10
Battery
    cables, replacement, 5-4
    check, maintenance and charging, 1-13
    disconnection, 5-3
    electrolyte fluid level check, 1-8
    removal and installation, 5-4
Blower motor resistor/module and blower motor,
    replacement, 3-10
Body repair
    major damage, 11-3
    minor damage, 11-2
Body, 11-1
Booster battery (jump) starting, 0-17
Brake
    Anti-lock Brake System (ABS), general information, 9-7
    caliper and mounting bracket, removal and
        installation, 9-10
    disc, inspection, removal and installation, 9-10
    fluid
        change, 1-19
        level check, 1-9
        type, 1-1
    general information, 9-2
    hoses and lines, inspection and replacement, 9-13
    hydraulic system, bleeding, 9-13
    master cylinder, removal and installation, 9-11
    pads, replacement, 9-8
    parking brake
        adjustment, 9-15
        shoes, replacement, 9-15
    power brake booster, check, removal and
        installation, 9-14
    system check, 1-17
    troubleshooting, 9-3
    vacuum auxiliary pump, 9-14

Brake Transmission Shift Interlock (BTSI) system, description, component replacement and adjustment, 7A-4
Bulb replacement, 12-11
Bumper covers, removal and installation, 11-7
Buying parts, 0-9

C

Cable replacement
  battery, 5-4
  hood release, 11-6
  shift, 7A-3
Caliper and mounting bracket, disc brake, removal and installation, 9-10
Camshaft Position (CMP) sensor, replacement, 6-13
Camshaft Position Actuator Solenoid Valve, replacement, 6-13
Camshafts, removal, inspection and installation, 2A-14
Capacities, fluids and lubricants, 1-2
Catalytic converter, replacement, 6-16
Center console, removal and installation, 11-17
Center support bearing, driveshaft (AWD models), replacement, 8-7
Charging system
  alternator, removal and installation, 5-5
  check, 5-2
  general information and precautions, 5-1
Chassis electrical system, 12-1
Chemicals and lubricants, 0-18
Circuit breakers, general information, 12-3
Clutch control module, replacement, 8-9
Coil spring (rear), removal and installation, 10-12
Coils, ignition, replacement, 5-5
Combination switches, steering column, replacement, 12-7
Compressor, air conditioning, removal and installation, 3-13
Condenser, air conditioning, removal and installation, 3-14
Constant velocity joints, general information and check, 8-8
Control arm, removal, inspection and installation, front, 10-9
Conversion factors, 0-19
Coolant
  level check, 1-8
  reservoir, 3-8
  temperature indicator, check, 3-7
  type, 1-1
Coolant Temperature (ECT) sensor, replacement, 6-14
Cooling fans, engine replacement, 3-7
Cooling system
  check, 1-15
  general information, 3-2

  servicing (draining, flushing and refilling), 1-22
  troubleshooting, 3-2
Cooling, heating and air conditioning systems, 3-1
Cowl cover, removal and installation, 11-17
Crankshaft front oil seal, removal and installation, 2A-9
Crankshaft Position (CKP) sensor, replacement, 6-13
Crankshaft pulley, removal and installation, 2A-8
Crankshaft, removal and installation, 2B-11
Cruise control system, description and check, 12-15
Cylinder compression check, 2B-4
Cylinder heads, removal and installation, 2A-7

D

Dashboard trim panels, removal and installation, 11-18
Data Link Communication system, description, 12-18
Daytime Running Lights (DRL), general information, 12-14
Defogger, rear window, check and repair, 12-14
Diagnosis, 0-22
Diagnostic Trouble Codes (DTCs), obtaining and clearing, 6-4
Differential (AWD models)
  assembly (rear), removal and installation, 8-8
  clutch drum, removal and installation, 8-8
  lubricant
    change, 1-22
    level check, 1-16
    type, 1-1
  pinion seal, replacement, 8-8
Disc brake
  caliper and mounting bracket, removal and installation, 9-10
  disc, inspection, removal and installation, 9-10
  pads, replacement, 9-8
Door
  latch, lock cylinder and handles, removal and installation, 11-13
  removal, installation and adjustment, 11-12
  trim panels, removal and installation, 11-9
  window glass regulator module, removal and installation, 11-14
  window glass, removal and installation, 11-14
Driveaxle
  boot check, 1-16
  boot, replacement, 8-3
  oil seals, replacement
    automatic transaxle, 7A-4
    rear (AWD models), 8-8
  removal and installation, 8-2
Drivebelt
  check and replacement, 1-14
  tensioner, replacement, 1-15
Driveline, 8-1

Driveplate, removal and installation, 2A-16
Driveshaft (AWD models)
    center support bearing, replacement, 8-8
    removal and installation, 8-7
    universal and constant velocity joints, general
        information and check, 8-8

E

Electric side view mirrors, description, 12-17
Electrical connectors, general information, 12-6
Electrical troubleshooting, general information, 12-1
Emissions and engine control systems, 6-1
    general information, 6-1
    information sensors, 6-3
Engine Coolant Temperature (ECT) sensor,
    replacement, 6-14
Engine coolant, level check, 1-8
Engine cooling fans, replacement, 3-7
Engine electrical systems, 5-1
    general information and precautions, 5-1
    troubleshooting, 5-2
Engine oil
    and oil filter change, 1-11
    level check, 1-7
    type and viscosity, 1-1
Engine overhaul
    disassembly sequence, 2B-8
    reassembly sequence, 2B-15
Engine rebuilding alternatives, 2B-5
Engine removal, methods and precautions, 2B-6
Engine, general overhaul procedures, 2B-1
    crankshaft, removal and installation, 2B-11
    cylinder compression check, 2B-4
    engine overhaul
        disassembly sequence, 2B-8
        reassembly sequence, 2B-15
    engine rebuilding alternatives, 2B-5
    engine removal, methods and precautions, 2B-6
    engine, removal and installation, 2B-7
    initial start-up and break-in after overhaul, 2B-15
    oil pressure check, 2B-4
    pistons and connecting rods, removal and
        installation, 2B-8
    vacuum gauge diagnostic checks, 2B-5
Engine, 3.6L V6, in-vehicle repair procedures, 2A-1
    camshafts, removal and installation, 2A-14
    crankshaft front oil seal, removal and
        installation, 2A-9
    crankshaft pulley, removal and installation, 2A-8
    cylinder heads, removal and installation, 2A-7
    driveplate, removal and installation, 2A-16
    exhaust manifolds, removal and installation, 2A-7
    intake manifold, removal and installation, 2A-6

    mounts, check and replacement, 2A-18
    oil pan, removal and installation, 2A-15
    oil pump, removal, inspection and
        installation, 2A-16
    rear main oil seal, replacement, 2A-17
    repair operations possible with the engine in the
        vehicle, 2A-3
    rocker arms and hydraulic lash adjusters, removal,
        inspection and installation, 2A-4
    timing chains and camshaft sprockets, removal,
        inspection and installation, 2A-9
    Top Dead Center (TDC) for number 1 piston,
        locating, 2A-3
    valve covers, removal and installation, 2A-3
    valve springs, retainers and seals,
        replacement, 2A-5
Engine, removal and installation, 2B-7
Evaporative emissions control (EVAP) system,
    component replacement, 6-17
Exhaust manifolds, removal and installation, 2A-7
Exhaust system
    check, 1-18
    servicing, general information, 4-5

F

Fans, engine cooling, replacement, 3-7
Fastener and trim removal, 11-5
Fault finding, 0-22
Fender, front, removal and installation, 11-8
Filter replacement
    engine air, 1-19
    engine oil, 1-11
Firing order, 1-2
Fluid level checks, 1-7
    automatic transaxle, 1-11
    battery electrolyte, 1-8
    brake fluid, 1-9
    differential (AWD models), 1-16
    engine coolant, 1-8
    engine oil, 1-7
    power steering, 1-10
    windshield washer, 1-8
Fluids and lubricants
    capacities, 1-2
    recommended, 1-1
Fog lights, adjustment, 12-9
Fraction/decimal/millimeter equivalents, 0-20
Front hub and bearing assembly, removal and
    installation, 10-11
Fuel
    disconnecting fuel line fittings, 4-4
    Fuel Pump Flow Control Module (FPFCM), removal
        and installation, 4-6

high pressure fuel pump, removal and installation, 4-6
lines and fittings, general information and
    disconnection, 4-5
pressure
    check, 4-3
    relief procedure, 4-3
    sensor, replacement, 4-6
pump module, removal and installation, 4-5
rail and injectors, removal and installation, 4-8
system check, 1-17
system warnings, 4-2
tank, removal and installation, 4-7
troubleshooting, 4-2
Fuel and exhaust systems, 4-1
Fuses, general information, 12-3
Fusible links, general information, 12-3

G

General engine overhaul procedures, 2B-1
crankshaft, removal and installation, 2B-11
cylinder compression check, 2B-4
engine overhaul
    disassembly sequence, 2B-8
    reassembly sequence, 2B-15
engine rebuilding alternatives, 2B-5
engine removal, methods and precautions, 2B-6
engine, removal and installation, 2B-7
initial start-up and break-in after overhaul, 2B-15
oil pressure check, 2B-4
pistons and connecting rods, removal and
    installation, 2B-8
vacuum gauge diagnostic checks, 2B-5

H

Hazard flashers, general information, 12-7
Headlight
adjustment, 12-9
bulb, replacement, 12-9
housing, removal and installation, 12-10
Heater core, replacement, 3-12
**Heater/air conditioning control assembly, removal
and installation, 3-11**
**Heating and air conditioning system, check and
maintenance, 3-5**
Hinges and locks, maintenance, 11-7
**Hood latch and release cable, removal and
installation, 11-6**
Hood, removal, installation and adjustment, 11-6

Horn, replacement, 12-14
Hub and bearing assembly, removal and installation
front, 10-11
rear, 10-13
**Hydraulic lash adjusters, removal, inspection and
installation, 2A-4**

I

Ignition system
coils, replacement, 5-5
switch and key lock cylinder, replacement, 12-7
Initial start-up and break-in after overhaul, 2B-15
Injectors, fuel, removal and installation, 4-8
Instrument cluster, removal and installation, 12-8
Instrument panel switches, replacement, 12-7
Intake manifold, removal and installation, 2A-6
**Intake manifold tuning valve (2007 and 2008 models),
removal and installation, 6-17**
Introduction, 0-5

J

Jacking and towing, 0-16
Jump starting, 0-17

K

Key lock cylinder and latch, removal and installation
door, 11-13
liftgate, 11-16
Key lock cylinder, ignition, replacement, 12-7
Keyless entry system, description and check, 12-16
Knock sensor, replacement, 6-14

L

Liftgate
latch and outside handle, removal and installation, 11-16
removal, installation and adjustment, 11-16
Lubricants and chemicals, 0-18
Lubricants and fluids
capacities, 1-2
recommended, 1-1

M

Maintenance
routine, 1-1
schedule, 1-6
techniques, tools and working facilities, 0-9
Manifold Absolute Pressure (MAP) sensor (2007 and 2008 models), replacement, 6-14
Mass Air Flow/Intake Air Temperature (MAF/IAT) sensor, replacement, 6-14
Master cylinder, brake, removal and installation, 9-11
Mirrors
electric side view, description, 12-17
removal and installation, 11-15
Mounts, engine, check and replacement, 2A-18
Multi-function switches, steering column, replacement, 12-7

O

Obtaining and clearing Diagnostic Trouble Codes (DTCs), 6-4
Oil, engine, level check, 1-7
Oil, engine, type and viscosity, 1-1
Oil life monitor, resetting, 1-12
Oil pan, removal and installation, 2A-15
Oil pressure check, 2B-4
Oil pump, removal, inspection and installation, 2A-16
On-Board Diagnostic (OBD) system, 6-2
Oxygen sensors, replacement, 6-15

P

Pads, disc brake, replacement, 9-8
Parking brake
adjustment, 9-15
shoes, replacement, 9-15
Parts, replacement, buying, 0-9
Pinion seal, differential (AWD models), replacement, 8-8
Pistons and connecting rods, removal and installation, 2B-8
Power brake booster, check, removal and installation, 9-14
Power door lock and keyless entry system, description and check, 12-16
Power seats, description, 12-17
Power steering
fluid
level check, 1-10
type, 1-1

pump, removal and installation, 10-17
system, bleeding, 10-17
Power window system, description and check, 12-15
Powertrain Control Module (PCM), removal and installation, 6-16

R

Radiator, removal and installation, 3-9
Radio and speakers, removal and installation, 12-8
Rear differential assembly (AWD models), removal and installation, 8-8
Rear hub and bearing assembly, removal and installation, 10-13
Rear knuckle, removal and installation, 10-13
Rear main oil seal, replacement, 2A-17
Rear quarter trim panel, removal and installation, 11-11
Rear window defogger, check and repair, 12-14
Recall information, 0-7
Receiver-drier, air conditioning, removal and installation, 3-14
Recommended lubricants and fluids, 1-1
Refrigerant, pressure sensor, replacement, 3-15
Relays, general information and testing, 12-4
Reminder light, engine oil change, resetting, 1-12
Repair operations possible with the engine in the vehicle, 2A-3
Repairing minor paint scratches, 11-2
Replacement parts, buying, 0-9
Rocker arms and hydraulic lash adjusters, removal, inspection and installation, 2A-4
Rotating the tires, 1-16
Rotor, brake, inspection, removal and installation, 9-10
Routine maintenance, 1-1
Routine maintenance schedule, 1-6

S

Safety first!, 0-21
Safety recall information, 0-7
Scheduled maintenance, 1-1
Seat belt check, 1-12
Seats, removal and installation, 11-20
Shift cable, replacement and adjustment, 7A-3
Shift lever, replacement, 7A-2
Shift lever knob, replacement, 7A-2
Shock absorber (rear), replacement, 10-11
Spare tire, installing, 0-16

Spark plug
    replacement, 1-19
    torque, 1-2
    type and gap, 1-2
Speakers, removal and installation, 12-8
Stabilizer bar and link(s), removal and installation
    front, 10-9
    rear, 10-13
Starter motor, removal and installation, 5-6
Steering
    column and housing covers, removal and
      installation, 11-20
    column multi-function switches, replacement, 12-7
    column, removal and installation, 10-14
    gear boots, replacement, 10-16
    gear, removal and installation, 10-16
    knuckle, removal and installation, 10-10
    wheel, removal and installation, 10-14
Strut/coil spring assembly (front)
    removal and installation, 10-7
    replacement, 10-8
Subframe (front), removal and installation, 10-18
Suspension and steering check, 1-20
Suspension and steering systems, 10-1
**Suspension arms (rear), removal and
    installation, 10-11**

T

Tensioner, drivebelt, replacement, 1-15
Thermostat, check and replacement, 3-3
**Thermostatic Expansion Valve (TXV), general
    information, 3-15**
Throttle body, removal and installation, 4-8
Tie-rod ends, removal and installation, 10-15
**Timing chains and camshaft sprockets, removal,
    inspection and installation, 2A-9**
Tire and tire pressure checks, 1-9
Tire rotation, 1-16
Tire, spare, installing, 0-16
Tools and working facilities, 0-9
**Top Dead Center (TDC) for number 1 piston,
    locating, 2A-3**
Torque specifications
    cylinder head bolts, 2A-2
    spark plugs, 1-2
    thermostat housing bolts, 3-1
    water pump bolts, 3-1
    wheel lug nuts, 1-2
    *Other torque specifications can be found in the
      Chapter that deals with the component being
      serviced.*
**Torque tube (AWD models), removal
    and installation, 8-8**
Towing, 0-16

Transaxle, automatic, 7A-1
    Brake Transmission Shift Interlock (BTSI) system,
      description, component replacement and
      adjustment, 7A-4
    diagnosis, general, 7A-2
    driveaxle oil seals, replacement, 7A-4
    fluid
      change, 1-21
      level check, 1-11
      type, 1-1
    mount, replacement, 7A-5
    overhaul, general information, 7A-5
    removal and installation, 7A-5
    shift cable, replacement and adjustment, 7A-3
    shift lever, replacement, 7A-2
    shift lever knob, replacement, 7A-2
Transfer case, 7B-1
    lubricant
      change, 1-22
      level check, 1-19
    removal and installation, 7B-1
**Transmission range switch, removal and
    installation, 6-16**
Transmission speed sensors, replacement, 6-16
Trim panels, door, removal and installation, 11-9
Trouble codes, obtaining and clearing, 6-4
Troubleshooting, 0-22
    cooling, heating and air conditioning systems, 3-2
    engine electrical systems, 5-2
    fuel and exhaust systems, 4-2
Tune-up and routine maintenance, 1-1
Tune-up general information, 1-7
**Turn signal and hazard flashers, general
    information, 12-7**

U

**Underhood fuse/relay box, removal
    and installation, 12-5**
Underhood hose check and replacement, 1-15
**Universal joints, general information and
    check, 8-8**
Upholstery and carpets, maintenance, 11-5

V

Vacuum gauge diagnostic checks, 2B-5
Valve covers, removal and installation, 2A-3
**Valve springs, retainers and seals,
    replacement, 2A-5**
Vehicle identification numbers, 0-6
Vinyl trim, maintenance, 11-5

W

Water pump replacement, 3-9
Wheel alignment, general information, 10-19
**Wheel bearing and hub assembly, removal and
  installation**
  front, 10-11
  rear, 10-13
Wheels and tires, general information, 10-18
Window glass
  door, removal and installation, 11-14
  regulator module, removal and installation, 11-14

Windshield
  and fixed glass, replacement, 11-7
  washer fluid, level check, 1-8
  wiper blade inspection and replacement, 1-12
  wiper motor, replacement, 12-13
Wiring diagrams, general information, 12-19
Working facilities, 0-9

Notes

Haynes Automotive Manuals

NOTE: If you do not see a listing for your vehicle, please visit haynes.com for the latest product information and check out our **Online Manuals!**

ACURA
- **12020** Integra '86 thru '89 & **Legend** '86 thru '90
- **12021** Integra '90 thru '93 & **Legend** '91 thru '95
  Integra '94 thru '00 - see HONDA Civic (42025)
  MDX '01 thru '07 - see HONDA Pilot (42037)
- **12050** Acura TL all models '99 thru '08

AMC
- **14020** Mid-size models '70 thru '83
- **14025** (Renault) Alliance & Encore '83 thru '87

AUDI
- **15020** 4000 all models '80 thru '87
- **15025** 5000 all models '77 thru '83
- **15026** 5000 all models '84 thru '88
  Audi A4 '96 thru '01 - see VW Passat (96023)
- **15030** Audi A4 '02 thru '08

AUSTIN-HEALEY
  Sprite - see MG Midget (66015)

BMW
- **18020** 3/5 Series '82 thru '92
- **18021** 3-Series incl. Z3 models '92 thru '98
- **18022** 3-Series incl. Z4 models '99 thru '05
- **18023** 3-Series '06 thru '14
- **18025** 320i all 4-cylinder models '75 thru '83
- **18050** 1500 thru 2002 except Turbo '59 thru '77

BUICK
- **19010** Buick Century '97 thru '05
  Century (front-wheel drive) - see GM (38005)
- **19020** Buick, Oldsmobile & Pontiac Full-size
  (Front-wheel drive) '85 thru '05
  Buick Electra, LeSabre and Park Avenue;
  Oldsmobile Delta 88 Royale, Ninety Eight
  and Regency; Pontiac Bonneville
- **19025** Buick, Oldsmobile & Pontiac Full-size
  (Rear wheel drive) '70 thru '90
  Buick Estate, Electra, LeSabre, Limited,
  Oldsmobile Custom Cruiser, Delta 88,
  Ninety-eight, Pontiac Bonneville,
  Catalina, Grandville, Parisienne
- **19027** Buick LaCrosse '05 thru '13
  Enclave - see GENERAL MOTORS (38001)
  Rainier - see CHEVROLET (24072)
  Regal - see GENERAL MOTORS (38010)
  Riviera - see GENERAL MOTORS (38030, 38031)
  Roadmaster - see CHEVROLET (24046)
  Skyhawk - see GENERAL MOTORS (38015)
  Skylark - see GENERAL MOTORS (38020, 38025)
  Somerset - see GENERAL MOTORS (38025)

CADILLAC
- **21015** CTS & CTS-V '03 thru '14
- **21030** Cadillac Rear Wheel Drive '70 thru '93
  Cimarron - see GENERAL MOTORS (38015)
  DeVille - see GENERAL MOTORS (38031 & 38032)
  Eldorado - see GENERAL MOTORS (38030)
  Fleetwood - see GENERAL MOTORS (38031)
  Seville - see GM (38030, 38031 & 38032)

CHEVROLET
- **10305** Chevrolet Engine Overhaul Manual
- **24010** Astro & GMC Safari Mini-vans '85 thru '05
- **24013** Aveo '04 thru '11
- **24015** Camaro V8 all models '70 thru '81
- **24016** Camaro all models '82 thru '92
- **24017** Camaro & Firebird '93 thru '02
  Cavalier - see GENERAL MOTORS (38016)
  Celebrity - see GENERAL MOTORS (38005)
- **24018** Camaro '10 thru '15
- **24020** Chevelle, Malibu & El Camino '69 thru '87
  Cobalt - see GENERAL MOTORS (38017)
- **24024** Chevette & Pontiac T1000 '76 thru '87
  Citation - see GENERAL MOTORS (38020)
- **24027** Colorado & GMC Canyon '04 thru '12
- **24032** Corsica & Beretta all models '87 thru '96
- **24040** Corvette all V8 models '68 thru '82
- **24041** Corvette all models '84 thru '96
- **24042** Corvette all models '97 thru '13
- **24044** Cruze '11 thru '19
- **24045** Full-size Sedans Caprice, Impala, Biscayne,
  Bel Air & Wagons '69 thru '90
- **24046** Impala SS & Caprice and Buick Roadmaster
  '91 thru '96
  Impala '00 thru '05 - see LUMINA (24048)
- **24047** Impala & Monte Carlo all models '06 thru '11
  Lumina '90 thru '94 - see GM (38010)
- **24048** Lumina & Monte Carlo '95 thru '05
  Lumina APV - see GM (38035)
- **24050** Luv Pick-up all 2WD & 4WD '72 thru '82
- **24051** Malibu '13 thru '19
- **24055** Monte Carlo all models '70 thru '88
  Monte Carlo '95 thru '01 - see LUMINA (24048)
- **24059** Nova all V8 models '69 thru '79

- **24060** Nova and Geo Prizm '85 thru '92
- **24064** Pick-ups '67 thru '87 - Chevrolet & GMC
- **24065** Pick-ups '88 thru '98 - Chevrolet & GMC
- **24066** Pick-ups '99 thru '06 - Chevrolet & GMC
- **24067** Chevrolet Silverado & GMC Sierra '07 thru '14
- **24068** Chevrolet Silverado & GMC Sierra '14 thru '19
- **24070** S-10 & S-15 Pick-ups '82 thru '93,
  Blazer & Jimmy '83 thru '94,
- **24071** S-10 & Sonoma Pick-ups '94 thru '04,
  including Blazer, Jimmy & Hombre
- **24072** Chevrolet TrailBlazer, GMC Envoy &
  Oldsmobile Bravada '02 thru '09
- **24075** Sprint '85 thru '88 & Geo Metro '89 thru '01
- **24080** Vans - Chevrolet & GMC '68 thru '96
- **24081** Chevrolet Express & GMC Savana
  Full-size Vans '96 thru '19

CHRYSLER
- **10310** Chrysler Engine Overhaul Manual
- **25015** Chrysler Cirrus, Dodge Stratus,
  Plymouth Breeze '95 thru '00
- **25020** Full-size Front-Wheel Drive '88 thru '93
  K-Cars - see DODGE Aries (30008)
  Laser - see DODGE Daytona (30030)
- **25025** Chrysler LHS, Concorde, New Yorker,
  Dodge Intrepid, Eagle Vision, '93 thru '97
- **25026** Chrysler LHS, Concorde, 300M,
  Dodge Intrepid, '98 thru '04
- **25027** Chrysler 300 '05 thru '18, Dodge Charger
  '06 thru '18, Magnum '05 thru '08 &
  Challenger '08 thru '18
- **25030** Chrysler & Plymouth Mid-size
  front wheel drive '82 thru '95
  Rear-wheel Drive - see Dodge (30050)
- **25035** PT Cruiser all models '01 thru '10
- **25040** Chrysler Sebring '95 thru '06, Dodge Stratus
  '01 thru '06 & Dodge Avenger '95 thru '00
- **25041** Chrysler Sebring '07 thru '10, 200 '11 thru '17
  Dodge Avenger '08 thru '14

DATSUN
- **28005** 200SX all models '80 thru '83
- **28012** 240Z, 260Z & 280Z Coupe '70 thru '78
- **28014** 280ZX Coupe & 2+2 '79 thru '83
  300ZX - see NISSAN (72010)
- **28018** 510 & PL521 Pick-up '68 thru '73
- **28020** 510 all models '78 thru '81
- **28022** 620 Series Pick-up all models '73 thru '79
  720 Series Pick-up - see NISSAN (72030)

DODGE
  400 & 600 - see CHRYSLER (25030)
- **30008** Aries & Plymouth Reliant '81 thru '89
- **30010** Caravan & Plymouth Voyager '84 thru '95
- **30011** Caravan & Plymouth Voyager '96 thru '02
- **30012** Challenger & Plymouth Sapporro '78 thru '83
- **30013** Caravan, Chrysler Voyager &
  Town & Country '03 thru '07
- **30014** Grand Caravan &
  Chrysler Town & Country '08 thru '18
- **30016** Colt & Plymouth Champ '78 thru '87
- **30020** Dakota Pick-ups all models '87 thru '96
- **30021** Durango '98 & '99 & Dakota '97 thru '99
- **30022** Durango '00 thru '03 & Dakota '00 thru '04
- **30023** Durango '04 thru '09 & Dakota '05 thru '11
- **30025** Dart, Demon, Plymouth Barracuda,
  Duster & Valiant 6-cylinder models '67 thru '76
- **30030** Daytona & Chrysler Laser '84 thru '89
  Intrepid - see CHRYSLER (25025, 25026)
- **30034** Neon all models '95 thru '99
- **30035** Omni & Plymouth Horizon '78 thru '90
- **30036** Dodge & Plymouth Neon '00 thru '05
- **30040** Pick-ups full-size models '74 thru '93
- **30042** Pick-ups full-size models '94 thru '08
- **30043** Pick-ups full-size models '09 thru '18
- **30045** Ram 50/D50 Pick-ups & Raider and
  Plymouth Arrow Pick-ups '79 thru '93
- **30050** Dodge/Plymouth/Chrysler RWD '71 thru '89
- **30055** Shadow & Plymouth Sundance '87 thru '94
- **30060** Spirit & Plymouth Acclaim '89 thru '95
- **30065** Vans - Dodge & Plymouth '71 thru '03

EAGLE
  Talon - see MITSUBISHI (68030, 68031)
  Vision - see CHRYSLER (25025)

FIAT
- **34010** 124 Sport Coupe & Spider '68 thru '78
- **34025** X1/9 all models '74 thru '80

FORD
- **10320** Ford Engine Overhaul Manual
- **10355** Ford Automatic Transmission Overhaul
- **11500** Mustang '64-1/2 thru '70 Restoration Guide
- **36004** Aerostar Mini-vans all models '86 thru '97
- **36006** Contour & Mercury Mystique '95 thru '00
- **36008** Courier Pick-up all models '72 thru '82

- **36012** Crown Victoria &
  Mercury Grand Marquis '88 thru '11
- **36014** Edge '07 thru '19 & Lincoln MKX '07 thru '18
- **36016** Escort & Mercury Lynx all models '81 thru '90
- **36020** Escort & Mercury Tracer '91 thru '02
- **36022** Escape '01 thru '17, Mazda Tribute '01 thru '11,
  & Mercury Mariner '05 thru '11
- **36024** Explorer & Mazda Navajo '91 thru '01
- **36025** Explorer & Mercury Mountaineer '02 thru '10
- **36026** Explorer '11 thru '17
- **36028** Fairmont & Mercury Zephyr '78 thru '83
- **36030** Festiva & Aspire '88 thru '97
- **36032** Fiesta all models '77 thru '80
- **36034** Focus all models '00 thru '11
- **36035** Focus '12 thru '14
- **36045** Fusion '06 thru '14 & Mercury Milan '06 thru '11
- **36048** Mustang V8 all models '64-1/2 '73
- **36049** Mustang II 4-cylinder, V6 & V8 '74 thru '78
- **36050** Mustang & Mercury Capri '79 thru '93
- **36051** Mustang all models '94 thru '04
- **36052** Mustang '05 thru '14
- **36054** Pick-ups & Bronco '73 thru '79
- **36058** Pick-ups & Bronco '80 thru '96
- **36059** F-150 '97 thru '03, Expedition '97 thru '17,
  F-250 '97 thru '99, F-150 Heritage '04
  & Lincoln Navigator '98 thru '17
- **36060** Super Duty Pick-ups & Excursion '99 thru '10
- **36061** F-150 full-size '04 thru '14
- **36062** Pinto & Mercury Bobcat '75 thru '80
- **36063** F-150 '15 thru '17
- **36064** Super Duty Pick-ups '11 thru '16
- **36066** Probe all models '89 thru '92
  Probe '93 thru '97 - see MAZDA 626 (61042)
- **36070** Ranger & Bronco II gas models '83 thru '92
- **36071** Ranger '93 thru '11 & Mazda Pick-ups '94 thru '09
- **36074** Taurus & Mercury Sable '86 thru '95
- **36075** Taurus & Mercury Sable '96 thru '07
- **36076** Taurus '08 thru '14, Five Hundred '05 thru '07,
  Mercury Montego '05 thru '07 & Sable '08 thru '09
- **36078** Tempo & Mercury Topaz '84 thru '94
- **36082** Thunderbird & Mercury Cougar '83 thru '88
- **36086** Thunderbird & Mercury Cougar '89 thru '97
- **36090** Vans all V8 Econoline models '69 thru '91
- **36094** Vans full size '92 thru '14
- **36097** Windstar '95 thru '03, Freestar & Mercury
  Monterey Mini-van '04 thru '07

GENERAL MOTORS
- **10360** GM Automatic Transmission Overhaul
- **38001** GMC Acadia '07 thru '16, Buick Enclave
  '08 thru '17, Saturn Outlook '07 thru '10
  & Chevrolet Traverse '09 thru '17
- **38005** Buick Century, Chevrolet Celebrity,
  Oldsmobile Cutlass Ciera & Pontiac 6000
  all models '82 thru '96
- **38010** Buick Regal '88 thru '04, Chevrolet Lumina
  '88 thru '04, Oldsmobile Cutlass Supreme
  '88 thru '97 & Pontiac Grand Prix '88 thru '07
- **38015** Buick Skyhawk, Cadillac Cimarron,
  Chevrolet Cavalier, Oldsmobile Firenza,
  Pontiac J-2000 & Sunbird '82 thru '94
- **38016** Chevrolet Cavalier & Pontiac Sunfire '95 thru '05
- **38017** Chevrolet Cobalt '05 thru '10, HHR '06 thru '11,
  Pontiac G5 '07 thru '09, Pursuit '05 thru '06
  & Saturn ION '03 thru '07
- **38020** Buick Skylark, Chevrolet Citation,
  Oldsmobile Omega, Pontiac Phoenix '80 thru '85
- **38025** Buick Skylark '86 thru '98, Somerset '85 thru '87,
  Oldsmobile Achieva '92 thru '98, Calais '85 thru '91,
  & Pontiac Grand Am all models '85 thru '98
- **38026** Chevrolet Malibu '97 thru '03, Classic '04 thru '05,
  Oldsmobile Alero '99 thru '03, Cutlass '97 thru '00,
  & Pontiac Grand Am '99 thru '03
- **38027** Chevrolet Malibu '04 thru '12, Pontiac G6
  '05 thru '10 & Saturn Aura '07 thru '10
- **38030** Cadillac Eldorado, Seville, Oldsmobile
  Toronado & Buick Riviera '71 thru '85
- **38031** Cadillac Eldorado, Seville, DeVille, Fleetwood,
  Oldsmobile Toronado & Buick Riviera '86 thru '93
- **38032** Cadillac DeVille '94 thru '05, Seville '92 thru '04
  & Cadillac DTS '06 thru '10
- **38035** Chevrolet Lumina APV, Oldsmobile Silhouette
  & Pontiac Trans Sport all models '90 thru '96
- **38036** Chevrolet Venture '97 thru '05, Oldsmobile
  Silhouette '97 thru '04, Pontiac Trans Sport
  '97 thru '98 & Montana '99 thru '05
- **38040** Chevrolet Equinox '05 thru '17, GMC Terrain
  '10 thru '17 & Pontiac Torrent '06 thru '09

GEO
  Metro - see CHEVROLET Sprint (24075)
  Prizm - '85 thru '92 see CHEVY (24060),
  '93 thru '02 see TOYOTA Corolla (92036)
- **40030** Storm all models '90 thru '93
  Tracker - see SUZUKI Samurai (90010)

(Continued on other side)

Haynes Automotive Manuals (continued)

*NOTE: If you do not see a listing for your vehicle, please visit **haynes.com** for the latest product information and check out our **Online Manuals!***

GMC
Acadia - *see GENERAL MOTORS (38001)*
Pick-ups - *see CHEVROLET (24027, 24068)*
Vans - *see CHEVROLET (24081)*

HONDA
42010 **Accord CVCC** all models '76 thru '83
42011 **Accord** all models '84 thru '89
42012 **Accord** all models '90 thru '93
42013 **Accord** all models '94 thru '97
42014 **Accord** all models '98 thru '02
42015 **Accord** '03 thru '12 **& Crosstour** '10 thru '14
42016 **Accord** '13 thru '17
42020 **Civic 1200** all models '73 thru '79
42021 **Civic 1300 & 1500 CVCC** '80 thru '83
42022 **Civic 1500 CVCC** all models '75 thru '79
42023 **Civic** all models '84 thru '91
42024 **Civic & del Sol** '92 thru '95
42025 **Civic** '96 thru '00, **CR-V** '97 thru '01
& **Acura Integra** '94 thru '00
42026 **Civic** '01 thru '11 **& CR-V** '02 thru '11
42027 **Civic** '12 thru '15 **& CR-V** '12 thru '16
42030 **Fit** '07 thru '13
42035 **Odyssey** all models '99 thru '10
Passport - *see ISUZU Rodeo (47017)*
42037 **Honda Pilot** '03 thru '08, **Ridgeline** '06 thru '14
& **Acura MDX** '01 thru '07
42040 **Prelude CVCC** all models '79 thru '89

HYUNDAI
43010 **Elantra** all models '96 thru '19
43015 **Excel & Accent** all models '86 thru '13
43050 **Santa Fe** all models '01 thru '12
43055 **Sonata** all models '99 thru '14

INFINITI
G35 '03 thru '08 - *see NISSAN 350Z (72011)*

ISUZU
Hombre - *see CHEVROLET S-10 (24071)*
47017 **Rodeo** '91 thru '02, **Amigo** '89 thru '94 & '98 thru '02
& **Honda Passport** '95 thru '02
47020 **Trooper** '84 thru '91 **& Pick-up** '81 thru '93

JAGUAR
49010 **XJ6** all 6-cylinder models '68 thru '86
49011 **XJ6** all models '88 thru '94
49015 **XJ12 & XJS** all 12-cylinder models '72 thru '85

JEEP
50010 **Cherokee, Comanche & Wagoneer Limited**
all models '84 thru '01
50011 **Cherokee** '14 thru '19
50020 **CJ** all models '49 thru '86
50025 **Grand Cherokee** all models '93 thru '04
50026 **Grand Cherokee** '05 thru '19
& **Dodge Durango** '11 thru '19
50029 **Grand Wagoneer & Pick-up** '72 thru '91
Grand Wagoneer '84 thru '91, Cherokee &
Wagoneer '72 thru '83, Pick-up '72 thru '88
50030 **Wrangler** all models '87 thru '17
50035 **Liberty** '02 thru '12 **& Dodge Nitro** '07 thru '11
50050 **Patriot & Compass** '07 thru '17

KIA
54050 **Optima** '01 thru '10
54060 **Sedona** '02 thru '14
54070 **Sephia** '94 thru '01, **Spectra** '00 thru '09,
Sportage '05 thru '20
54077 **Sorento** '03 thru '13

LEXUS
ES 300/330 - *see TOYOTA Camry (92007, 92008)*
ES 350 - *see TOYOTA Camry (92009)*
RX 300/330/350 - *see TOYOTA Highlander (92095)*

LINCOLN
MKX - *see FORD (36014)*
Navigator - *see FORD Pick-up (36059)*
59010 **Rear-Wheel Drive Continental** '70 thru '87,
Mark Series '70 thru '92 **& Town Car** '81 thru '10

MAZDA
61010 **GLC (rear-wheel drive)** '77 thru '83
61011 **GLC (front-wheel drive)** '81 thru '85
61012 **Mazda3** '04 thru '11
61015 **323 & Protogé** '90 thru '03
61016 **MX-5 Miata** '90 thru '14
61020 **MPV** all models '89 thru '98
Navajo - *see Ford Explorer (36024)*
61030 **Pick-ups** '72 thru '93
Pick-ups '94 thru '09 - *see Ford Ranger (36071)*
61035 **RX-7** all models '79 thru '85
61036 **RX-7** all models '86 thru '91
61040 **626 (rear-wheel drive)** all models '79 thru '82
61041 **626 & MX-6 (front-wheel drive)** '83 thru '92
61042 **626** '93 thru '01 **& MX-6/Ford Probe** '93 thru '02
61043 **Mazda6** '03 thru '13

MERCEDES-BENZ
63012 **123 Series Diesel** '76 thru '85
63015 **190 Series** 4-cylinder gas models '84 thru '88
63020 **230/250/280** 6-cylinder SOHC models '68 thru '72
63025 **280 123 Series** gas models '77 thru '81
63030 **350 & 450** all models '71 thru '80
63040 **C-Class:** C230/C240/C280/C320/C350 '01 thru '07

MERCURY
64200 **Villager & Nissan Quest** '93 thru '01
All other titles, see FORD Listing.

MG
66010 **MGB** Roadster & GT Coupe '62 thru '80
66015 **MG Midget, Austin Healey Sprite** '58 thru '80

MINI
67020 **Mini** '02 thru '13

MITSUBISHI
68020 **Cordia, Tredia, Galant, Precis & Mirage** '83 thru '93
68030 **Eclipse, Eagle Talon & Plymouth Laser** '90 thru '94
68031 **Eclipse** '95 thru '05 **& Eagle Talon** '95 thru '98
68035 **Galant** '94 thru '12
68040 **Pick-up** '83 thru '96 **& Montero** '83 thru '93

NISSAN
72010 **300ZX** all models including Turbo '84 thru '89
72011 **350Z & Infiniti G35** all models '03 thru '08
72015 **Altima** all models '93 thru '06
72016 **Altima** '07 thru '12
72020 **Maxima** all models '85 thru '92
72021 **Maxima** all models '93 thru '08
72025 **Murano** '03 thru '14
72030 **Pick-ups** '80 thru '97 **& Pathfinder** '87 thru '95
72031 **Frontier** '98 thru '04, **Xterra** '00 thru '04,
& Pathfinder '96 thru '04
72032 **Frontier & Xterra** '05 thru '14
72037 **Pathfinder** '05 thru '14
72040 **Pulsar** all models '83 thru '86
72042 **Roque** all models '08 thru '20
72050 **Sentra** all models '82 thru '94
72051 **Sentra & 200SX** all models '95 thru '06
72060 **Stanza** all models '82 thru '90
72070 **Titan pick-ups** '04 thru '10, **Armada** '05 thru '10
& Pathfinder Armada '04
72080 **Versa** all models '07 thru '19

OLDSMOBILE
73015 **Cutlass** V6 & V8 gas models '74 thru '88
*For other OLDSMOBILE titles, see BUICK,
CHEVROLET or GENERAL MOTORS listings.*

PLYMOUTH
For PLYMOUTH titles, see DODGE listing.

PONTIAC
79008 **Fiero** all models '84 thru '88
79018 **Firebird** V8 models except Turbo '70 thru '81
79019 **Firebird** all models '82 thru '92
79025 **G6** all models '05 thru '09
79040 **Mid-Size Rear-wheel Drive** '70 thru '87
Vibe '03 thru '10 - *see TOYOTA Corolla (92037)*
*For other PONTIAC titles, see BUICK,
CHEVROLET or GENERAL MOTORS listings.*

PORSCHE
80020 **911 Coupe & Targa** models '65 thru '89
80025 **914** all 4-cylinder models '69 thru '76
80030 **924** all models including Turbo '76 thru '82
80035 **944** all models including Turbo '83 thru '89

RENAULT
Alliance & Encore - *see AMC (14025)*

SAAB
84010 **900** all models including Turbo '79 thru '88

SATURN
87010 **Saturn** all S-series models '91 thru '02
Saturn Ion '03 thru '07- *see GM (38017)*
Saturn Outlook - *see GM (38001)*
87020 **Saturn L-series** all models '00 thru '04
87040 **Saturn VUE** '02 thru '09

SUBARU
89002 **1100, 1300, 1400 & 1600** '71 thru '79
89003 **1600 & 1800** 2WD & 4WD '80 thru '94
89080 **Impreza** '02 thru '11, **WRX** '02 thru '14,
& WRX STI '04 thru '14
89100 **Legacy** all models '90 thru '99
89101 **Legacy & Forester** '00 thru '09
89102 **Legacy** '10 thru '16 **& Forester** '12 thru '16

SUZUKI
90010 **Samurai/Sidekick & Geo Tracker** '86 thru '01

TOYOTA
92005 **Camry** all models '83 thru '91
92006 **Camry** '92 thru '96 **& Avalon** '95 thru '96
92007 **Camry, Avalon, Solara, Lexus ES 300** '97 thru '01

92008 **Camry, Avalon, Lexus ES 300/330** '02 thru '06
& Solara '02 thru '08
92009 **Camry, Avalon & Lexus ES 350** '07 thru '17
92015 **Celica Rear-wheel Drive** '71 thru '85
92020 **Celica Front-wheel Drive** '86 thru '99
92025 **Celica Supra** all models '79 thru '92
92030 **Corolla** all models '75 thru '79
92032 **Corolla** all rear-wheel drive models '80 thru '87
92035 **Corolla** all front-wheel drive models '84 thru '92
92036 **Corolla & Geo/Chevrolet Prizm** '93 thru '02
92037 **Corolla** '03 thru '19, **Matrix** '03 thru '14,
& Pontiac Vibe '03 thru '10
92040 **Corolla Tercel** all models '80 thru '82
92045 **Corona** all models '74 thru '82
92050 **Cressida** all models '78 thru '82
92055 **Land Cruiser FJ40, 43, 45, 55** '68 thru '82
92056 **Land Cruiser FJ60, 62, 80, FZJ80** '80 thru '96
92060 **Matrix** '03 thru '11 **& Pontiac Vibe** '03 thru '10
92065 **MR2** all models '85 thru '87
92070 **Pick-up** all models '69 thru '78
92075 **Pick-up** all models '79 thru '95
92076 **Tacoma** '95 thru '04, **4Runner** '96 thru '02
& T100 '93 thru '08
92077 **Tacoma** all models '05 thru '18
92078 **Tundra** '00 thru '06 **& Sequoia** '01 thru '07
92079 **4Runner** all models '03 thru '09
92080 **Previa** all models '91 thru '95
92081 **Prius** all models '01 thru '12
92082 **RAV4** all models '96 thru '12
92085 **Tercel** all models '87 thru '94
92090 **Sienna** all models '98 thru '10
92095 **Highlander** '01 thru '19
& Lexus RX330/330/350 '99 thru '19
92179 **Tundra** '07 thru '19 **& Sequoia** '08 thru '19

TRIUMPH
94007 **Spitfire** all models '62 thru '81
94010 **TR7** all models '75 thru '81

VW
96008 **Beetle & Karmann Ghia** '54 thru '79
96009 **New Beetle** '98 thru '10
96016 **Rabbit, Jetta, Scirocco & Pick-up**
gas models '75 thru '92 & Convertible '80 thru '92
96017 **Golf, GTI & Jetta** '93 thru '98, **Cabrio** '95 thru '02
96018 **Golf, GTI, Jetta** '99 thru '05
96019 **Jetta, Rabbit, GLI, GTI & Golf** '05 thru '11
96020 **Rabbit, Jetta & Pick-up** diesel '77 thru '84
96021 **Jetta** '11 thru '18 **& Golf** '15 thru '19
96023 **Passat** '98 thru '05 **& Audi A4** '96 thru '01
96030 **Transporter 1600** all models '68 thru '79
96035 **Transporter 1700, 1800 & 2000** '72 thru '79
96040 **Type 3 1500 & 1600** all models '63 thru '73
96045 **Vanagon Air-Cooled** all models '80 thru '83

VOLVO
97010 **120, 130 Series & 1800 Sports** '61 thru '73
97015 **140 Series** all models '66 thru '74
97020 **240 Series** all models '76 thru '93
97040 **740 & 760 Series** all models '82 thru '88
97050 **850 Series** all models '93 thru '97

TECHBOOK MANUALS
10205 **Automotive Computer Codes**
10206 **OBD-II & Electronic Engine Management**
10210 **Automotive Emissions Control Manual**
10215 **Fuel Injection Manual** '78 thru '85
10225 **Holley Carburetor Manual**
10230 **Rochester Carburetor Manual**
10305 **Chevrolet Engine Overhaul Manual**
10320 **Ford Engine Overhaul Manual**
10330 **GM and Ford Diesel Engine Repair Manual**
10331 **Duramax Diesel Engines** '01 thru '19
10332 **Cummins Diesel Engine Performance Manual**
10333 **GM, Ford & Chrysler Engine Performance Manual**
10334 **GM Engine Performance Manual**
10340 **Small Engine Repair Manual,** 5 HP & Less
10341 **Small Engine Repair Manual,** 5.5 thru 20 HP
10345 **Suspension, Steering & Driveline Manual**
10355 **Ford Automatic Transmission Overhaul**
10360 **GM Automatic Transmission Overhaul**
10405 **Automotive Body Repair & Painting**
10410 **Automotive Brake Manual**
10411 **Automotive Anti-lock Brake (ABS) Systems**
10420 **Automotive Electrical Manual**
10425 **Automotive Heating & Air Conditioning**
10435 **Automotive Tools Manual**
10445 **Welding Manual**
10450 **ATV Basics**

Over a 100 Haynes
motorcycle manuals
also available

1 0/22